TRENDS IN RNA RESEARCH

TRENDS IN RNA RESEARCH

PETER A. MCNAMARA
EDITOR

Nova Science Publishers, Inc.
New York

For permission to use material from this book please contact us:
Telephone 631-231-7269; Fax 631-231-8175
Web Site: http://www.novapublishers.com

Library of Congress Cataloging-in-Publication Data

Trends in RNA research / Peter A. McNamara (editor).
p. ; cm.
Includes index.
ISBN 1-59454-506-5
1. RNA--Research. 2. Genetic regulation--Research. 3. Molecular dynamics. I. McNamara, Peter A.
[DNLM: 1. Genetic Techniques--trends. 2. RNA--genetics. QU 58.7 T794 2006]
QP623.T74 2006
572.8'8--dc22 2005031861

Published by Nova Science Publishers, Inc. ✦New York

Contents

Preface

RNA is a chemical found in the nucleus and cytoplasm of cells. RNA plays an important role in protein synthesis and other chemical activities of the cell. The structure of RNA is similar to that of DNA, although RNA is single stranded whereas DNA has a characteristic "double helix." Genetic information is stored by DNA in the nucleus of cells, and RNA carries that information to other parts of the cell where it is converted into protein. Three types of RNA are; mRNA (messenger RNA, which contains the specific sequence of nucleotides necessary to dictate amino acid sequence in proteins), tRNA (transfer RNA, which serves as the "adaptor" to position the appropriate amino acid next to a growing polypeptide chain during protein synthesis), and rRNA (ribosomal RNA, which is the RNA component of ribosomes). This new book presents state-of-the-research from throughout the world.

Targeting RNA with small molecules is gaining interest in the drug discovery industry as an alternative approach to traditional drug discovery focusing on proteins. The targeting of bacterial 16S ribosomal RNA (rRNA) with aminoglycoside antibiotics provides a clear precedence for the targeting of RNA with small molecule drugs. The mechanism how aminoglycosides, through binding to rRNA, interfere with the fidelity of the translation machinery is understood in great detail, due to the resolution of the X-ray crystal structure of the 30S ribosomal subunit in complex with paromomycin. Binding of aminoglycosides to rRNA fragments is predominantly determined by electrostatic interactions: The negatively charged RNA backbone attracts the positively charged aminoglycosides. Binding is relatively non-specific. Aminoglycosides also bind to human 18S rRNA, with affinities that are only two- to six-fold lower than for binding to bacterial 16S rRNA. However, the specific anti-bacterial effect of aminoglycosides can also be subscribed to other mechanisms, such as the occurrence in prokaryotes of tranporter proteins that actively take up aminoglycosides. Aminoglycosides can suppress premature termination codons in human messenger RNAs (mRNAs), thus permitting translation to continue to the normal end of the transcript. This suppression activity of aminoglycosides is explored in drug discovery for the development of novel therapy for genetic disorders that are caused by stop codon mutations, such as cystic fibrosis and Duchenne muscular dystrophy. While aminoglycosides act on rRNA, certain small molecule metabolites, including vitamins, lysine and purines, can regulate gene expression by binding to mRNA. Binding occurs at defined target sites, which are called 'riboswitches'. Riboswitches consist of complex three-dimensional structures, which are located in the 5'-untranslated region (UTR) of bacterial and fungal mRNAs, coding for proteins involved in the uptake and biosynthesis of these metabolites. It is unknown whether riboswitches also occur in human cells. Riboswitches provide yet another proof that small molecules can influence biological events by binding directly to RNA. The targeting of human mRNAs with small synthetic chemical molecules is a relatively new approach, but is expected to create new and unique opportunities for drug development. Recently, it was shown that analogues of thalidomide, a sedative that has been withdrawn from the market in the early 1960s, can suppress the synthesis of the cytokine Tumor Necrosis Factor-α (TNFα) at the post-transcriptional level, through the involvement of sequences in the 3'-UTR of TNFα mRNA. Targeting RNA is in particular attractive for the development of new male contraceptives, as several RNA binding proteins and their cognate targets in mRNA have shown play critical roles in spermatogenesis. These and other opportunities of RNA as a target for small molecule drugs will be discussed in chapter I.

Alzheimer's disease (AD) is a complex neurodegenerative disorder characterized by the progressive accumulation of β-amyloid (Aβ) in specific brain regions. Aβ is derived from the amyloid precursor protein (APP) after β- and γ-secretase cleavages. β-secretase was recently identified as a membrane-anchored aspartyl protease, but the identity of γ-secretase has remained elusive. Recent studies have demonstrated that γ-secretase is a multiprotein complex consisting of at least four proteins: presenilin (PS), which may harbor the catalytic site, nicastrin (NCT), Aph-1 and Pen-2. γ-Secretase cleaves not only APP but also Notch and several other transmembrane receptors indicating that it serves as a mediator of diverse signaling pathways with multiple biological functions. The discovery that double-stranded RNA is able to silence gene expression was made only a few years ago but methods for experimentally silencing genes have already been extended to a broad diversity of organisms and cell types, including neurons. Chapter II reviews recent studies that have employed RNA-mediated gene silencing techniques to dissect the roles of the different factors involved in β- and γ-secretase-mediated amyloidogenic cleavage of APP in AD.

CRD-BP/IMP1 belongs to a family of highly conserved RNA binding proteins which contain four hnRNP K homology (KH) domains and two RNA recognition motifs (RRM) and whose members appear to be involved in the localization, stability and/or translational control of their RNA targets as described in chapter III. In humans, besides CRD-BP/IMP1 this family is represented by two other members, KOC/IMP3 and IMP-2 isolated in screens designed to identify genes or tumor-associated antigens specifically expressed in pancreatic and hepatocellular carcinomas respectively. CRD-BP/IMP-1 was initially isolated by means of its high specific affinity to the *c-myc* mRNA Coding Region Determinant (CRD) instability element. In a cell-free mRNA decay system CRD-BP/IMP1 binding protects the message of this proto-oncogene from rapid destruction *via* endonucleolytic cleavage. CRD-BP/IMP-1, IMP-2 and KOC/IMP-3 were also isolated independently, as members of the protein family that binds to the 5' UnTranslated Region (UTR) of the Insulin-like Growth Factor II (*IGF-II*) leader 3' mRNA species and postulated to play a role in its translational regulation. This mRNA species represents the major embryonic form of IGF-II transcripts. Besides leader 3' *IGF-II* mRNA, IMP-1, 2 and 3, bind multiple sites of the *H19* RNA as well, a transcript with no open reading frame that exhibits an oncofetal pattern of expression and whose role remains enigmatic. Recently was shown that CRD-BP/IMP1 co-associates with the G3BP and HuD proteins to the *Tau* mRNA that is localized primarily in the cell body and axon of neurons. CRD-BP/IMP1 is able to translocate to the nucleus and also contains two nuclear export signals. In the cytoplasm CRD-BP/IMP1 is distributed along with microtubules and transports towards the leading edge in migrating cells. It has been postulated that CRD-BP/IMP-1 initially binds to its targets in the nucleus and subsequently plays a role in the polarization of genetic information *via* specific cytoplasmic RNA localization, as it has been shown for two other members of this protein family, the Xenopus Vg1RBP/Vera and chicken ZBP-1 that promote the localization of *Vg1* mRNA to the vegetal pole of the oocyte and beta-actin mRNA to the leading edge of motile cells, respectively. Normally, CRD-BP/IMP-1 is expressed only during fetal life and its expression becomes undetectable shortly after birth. CRD-BP/IMP1 null mice exhibit high perinatal lethality, dwarfism and impaired gut development. Growing evidence demonstrates the *de novo* activation of this gene in high percentages of tumor specimens from different types of human malignancies, including some of the most common, such as breast and colon carcinomas. Furthermore, transgenic mice exhibiting targeted expression of CRD-BP/IMP-1 in adult mammary tissue develop tumors. These findings imply that the *de novo* activation of CRD-BP/IMP-1 in adult tissue, whatever its precise role in neoplasia, may be causally related to the onset or progression of tumorigenesis.

Human Papillomavirus (HPV) is related to more than 90% of cervical cancer. The virus is shown to be essential for the induction and maintenance of the malignant phenotype in cervical cancer. In chapter IV, the authors designed a hammerhead ribozyme to specifically target the HPV16 E6 mRNA, and our results demonstrated that the ribozyme can cleave HPV16 E6 mRNA effectively and with high specificity. When transfected into a HPV16 positive cervical cancer cell, CaSKi, the ribozyme reduced the expression of HPV16 E6 mRNA, and inhibited cell growth both *in vitro* and *in vivo*. The percentage of apoptosis cells was also increased. They found that the anti-HPV16E6 ribozyme reduced the expression of the viral E6 gene, and cellular c-myc, bcl-2 genes, but increased the expression of p53. It is likely that anti-HPV16E6 ribozyme inhibited cervical cancer cell growth by reducing the expression of the HPV16 E6 gene, which may alter the expression of p53, c-myc and bcl-2, and led to apoptosis in cancer cells. They also found that CaSKi cells transfected with the anti-HPV16E6 ribozyme showed increased sensitivity to cisplatin and radiation. Furthermore, the anti-HPV16E6 ribozyme increased expression of HLA-2, B7-1 and B7-2 molecules on CaSKi cell, and enhanced its antigenic property to immune therapy. Our study demonstrated the potential of anti-HPV16E6 ribozyme for treating cervical cancer, and the possibility of combining ribozyme approach with anti-tumor therapies including chemotherapy, radiotherapy or immunotherapy.

Dendritic cells (DCs) are potent antigen presenting cells as reported in chapter V. They initiate and tailor the immune response by sensing immunostimulatory signals and foreign molecular structures. DCs play the role of central processing unit of the immune system, whereby different stimuli lead to a qualitatively different maturation outcome. There is growing evidence that DC subsets can differ in their response based on microenvironmental changes or pathogens and their products. Many of the functional characteristics of these tailored responses have been defined, but the underlying molecular pathways involved in DC maturation are not fully characterized. In the past few years, highly parallel expression analyses by microarrays have begun to elucidate the genetic pathways underlying DC maturation as well as DC subset differentiation. These experiments have identified unexpected cytokine pathways, which play an important role in DC-mediated T cell proliferation, as well as transcription regulators which proved to be critical for the development of certain DCs lineages *in vivo*. Furthermore, global gene expression profiling has revealed that the DC maturation program contains a set of genes that is commonly induced in response to various stimuli as well as stimulus-specific genes, indicating a certain plasticity.

Cyclosporine (CsA) introduction is a milestone in immunosuppresive therapy improving morbidity and mortality of transplanted patients. However, the use of CsA is complicated by side-effects such as hypertension and nephrotoxicity through effects on factors controlling vasomotor tone (endothelin, thromboxane and NO). Nitric oxide (NO) is crucial for maintaining a state of basal vasodilation and inhibition of platelet activation and its impairment by CsA has been advanced as a mechanism for CsA-induced hypertension. However, direct effect of CsA on NO synthesis is not supported as studies in normal human volunteers showed that acute CsA administration produced vasoconstriction and simultaneous increase in endothelial NO release and that CsA produced vasoconstriction in the presence of a simultaneous inhibition of nitric oxide synthase (NOS). Furthermore, incubation of endothelial cells with CsA showed an increased gene expression of endothelial NOS (ecNOS). It is therefore unlikely that CsA is a direct inhibitor of the NO system and decreased NO activity is unlikely to mediate the vasoconstrictive actions of CsA. On the other hand, CsA has been shown to induce superoxide anions production which increase NO metabolism and could represent an link between NO increased metabolism and reduced activity and CsA induced vasoconstriction and hypertension. Chapter VI has evaluated in mononuclear cells of kidney transplanted patients chronically treated with CsA or Tracolimus (FK-506) and post-transplant hypertension, NO system status (ecNOS gene expression and NO metabolites in plasma) and the presence of oxidative stress using a molecular biology approach (gene expression of oxidative stress related proteins such as $p22^{phox}$, heme oxygenase (HO-1), and TGF-β and markers of oxidative stress such has hydroperoxides, peroxynitrites and plasma antioxidant power). In kidney transplanted patients chronically treated with calcineurin inhibitors and post-transplant hypertension, NO system is upregulated (ecNOS mRNA and NO metabolites are increased in patients). CsA or FK-506 induced oxidative stress as shown by the induction of $p22^{phox}$ and TGF-β expression and of hydroperoxides and peroxynitrites. CsA and FK-506 induced comparable hypertension and degree of oxidative stress with no difference between these two treatments in terms of induction of oxidative stress related proteins, therefore supporting the role of calcineurin inhibitors-induced oxidative stress for the endothelial dysfunction and posttransplant hypertension. The authors have also evaluated in kidney transplanted patients chronically treated with CsA or FK-506 and post-transplant hypertension the effect of antihypertensive drugs on oxidative stress related biochemical and molecular markers. They have considered the ACE inhibitor Ramipril, the calcium antagonist Nifedipine and the α1-β blocker Carvedilol. These drugs are equally effective in the reduction of the blood pressure and leave unchanged the upregulation of NO system observed in kidney transplanted patients under chronic CsA or FK-506 treatment. Carvedilol and Nifedipine leave unchanged NO upregulation but only Carvedilol reduced TGF-β and increased HO-1 mRNA production. Ramipril reduced $p22^{phox}$ and TGF-β gene expression while was ineffective on HO-1 mRNA. The demonstration that in kidney transplant patients chronically treated with CsA or FK-506 and posttransplant hypertension, different antihypertensive drugs induce a comparable blood pressure reduction but different effects on proteins induced by and/or protective from oxidative stress, closely correlated with endothelial dysfunction and fibrogenesis, could give further suggestions on the antihypertensive therapeutic approach to the transplant patient chronically treated with calcineurin inhibitors and with calcineurin inhibitor-induced hypertension besides the need of blood pressure reduction.

Detection of hepatitis C virus (HCV) RNA by AMPLICOR system (Roche Diagnostic Systems) has been improved in assay quality assurance through placement of new reagents and automated instruments as well as quality control methods. Chapter VII correlated laboratory test results of consecutive 5395 serum samples received for routine detection of HCV by the qualitative AMPLICOR HCV test (AMPLICOR, version 1.0), and investigated causes of the inconsistency. After excluding patients under treatment with anti-viral agents, 519 and 866 of 1133 samples were positive for HCV RNA by the qualitative AMPLICOR HCV test and HCV antibodies by a third-generation enzyme immunoassay test, respectively. Titers of positive HCV antibodies were widely distributed with two groups: low (<40.0) and high (≥ 40.0), which showed distinctive positive rates for HCV RNA in 19/371 (5.1%) and 488/495 (98.6%), respectively ($p<0.0001$). Patterns of laboratory test results suggested false-negative results for HCV RNA in a total of 8 samples. Seven of them had high-titers of HCV antibodies. One sample with negative HCV antibodies and increased levels of aminotransferases was considered a possibility of acute hepatitis C infection. Four of the 8 samples were suggested the presence of an inhibitor by a dilution study. Results for HCV RNA of another 3 samples turned to be positive, when assayed by a second version of the AMPLICOR HCV test (AMPLICOR, version 2.0), the reagent with more equivalent detection sensitivity of all HCV genotypes than the first version. All of the 3 samples were revealed to have genotype 2a. The last one sample was suggested a negative PCR result under the detection limit caused by RNA degradation. False-positive results were not observed. In summary, individual correlation of laboratory test results of each sample followed by further assessment in detection of HCV RNA by the qualitative AMPLICOR HCV test would support the quality assurance of each result as well as the system of assay. The 5′-untranslated region of the human immunodeficiency virus type 1 (HIV-1) genomic RNA contains several stem-loops that play important roles in the viral replication cycle. Amongst these are the dimerization initiation site (DIS), the hairpin containing the major splice donor site (SD) and the packaging signal (ψ), all three hairpins are sequentially located in the region of approximately 100 nt length preceding GAG start codon. In this chapter, the authors analyzed the primary and the secondary structures of this region in all HIV-1 isolates for which this region was submitted in NCBI database, 350 isolates in total. The secondary structures were predicted by the mfold program (Zuker, 2003). The authors found that 94% of viral isolates (groups M and N) contain three hairpins (DIS, SD and ψ) in the region preceding GAG start codon. In the rest of HIV-1 isolates (group O) this region has four hairpins. Two hairpins are DIS- and ψ-like hairpins, a hairpin located between them being different from SD-like hairpin. The additional fourth hairpin is located immediately upstream of GAG start codon. Such structure of the region preceding GAG start codon has first been reported by Harrison et al. (1998). They determined the most frequently occurring secondary structures of DIS, SD and ψ hairpins in HIV-1 isolates of M group, and compared our phylogenetic findings with literature data on high-resolution structures for these hairpins. In particular, they found that 73% of HIV-1 group M isolates surveyed contain DIS hairpins with the AA[GCGCGC]A (Lai variant), AG[GUGCAC]A (Mal variant) or A[GUGCAC] apical loops. The last apical loop is typical for HIV-1 isolates (subtype C) from Botswana, South Africa and India. Contrary to the well-known structures of DIS hairpins in Lai and Mal isolates, the structure of the DIS hairpin with the A[GUGCAC] apical loop is not well studied yet. They observed that HIV-1 genomic RNAs with this loop often contain SD and ψ hairpins with the specific base changes (A^5 to G^5 in SD and U^1 to A^1 in ψ hairpins). This leads to alterations in the hairpins secondary structures. The stem of the SD hairpin in these isolates has no A-U-A base triple platform and the region corresponding to ψ hairpin is folded mainly in two alternative structures, including one with a truncated stem. Phylogenetic analysis data provide evidence that DIS, SD and ψ motifs are essential for HIV-1 replication cycle.

The 5′-untranslated region of the human immunodeficiency virus type 1 (HIV-1) genomic RNA contains several stem-loops that play important roles in the viral replication cycle. Amongst these are the dimerization initiation site (DIS), the hairpin containing the major splice donor site (SD) and the packaging signal (ψ), all three hairpins are sequentially located in the region of approximately 100 nt length preceding GAG start codon. In chapter VIII, the authors analyzed the primary and the secondary structures of this region in all HIV-1 isolates for which this region was submitted in NCBI database, 350 isolates in total. The secondary structures were predicted by the mfold program (Zuker, 2003). They found that 94% of viral isolates (groups M and N) contain three hairpins (DIS, SD and ψ) in the region preceding GAG start codon. In the rest of HIV-1 isolates (group O) this region has four hairpins. Two hairpins are DIS- and ψ-like hairpins, a hairpin located between them being different from SD-like hairpin. The additional fourth hairpin is located immediately upstream of GAG start codon. Such structure of the region preceding GAG start codon has first been reported by Harrison et al. (1998). They determined the most frequently occurring secondary structures of DIS, SD and ψ hairpins in HIV-1 isolates of M group, and compared our phylogenetic findings with literature data on high-resolution structures for these hairpins. In particular, they found that 73% of HIV-1 group M isolates surveyed contain DIS hairpins with the AA[GCGCGC]A (Lai variant), AG[GUGCAC]A (Mal variant) or A[GUGCAC] apical loops. The last apical loop is typical for HIV-1 isolates (subtype C) from Botswana, South Africa and India. Contrary to the well-known structures of DIS hairpins in Lai and Mal isolates, the structure of the DIS hairpin with the A[GUGCAC] apical loop is not well studied yet. They observed that HIV-1 genomic RNAs with this loop often contain SD and ψ hairpins with the specific base changes (A^5 to G^5 in SD and U^1 to A^1 in ψ hairpins). This leads to alterations in the hairpins secondary structures. The stem of the SD hairpin in these isolates has no A-U-A base triple platform and the region corresponding to ψ hairpin is folded mainly in two alternative structures, including one with a truncated stem. Phylogenetic analysis data provide evidence that DIS, SD and ψ motifs are essential for HIV-1 replication cycle.

Sodium hydrogen exchanger [NHE] plays an important role in the regulation of intracellular pHi and net uptake of NaCl and HCO_3^- from intestine and renal tubules. Several pharmacologically and biochemically distinct NHE isoforms have been identified, and classified broadly as amiloride sensitive or resistant ones. Sardet et al cloned an amiloride sensitive human NHE-1 isoform for the first time in 1989. Using this cDNA as a probe, NHE-1 isoform was cloned from several other species including chinese hamster, mouse, rabbit, rat and swine. So far nine distinct NHE genes [NHE-1–NHE-9], and an alternatively spliced variant of NHE-1 have been reported. These isoforms show considerable heterogeneity in their tissue/cell expression and membrane localization, their sensitivity to pharmacological agents, kinetic properties, and regulation by hormonal and physical stimuli. Although the primary amino acid sequence of the human NHE isoforms shares only limited [21-44%] identity, they show a common secondary structure consisting of 10-13 conserved trans membrane domains in their N-terminus, and a regulatory C-terminus located in the cytoplasm. Most of the NHE isoforms are located in the plasma membrane, but the NHE-5, NHE-6 and NHE-7 are reported to be present intracellularly. Analysis also predicts strong probability of a mitochondrial target sequence in the NHE-1 and NHE-3 which are normally located on the plasma membrane. Among the known isoforms, NHE-1 and NHE-3 are the most extensively studied isoforms. The authors of chapter IX and others have shown that NHE-1 is localized on basolateral domain, and NHE-3 on the apical domain of epithelial cells. Their distinct location on the epithelial cell suggests that the main function of NHE-1 is to regulate pHi, while that of NHE-3 is to regulate absorption of NaCl. Alterations in the NHE-1 and NHE-3 activity and expression of mRNA and protein have been linked to pathogenesis of several diseases including colitis, essential hypertension, congenital secretory diarrhea, gastroesophageal reflux disease, diabetes, and tissue damage in ischemia/reperfusion. Chapter VIII documents recent developments pertaining to the molecular diversity and the regulation of NHE isoforms in pathogenic conditions. A comprehensive analysis based physical properties, predictions on the primary and secondary structures, topology, and functional relevance of NHE in disease pathogenesis will also be presented.

The most useful molecular chronometers used today for the determination of phylogenetic relationships are the 16S ribosomal RNA genes (16S rDNA genes) and their gene products, the 16S rRNA. Genotypic identification procedures have increasingly received attention as an alternative or complement to conventional phenotypic methods in a polyphasic approach. Broad-range primers that recognize 16S rDNA sequences conserved among a wide variety of bacteria are used to amplify species-specific variable regions of interest. Sequence determination and comparative database searches allow the unknown isolates to be assigned to a bacterial taxon. In this chapter, application of 16S rRNA gene sequencing to the members of the family *Acidaminococcaceae*, formerly *Sporomusa* sub-branch, will be described. First of all, the family *Acidaminococcaceae* and its anaerobic, Gram-negative members will be presented as well as the particular taxonomic position of this family in the bacterial phylogenetic tree within the Gram-positive bacteria phylum. Chapter X will focus on the genera recovered in humans in a medical practice, i.e., *Acidaminococcus*, *Anaeroglobus*, *Megasphaera*, *Veillonella*, *Dialister* and *Selenomonas*. Based on recent studies, two main interesting aspects of sequencing 16S rDNA genes will be developed. First, they underlined the interest in identification of the members of the family *Acidaminococcaceae* when phenotypic characteristics are poor or not discriminating enough to affiliate a bacterial isolate to a definite species. This part is supported by a study conducted over a 3-year period in a teaching hospital demonstrating the powerful capacity of identification of this method. Secondly, the authors highlighted the interest in detection and characterization of hitherto unknown taxa. The underestimated biodiversity of this family will be underlined here based on culture-dependent and culture-independent studies. The latter part will be illustrated by examples of polyphasic characterization of novel taxa within the family *Acidaminococcaceae* like the genus *Anaeroglobus* and the species *Veillonella montpellierensis* and *Megasphaera micronuciformis*. Despite the efficiency of the 16S rDNA-based methods, it is important to know their limits, particularly when its discriminating power is not high enough in some genera like *Selenomonas* or impaired by a high level of heterogeneity between *rrn* copies in a genome like in the genus *Veillonella*. To avoid pitfalls of 16S rRNA gene sequencing, developing alternative molecular markers, like *dnaK* gene sequencing in the genus *Veillonella*, is still needed as well as constituting larger computerized databases for sequences of these markers. In a more global approach, the considerations presented herein could be extended to all bacterial species.

In 1987, rRNAs were coined by Woese as the ultimate molecular chronometers for determining the phylogenetic relationships between bacteria. Therefore, the primary aim of 16S rRNA gene sequencing was to assess bacterial evolution (with the building of phylogenetic trees). Ever since, other applications have been derived from 16S rDNA sequence knowledge such as identification tools (mainly probes and primers for polymerase chain reactions), description of uncultured bacterial ecosystems through the variety of 16S rDNA they contain or, in taxonomy, the definition of a new species on the sole basis of its 16S rRNA gene sequence being less than 99% similar to any known one. After a brief overview of 16S rRNA structure and its many uses, chapter XI will illustrate them by taking the genus *Bifidobacterium* as an example for the determination of a global phylogenetic position, phylogenetic studies of a genus and identification techniques derived from 16S rRNA sequences. Special attention will be devoted to a multiplex PCR approach developed for the identification of bifidobateria from human origin. This technique has successfully been applied to the identification of isolated bifidobacterial strains as well as to the direct detection of bifidobacterial human species in stool samples. The authors will also point out some of the limits that arose from 16S rRNA almost universal use, i.e. the existing *continua* within the genus *Bifidobacterium* (limits in the species definition) and the presence of multiple and non similar copies of 16S rRNA genes within a single bacterial cell.

RNA amplification is a series of somewhat complicated methods to amplify genetic signals from minute quantities of starting materials for complementary (cDNA) array analysis and other downstream genetic methodologies. Essentially, RNA amplification produces quantities of RNA through *in vitro* transcription (IVT). The present review will identify and provide historical perspective for each step in conventional and newly developed RNA amplification schemes, starting from isolating the RNAs to be amplified though the generation of double stranded (ds) cDNA template, to labeling the amplified RNA species for use in microarray analysis and other related experiments. A novel methodology has been developed in this laboratory to amplify RNA from minute amounts of starting material.

Briefly, an efficient means of second strand (ss) cDNA synthesis using a sequence-specific "terminal continuation" (TC) method is demonstrated. An RNA synthesis promoter is attached to the 3' and/or 5' region of cDNA utilizing the TC mechanism. The orientation of amplified RNAs is 'antisense', or a novel 'sense' orientation. TC RNA amplification is utilized for many downstream applications including gene expression profiling, microarray analysis, and cDNA library/subtraction library construction. Biological samples harvested for this procedure can originate from a myriad of *in vivo* and *in vitro* tissue sources. Moreover, a variety of fixations can be employed, and tissues can be processed for histochemistry and/or immunocytochemistry prior to microdissection for TC RNA amplification, allowing for tremendous cell type and tissue specificity of downstream genetic applications. A likely mechanism of action is that the TC primer binds preferentially to GC-rich, CpG islands flanking 5' regions of DNA that contain promoter sequences. In summary, chapter XII will focus on the development and practical usage of RNA amplification technologies, including T7-based RNA amplification methodologies and the mechanism(s) of action of each amplification step.

In: Trends in RNA Research
Editor: Peter A. McNamara, pp. 1-21

ISBN: 1-59454-506-5

Chapter I

Targeting RNA with Small Molecule Drugs

Guido J. R. Zaman and Paulus J. A. Michiels
N.V. Organon, Molecular Pharmacology, P.O. Box 20, 5340 BH Oss, the Netherlands

Abstract

Targeting RNA with small molecules is gaining interest in the drug discovery industry as an alternative approach to traditional drug discovery focusing on proteins. The targeting of bacterial 16S ribosomal RNA (rRNA) with aminoglycoside antibiotics provides a clear precedence for the targeting of RNA with small molecule drugs. The mechanism how aminoglycosides, through binding to rRNA, interfere with the fidelity of the translation machinery is understood in great detail, due to the resolution of the X-ray crystal structure of the 30S ribosomal subunit in complex with paromomycin. Binding of aminoglycosides to rRNA fragments is predominantly determined by electrostatic interactions: The negatively charged RNA backbone attracts the positively charged aminoglycosides. Binding is relatively non-specific. Aminoglycosides also bind to human 18S rRNA, with affinities that are only two- to six-fold lower than for binding to bacterial 16S rRNA. However, the specific anti-bacterial effect of aminoglycosides can also be subscribed to other mechanisms, such as the occurrence in prokaryotes of tranporter proteins that actively take up aminoglycosides. Aminoglycosides can suppress premature termination codons in human messenger RNAs (mRNAs), thus permitting translation to continue to the normal end of the transcript. This suppression activity of aminoglycosides is explored in drug discovery for the development of novel therapy for genetic disorders that are caused by stop codon mutations, such as cystic fibrosis and Duchenne muscular dystrophy. While aminoglycosides act on rRNA, certain small molecule metabolites, including vitamins, lysine and purines, can regulate gene expression by binding to mRNA. Binding occurs at defined target sites, which are called 'riboswitches'. Riboswitches consist of complex three-dimensional structures, which are located in the 5'-untranslated region (UTR) of bacterial and fungal mRNAs, coding for proteins involved in the uptake and biosynthesis of these metabolites. It is unknown whether riboswitches also occur in human cells. Riboswitches provide yet another proof that small molecules can influence biological events by binding directly to RNA. The targeting of human mRNAs with small synthetic chemical molecules is a relatively new approach, but is expected to create new and unique opportunities for drug development. Recently, it was shown that analogues of thalidomide, a sedative that has been withdrawn from the market in the early 1960s, can suppress the synthesis of the cytokine Tumor Necrosis Factor-α (TNFα) at the post-transcriptional level, through the involvement of sequences in the 3'-UTR of TNFα mRNA. Targeting RNA is in particular attractive for the development of new male contraceptives, as several RNA binding proteins and their cognate targets in mRNA have shown play critical roles in spermatogenesis. These and other opportunities of RNA as a target for small molecule drugs will be discussed in this chapter.

Why Consider RNA as a Drug Target?

Modern drug discovery, as it is carried out in larger and most mid-sized pharmaceutical companies, is a highly industrialized process, in which the chance of finding good drug leads is enhanced by screening high numbers of protein targets against large collections of synthetic chemical molecules. These compounds usually have a molecular weight < 500 Dalton. The drug discovery process is schematically represented in Figure 1.

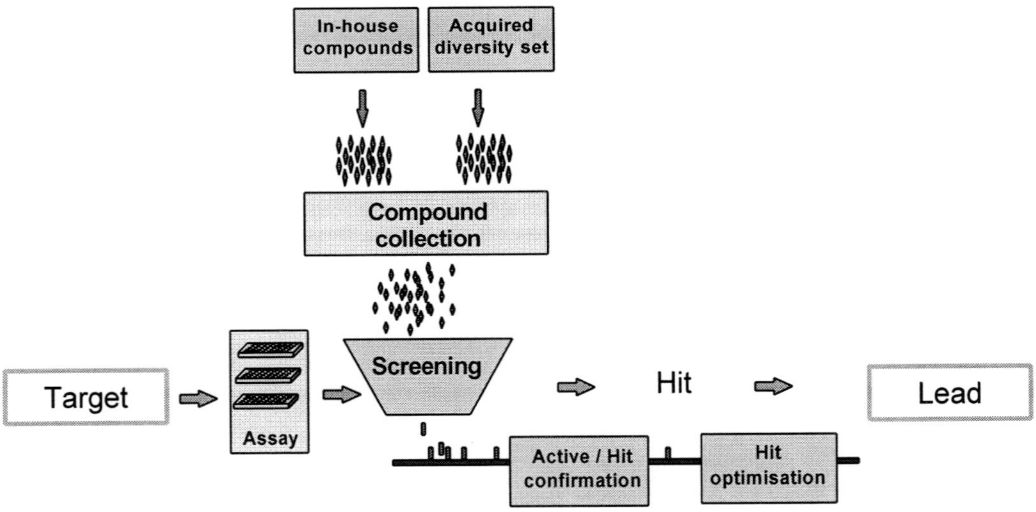

Figure 1. The discovery of novel lead compounds in a drug discovery company (N.V. Organon; www.organon.com).

To measure the activity of bio-molecular targets and their modulation with small molecules, biochemical and cellular assays are developed. These assays are automated and miniaturized, to allow screening at sufficiently high throughput (usually ≥ 50,000 compounds per day). Compound collections usually are composed of historical collections, *i.e.* compounds collected from previous projects, and compound files that have been acquired or synthesized specifically for the purpose of screening. These can be dedicated libraries around structures that are expected to yield high hit rates on certain target classes, or random libraries, covering a large proportion of the possible drug-like chemical structures that are available through commercial suppliers. High-throughput screening (HTS) initially yields wells containing active compounds. After confirmation of the structure of the compound, and the activity of the compound in the assay, the hits are optimized for potency and selectivity by repeated cycles of organic chemistry and molecular pharmacological testing, in a process called hit optimization. Finally, a lead compound is identified that fulfills certain pre-defined criteria with respect to potency, selectivity, structure-activity-relationship, and usually has shown activity in a proof-of-principle *in vivo* animal model.

To further reduce the risk of failure, many companies confine themselves more and more to certain families of targets that have proven to be particularly successful, as evidenced by the number of marketed medicines interacting with these protein classes. These 'druggable' protein families (Hopkins and Groom, 2002) include G-protein coupled receptors, nuclear receptors, ion channels, transporters and enzymes. Generic technologies to assay these druggable proteins are available, and greatly speed up the process from gene-to-screen (see *e.g.* Zaman *et al.*, 2003a). Since the start of high-throughput screening (HTS) in the early 1990s, this factory-like approach for drug discovery has yielded a limited number of new drugs and several tens of new drug candidates (Fox *et al.*, 2002) (Fig. 2). Sequencing of the human genome has revealed the DNA code of many new members of the druggable protein classes (Venter *et al.*, 2001). However, the human genome project has also identified many proteins that are not amenable to HTS, because assay technologies are not available, or because the biochemical function of these proteins are not yet known (Zaman *et al.*, 2003b) (Fig. 3). On the other hand, in more than half of all instances, the screening of druggable proteins does not deliver a good lead compound (Fox *et al.*, 2002). It is therefore not surprising that targeting at the RNA level is gaining interest in pharmaceutical industries as an alternative approach to traditional drug discovery focusing on proteins.

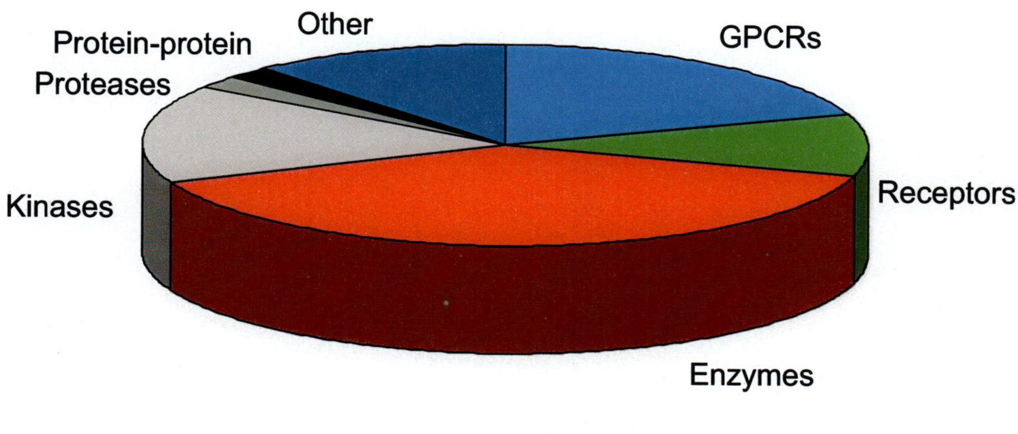

Figure 2. Since the start of HTS in the early 1990s, more than 60 lead compounds have been developed further to drug candidates that were actually tested in humans (Fox *et al.*, 2002). The figure shows the distribution of these novel drug candidates over the different druggable protein target classes.

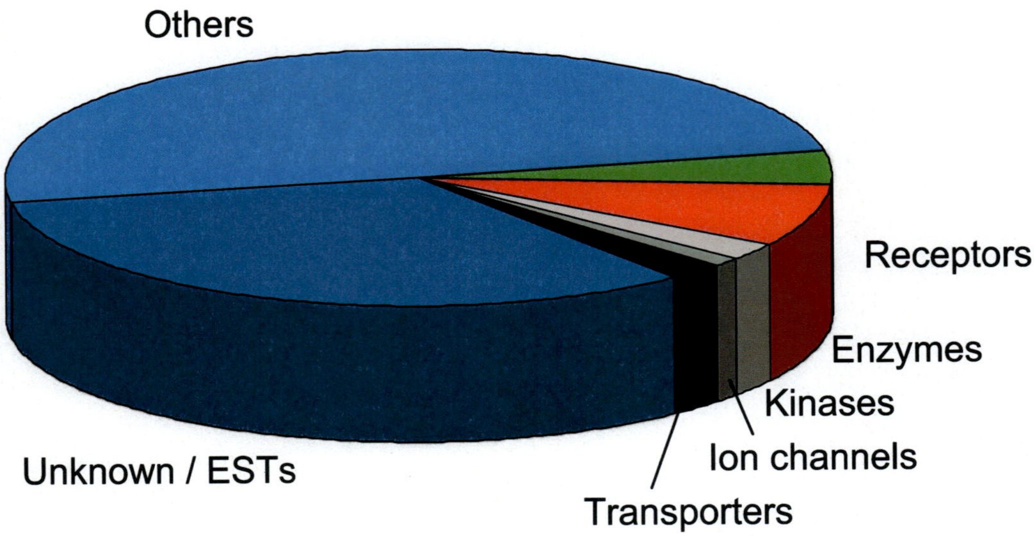

Figure 3. The major drug target classes (i.e. receptors, kinases, enzymes, ion channels and transporters) cover only 17 % of the human genome. The remaining part encodes proteins that are not amenable to HTS (50 %), such as structural proteins, or are genes with unknown functions or expressed sequence tags (ESTs) (33 %) (Venter *et al.*, 2001).

A characteristic feature of RNA is that it can fold into complex three-dimensional structures comprising loops, bulges, pseudoknots and turns (Fig. 4). These structures are responsible for the diverse actions of RNA molecules within cells. In this respect, RNA resembles more a protein than DNA, which is less flexible and has a less diverse tertiary structure. The unique shapes in various target RNAs create potential binding sites for small molecules. A property of messenger RNA (mRNA) that makes it an attractive target for drug discovery is that mRNA molecules are usually present in cells at relatively low copy numbers, as compared to proteins. mRNA is therefore highly sensitive to small changes in the concentrations of regulatory signals, such as proteins and small molecules.

In summary, targeting at the RNA level provides new opportunities to address proteins from druggable families, such as receptors and enzymes that have failed thus far to give good leads by HTS. RNA also creates strategies for drug discovery on targets that have thus far not been amenable to HTS.

Proof-of-Concept

Aminoglycoside Antibiotics

Proof-of-concept that RNA can function as a target for small molecule drugs has been provided by the aminoglycoside antibiotics, which bind to the RNA component of the small ribosomal subunit, the 16S ribosomal RNA (rRNA) (Moazed and Noller, 1987). Aminoglycosides are oligosaccharides containing several amino groups (Fig. 5). They exert their antimicrobial activity by inducing misreading of the genetic code. This results in the loss of functional proteins and the accumulation of abnormal proteins in the bacterial membrane (Davis *et al.,* 1986). This compromises the barrier function of the membrane and finally results in cell lysis (Davis *et al.,* 1986).

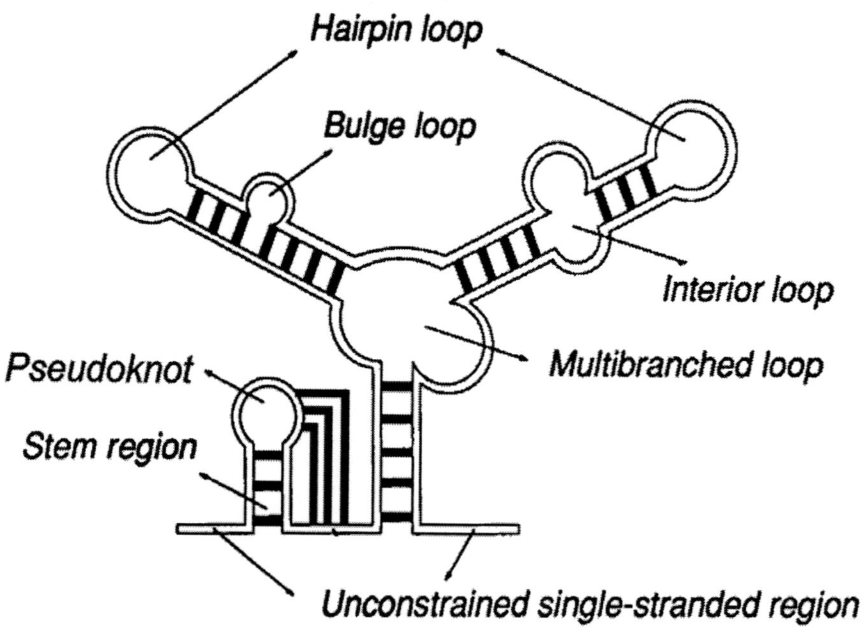

Figure 4. Possible secondary structures in RNA.

In 1943, streptomycin was the first member of the aminoglycoside family discovered to have antimicrobial activity against gram negative bacteria and *Mycobacterium tuberculosis*, the causative agent of tuberculosis. In contrast to the discovery of penicillin by Alexander Fleming in 1929, which was largely due to a matter of chance, the isolation of streptomycin was the result of a long-term, systematic effort by Selman A. Waksman and co-workers, who tested the antimicrobial activity of 10,000 different soil microbes for their antibacterial activity. In fact, streptomycin is the first drug that was discovered as the result of 'screening', as opposed to 'rational design'. Since this breakthrough discovery, which was awarded with the Nobel Prize to Dr Waksman in 1952, more than 150 aminoglycosides with highly diverse chemical structures have been isolated from the culture filtrates of bacteria. In addition, many semi-synthetic variants have been derived from these natural products, with the aim to develop more selective RNA targeting drugs, or to combat drug resistance.

Evidence for the involvement of the 16S rRNA in aminoglycoside action was provided by *in vitro* reconstitution experiments performed in the 1980s (Cundliffe, 1989). It was shown that aminoglycoside-resistant ribosomes could be constructed by combination of the ribosomal proteins of sensitive bacteria and the 16S rRNA of resistant bacterial strains. These resistant species, such as *Streptomyces* and *Micromonospora*, actually produce aminoglycosides. To prevent their own destruction, they express methylases that modify the 16S rRNA at specific bases in a region that is involved in the decoding of the mRNA (the decoding loop). This region is located in the A-site of the 30S ribosomal subunit. It was shown that the expression of 16S rRNA mutants in *Escherichia coli* could confer resistance to aminoglycosides, and that resistance in cells correlated with a decreased binding of aminoglycosides to the 16S rRNA mutants (De Stasio *et al.*, 1989). Direct binding of aminoglycosides to 16S rRNA was shown first in chemical foot printing experiments (Moazed and Zoller, 1987; De Stasio *et al.*, 1989). Since then, the binding of aminoglycosides to 16S rRNA has been demonstrated by various techniques, including nuclear magnetic resonance (NMR) (Fourmy *et al.*, 1996), surface plasmon resonance (Wong *et al.*, 1998), mass spectroscopy (Griffey *et al.*, 1999), and X-ray crystallography (Carter *et al.*, 2000).

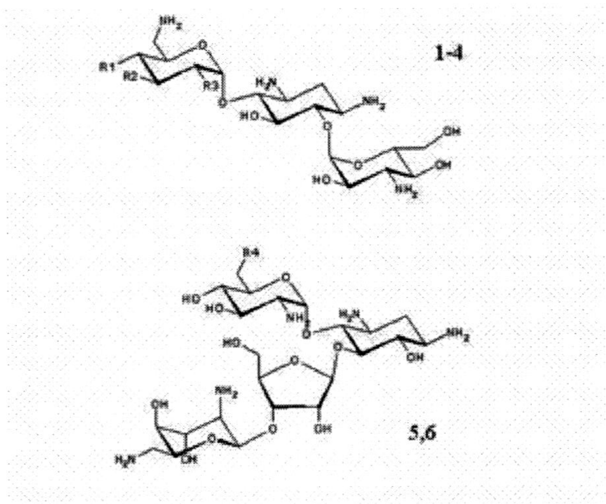

Antibiotic	R1	R2	R3	R4
1. Tobramyacin	OH	H	NHz	
2. Kanamycin A	OH	OH	NHz	
3. Kanamycin B	OH	OH	OH	
4. DIbekacin	H	H	NHz	
5. Neomycin				NHz
6. Paromomycin				OH

Figure 5. Structure of the aminoglycoside antibiotics tobramycin, kanamycin A and B, dibekacin, neomycin and paromomycin.

Riboswitches

A second proof that small molecules can influence biological events by binding to RNA is the recent discovery of 'riboswitches' in the 5'-UTR of certain bacterial mRNAs (Mironov *et al.*, 2002; Winkler *et al.*, 2002). Riboswitches function as high affinity binding sites for small molecule metabolites, such as vitamins (*e.g.* thiamine pyrophosphate and flavin mononucleotide), S-adenosylmethionine, lysine, guanine and adenine (Winkler and Breaker, 2003). These metabolites control their own synthesis by binding directly to mRNAs of genes involved the uptake and biosynthesis of these molecules. Binding of a ligand to the riboswitch induces a conformational change in the mRNA that occludes the Shine-Dalgarno sequence. In *Bacillus subtilus*, at least 69 genes are predicted to be under the control of riboswitches (Mandal *et al.*, 2003). This comprises approximately 2% of the total genome of this bacterium, and illustrates the extensive use of riboswitches for genetic control in prokaryotes. Riboswitches also occur in fungal mRNAs, and have been predicted to occur in the mRNA of certain lower eukaryotes and plants (Sudarsan *et al.,* 2003). However, most of the genes in which riboswitches have been found in prokaryotes are absent in the genome of 'higher' species, such as man. It is therefore unknown whether riboswitches, or similar mechanisms occur in human cells.

Principles of Aminoglycoside - RNA Interactions

Structure-Activity-Relationship Study

The study of the binding of aminoglycoside antibiotics to rRNA has revealed the basic principles of RNA – drug interactions. Small RNA oligonucleotides derived from the decoding A-site bind aminoglycosides similar to the whole ribosome (Purohit and Stern, 1994) (Fig. 6). Under physiological conditions, most amino functions on aminoglycosides are protonated and binding is primarily mediated through electrostatic interactions: The negatively charged phosphate backbone of the RNA attracts and binds the positively charged amino-groups of the aminoglycoside. Not surprisingly, there is a correlation between the number of protonated amino groups and the dissociation constant (Wong *et al.,* 1998; Verhelst *et al.*, 2004). We have used the binding of aminoglycoside antibiotics to the decoding region of rRNA as a model system to study the relationship between the affinity and the selectivity of binding of small molecules to RNA. As with compounds that address protein targets, compounds that act through RNA should bind either selectively or have a specific biological effect. The three-dimensional X-ray crystal structure of the 30S ribosome co-crystallized with paromomycin has shown that five sites in the neamine scaffold of the antibiotic have specific interactions with the ribosomal RNA (Fig. 7) (Carter *et al.*, 2000). In our studies, the binding of a set of six aminoglycoside antibiotics (Fig. 5) with either four (kanamycin A), five (tobramycin, kanamycin B, dibekacin, paromomycin), or six amino-groups (neomycin), to the decoding region A-site of bacterial 16S and human 18 S rRNA stem-loop structures (Fig. 6) were compared by means of surface plasmon resonance. At physiological pH, the pK_a of the amino-groups differ due to neighboring residues, such as hydroxyl-groups, within the aminoglycoside. Since the protonation state determines the affinity towards negatively charged RNA targets (Kaul *et al.*, 1997), the comparison between the aminoglycosides at physiological pH is complicated. Therefore, two buffers were used in the analysis, one at pH 7.4 and the other at pH 5.0. In the more acidic buffer of pH of 5.0, all amino groups are protonated. As can be seen in Table 1, the aminoglycosides bind with higher affinity to the 16S than to the 18S rRNA fragment at both pHs. The differences are two- to six-fold. The overall affinity of the aminoglycoside antibiotics for the two RNA fragments is determined by the number of amino-groups: Neomycin, comprising six amino-groups has the highest positive charge and binds with the highest affinity to both RNA fragments. The four aminoglycosides with five amino-groups, *i.e.* tobramycin, kanamycin B, dibekacin and paromomycin, have very similar K_d values. Lowest affinity was found for kanamycin A that contains only four amino-groups. These results show that there is almost a linear correlation between the number of (protonated) amino-groups and the dissociation constant. There are no significant differences in the affinities of the four aminoglycosides with five amino-groups, *i.e.* tobramycin, kanamycin B, dibekacin and paromomycin, but containing differences in the number of hydroxyl groups. Based on the X-ray crystal structure of the 30S rRNA and paromomycin (Carter *et al.*, 2000) (Fig. 6), the hydroxyl groups R1 and R2 can be predicted to interact directly with the rRNA.

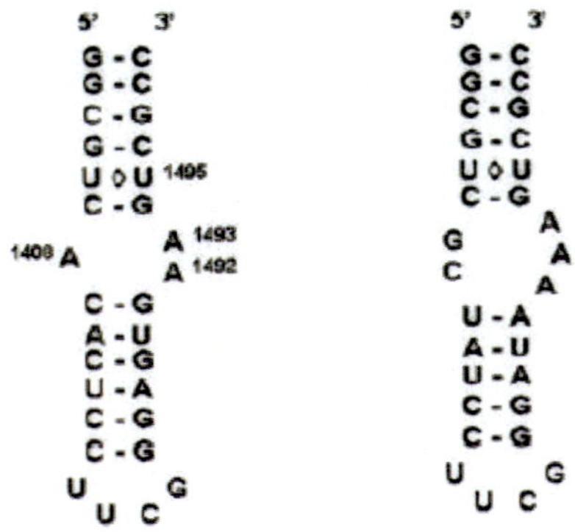

Figure 6. Target sites of aminoglycosides in bacterial 16S and human cytoplasmic 18S rRNA. 5'-biotinylated RNA fragments with these sequences were used in the surface plasmon resonance experiments of which the data are presented in Table 1.

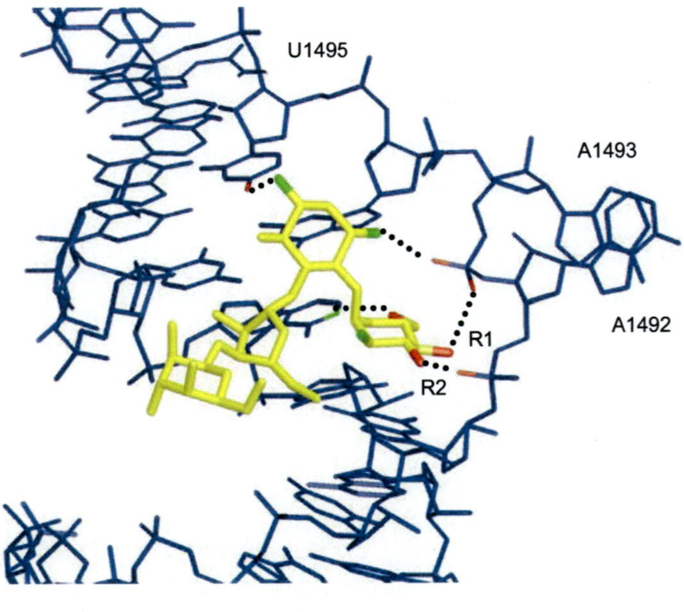

Figure 7. Model of the neomycin analogue paromomycin (yellow) complexed to the decoding A-site of 16S rRNA (shown in blue), based on the X-ray crystal published by Carter *et al.* (2000). Some nitrogen and oxygen atoms of paromomycin and the RNA are color coded in green and red, respectively. Out of several hydrogen bonds between the antibiotic and RNA, five are marked with dashed lines. The picture was created with the program MOLMOL (Koradi *et al.*, 1996).

Apparently they do not, however, contribute significantly to the affinity of binding (Table 1). Therefore, there is also no structure-activity-relationship observed for the hydroxyl-groups of tobramycin, kanamycin B and dibekacin, which differ at position R1 and R2 (Fig. 5).

Table 1. Affinities of aminoglycoside antibiotics for binding to 16S or 18S rRNA fragments at pH 7.4 and pH 5.0, as measured by surface plasmon resonance.*

aminoglycoside	amino groups	16S Kd (μM) pH 7.4	18S Kd (μM) pH 7.4	18S Kd (μM) pH 5.0	18S Kd (μM) pH 5.0
kanamycin A	4	70 ± 21	110 ± 13	2.9 ± 0.43	6.6 ± 0.57
kanamycin B	5	5.2 ± 0.66	25 ± 18	0.33 ± 0.11	0.79 ± 5.9
tobramycin	5	9.0 ± 1.0	32 ± 11	0.55 ± 0.27	0.97 ± 1.2
dibekacin	5	5.8 ± 0.67	20 ± 6.2	0.42 ± 0.11	0.75 ± 0.20
paromomycin	5	5.0 ± 0.61	31 ± 8.6	0.53 ± 0.17	1.4 ± 2.1
neomycin	6	0.32 ± 0.031	1.2 ± 0.34	0.12 ± 0.15	0.42 ± 0.014

* Surface plasmon resonance measurements were conducted on a BIAcore 3000 system from Biacore AB, essentially as described by Alper *et al.* (1998). Aminoglycosides were dissolved at concentrations ranging from 2 nM to 100 μM in autoclaved buffer with pH 7.4 (10 mM HEPES, 150 mM NaCl, 3 mM EDTA, 0.005 % surfactant P20) or pH 5.0 (10 mM NaAc pH 5.0, 150 mM NaCl, 3 mM EDTA). The two 5'-biotin-labeled RNA fragments, derived from the 16S and 18S rRNA decoding A-site (Fig. 6), were immobilized on a streptavidin-coated sensor chip at approximately 1000 response units (RU). Double referencing was applied to correct for the dextran matrix and to correct for buffer effects.

Mechanism of Drug Action and Side Effects

In view of the success of the aminoglycosides as antibacterial agents for over half a century, it may seem odd that these compounds also bind to human rRNA with affinities that are only two- to six-fold lower than for bacterial rRNA (Table 1). It should be noted, however, that aminoglycoside therapy can cause severe side-effects. Of these, renal toxicity is the most common. Despite changes in treatment regimens and monitoring procedures, still 10 % of patients suffer from this side-effect. The second main side-effect is ear toxicity, which can result in reversible or irreversible deafness. Both side-effects are explained by the specific accumulation of aminoglycosides in renal tubular epithelial cells or in hair cells of the inner ear, which express specific membrane receptors for aminoglycosides (Schmitz *et al.*, 2002).

A point mutation in the 12S rRNA of human mitochondrial ribosomes has been associated with aminoglycoside hypersensitivity (Hutchin and Cortopassi, 1994) (Fig. 8). It has been reported that a fluorescently-labeled aminoglycoside bound to an RNA fragment containing the mutation, but not to a fragment containing the wild-type mitochondrial RNA sequence (Ryu and Rando, 2002). In contrast, by using surface plasmon resonance, measuring the direct binding of various aminoglycosides to the two RNA fragments, we could not confirm these differences in affinity (data not shown). Both RNA variants were bound with affinities similar to those of human cytoplasmic and bacterial rRNA fragments (Table 1).

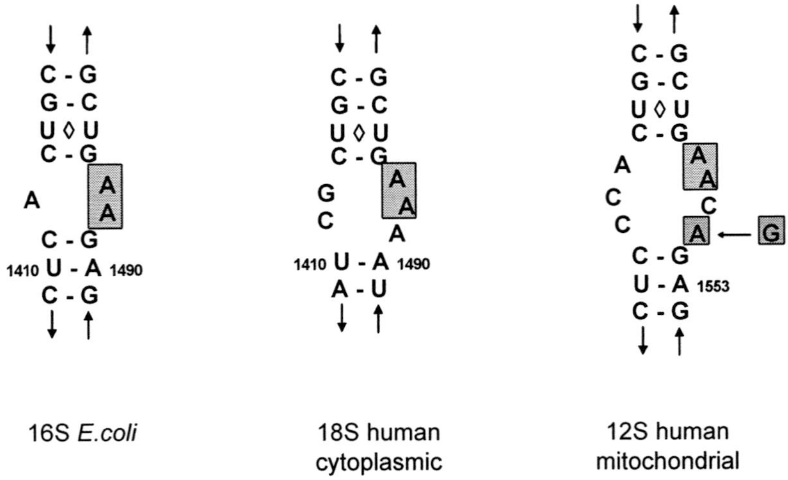

Fig. 8. Comparison of the sequences of the aminoglycoside target sites in different rRNA species. The universally conserved adenine bases, located at positions 1492 and 1493 of the Escherichia coli 16S rRNA are highlighted in yellow. Mutation of the adenine at position 1555 of the human 12S rRNA to guanine is associated with aminoglycoside hypersensitivity. This base and the substituent are highlighted in blue.

There are two main factors that determine the selective cytotoxic action of aminoglycosides on bacterial cells. The first is the sequence differences between bacterial and human rRNA. This is illustrated by the fact that a 'pseudo-eukaryotic' rRNA can be created by mutation of *Escherichia coli* rRNA at one specific residue in the A-site (Recht *et al.*, 1999). Expression of this 'pseudo-eukaryotic' rRNA renders bacteria 100 times more resistant to neomycin. The second important factor is the occurrence in prokaryotes of transporter proteins that actively take up and concentrate aminoglycosides in the cytoplasm.

In conclusion, while aminoglycoside antibiotics act by binding to rRNA, both their therapeutic and their side-effects are determined by other factors as well, such as specific drug transport mechanisms. Aminoglycosides are not the only antibiotics that act by interfering with the translation machinery of the cell, the ribosome. For instance, macrolide antibiotics inhibit protein synthesis by binding to the 50S ribosomal subunit, where the drugs interact with bases in the peptidyl transferase loop of 23S rRNA (Rodriguez-Fonseca *et al.*, 1995). Tetracyclines bind to the 30S ribosomal subunit where they inhibit the binding of aminoacylated tRNA to the A-site (Brodersen *et al.*, 2000).

The Necessity of a Switch

Aminoglycosides

A model to explain the induction of translational misreading by aminoglycosides has been inferred from the crystal structure of the 30S ribosomal subunit in complex with a 3 nucleotide piece of mRNA and paromomycin (Carter *et al.*, 2000). Critical in this model are two universally conserved adenine residues, located at positions 1492 and 1493 in the bacterial rRNA (Fig. 8). In the process of decoding of the mRNA, A1492 and A1493 will protrude from the RNA helix, where they previously were stacked within the helical context, to interact with the codon-anticodon duplex formed between cognate aminoacyl-tRNA and mRNA. There is equilibrium between the two states. Paromomycin shifts the equilibrium to the structure where the two adenines are bulged out. This results in an increased error rate of the ribosome, and eventually cell death.

Riboswitches

The mechanism of action of riboswitches is in fact very similar to that of aminoglycosides acting on rRNA. In the case of the riboswitch, a small molecule, such as flavin mononucleotide (FMN) or thiamine pyrophosphate (TTP), causes a conformational change in the structure of the 5'-UTR, allowing a hairpin to be formed by which mRNA transcription is interrupted (Winkler and Breaker, 2003). For FMN and TTP almost 1000-fold differences in binding constants were observed compared to riboflavine and thiamine, which lack the phosphate groups (Winkler *et al.*, 2002; Winkler and Breaker, 2003).

Opportunities for Drug Discovery

Antibacterials

Bacterial rRNA is a target that is still being explored in drug discovery today. Research is ongoing to design novel rRNA-targeting drugs that are active against a broad spectrum of pathogenic bacteria (Dutton, 2000). These drugs may be used in the defense of bioterrorism attacks. To look for new targets for broad-spectrum antibiotics, the rRNA genes of many bacteria are searched for conserved secondary structures (Carter *et al.*, 2001). It is assumed that the more frequently a predicted structure is found across pathogenic bacterial species, the more likely it is real, and the better it should be as a drug target for broad-spectrum antibiotics.

The recent discovery of riboswitches creates entirely new opportunities for the development of novel anti-microbials (Winkler and Breaker, 2003). Since many of the genes in bacteria that are regulated by riboswitches are involved in fundamental biochemical processes, metabolite analogs that interfere with riboswitch function can be expected to impair vital metabolic functions and cause cessation of growth. If the phylogenetic distribution of riboswitches indeed is limited to prokaryotes, riboswitches are attractive targets for the development of specific bacterial anti-metabolites.

Human Genetic Disorders

Not only bacterial, but also human rRNA is being explored as a target for drug discovery. The antibiotics G-418 and gentamicin have been shown to cause read-through of premature stop codons in genes involved in human genetic disorders, such as cystic fibrosis (Howard *et al.*, 1996; Bedwell *et al.*, 1997), Duchenne muscular dystrophy (Barton-Davis *et al.*, 1999), Hurler syndrome (a lysosomal storage disease) (Keeling *et al.*, 2001), cystinosis (Helip-Wooley *et al.*, 2002), late infantile neuronal ceroid lipofuscinosis (Sleat *et al.*, 2001), and disorders involving the p53 gene (Keeling *et al.*, 2002). Read-through results in increased levels of full-length proteins. Topical administration of gentamicin to the nasal epithelium of patients with cystic fibrosis resulted in the expression of functional cystic fibrosis transmembrane conductance regulator channels (Wilschanski *et al.*, 2003). Promising data were also obtained with a mouse model of Duchenne's muscular dystrophy: Aminoglycosides increased the expression of full-length protein from a dystrophin allele carrying a stop codon mutation (Barton-Davis *et al.*, 1999). However, these results could not be reproduced in patients with molecular dystrophy (Wagner *et al.*, 2001).

AIDS

Besides rRNA, aminoglycosides can bind in a saturable fashion and with similar micro molar affinities to a variety of other RNA structures containing non-duplex elements. Aminoglycosides have been used as lead compounds for the development of more selective RNA targeting drugs. In human immunodeficiency virus-1 (HIV) RNA, there are two structures that are bound by aminoglycosides: *i.e.* the Rev Protein response element (RRE) and the transactivation-responsive element (TAR) (Fig. 9). RRE is the binding site of the viral Rev protein in the *envelope* gene of HIV, and plays a role in the transport of unspliced viral RNA from the nucleus to the cytoplasm. The TAR RNA hairpin is present in all HIV mRNAs and plays and important role in the viral replication cycle. TAR is bound by the viral Tat protein, which is essential for transcriptional activation and elongation of HIV mRNA synthesis. This protein – RNA interaction has been recognized as an important target for the development of new therapies against AIDS and has been studied extensively (Krebs *et al.*, 2003). A number of molecules, varying from heterocyclic compounds, peptides and peptidomimetics have been reported to interfere with the Tat – TAR interaction (Wilson and Li, 2000; Du *et al.*, 2002). Some of these have been shown to inhibit HIV replication in cells (Mei *et al.*, 1998).

Targets in Human mRNAs

Targeting human mRNAs with small molecule compounds for drug discovery is a relatively new approach. Tertiary structures in mRNA, such as hairpins or pseudoknots, are expected to provide more selective binding sites than single-stranded regions or stable double-stranded domains (Zaman *et al.*, 2003b).

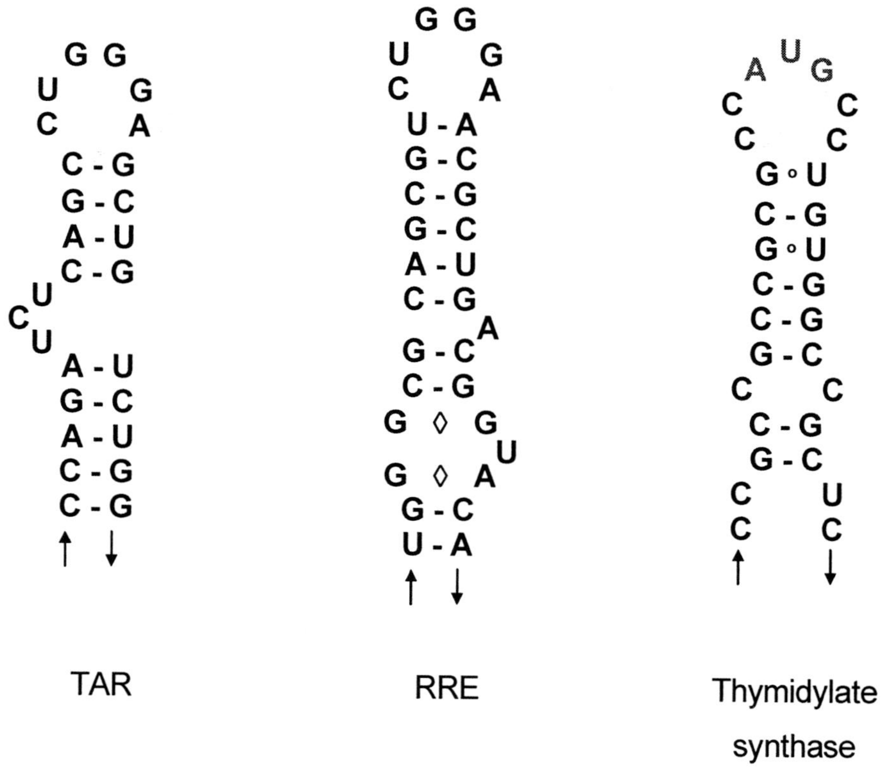

Fig. 9. Target sites of aminoglycosides in eukaryotic viral and cellular mRNAs. The TAR hairpin in HIV mRNA is bound by the viral Tat protein. The RRE hairpin in HIV mRNA is the target site of the viral Rev protein. The stem-loop structure in the thymidylate synthase mRNA includes the start codon, which is indicated in red.

Although the structure of mRNA is considered to be less complex than that of rRNA, there are a few well-defined secondary structure elements in mRNA and drug target sites described in literature. One is the autoregulatory stem-loop structure in the 5' end of the thymidylate synthase mRNA (Tok *et al.*, 1999) (Fig. 9). Thymidylate synthase regulates its own synthesis by binding to and stabilizing this stem-loop structure, which includes the AUG start codon. Aminoglycosides have been shown to bind to this stem-loop structure with affinities of approximately 1 μM affinity (Tok *et al.*, 1999). Thymidylate synthase is the chemotherapeutic target of the anti-cancer drug 5-fluorouracil and an example of an established drug target that is being re-addressed at the mRNA level.

Another example of a well-defined secondary structure in mRNA is the iron response element (IRE) (Fig. 10), which is contained in several mRNAs involved in iron homeostasis (Theil, 2000). At low cytosolic concentrations of iron, the IRE is bound by the iron sensor protein aconitase. When the concentration of iron in the cell increases, the protein dissociates from the IRE. In some mRNAs, such as the mRNA that codes for the iron-storage protein ferritin, the IRE is located in the 5' leader. The IRE functions here as a translational silencer. In other mRNAs, such as the mRNA that codes for the transferrin receptor, the IRE is located in the 3' untranslated region. Here the function of the IRE is to increase the stability of the mRNA.

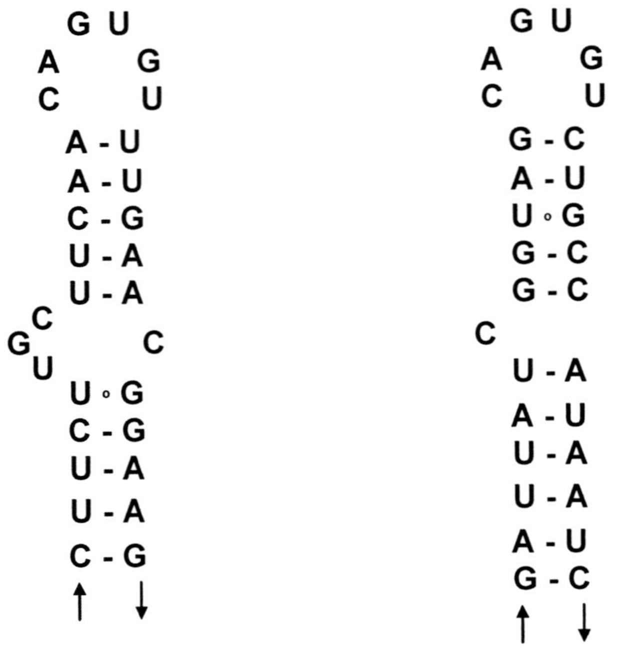

Ferritin

Transferrin receptor

Fig. 10. Iron response element in the 5' leader of ferritin mRNA and 3'-UTR of transferrin receptor mRNA. Ferritin is an iron-storage protein. The transferrin receptor is involved in the cellular uptake of iron. At low cytosolic concentrations of iron, aconitase binds to the IRE in both the ferritin and the transferring receptor mRNA. This results in inhibition of ferritin translation. At the same time, the stability, and consequently the translation, of the transferring receptor mRNA is increased. In response to increased levels of iron, aconitase dissociates from both mRNAs. This results in increased synthesis of ferritin and iron storage, while the synthesis of the transferring receptor and the cellular uptake of iron are decreased.

At low cytosolic concentrations of iron, aconitase binds to the IRE in both the ferritin and transferrin receptor mRNA. This results in inhibition of ferritin translation. At the same time, the stability, and consequently the translation, of the transferrin mRNA is increased. In response to increased levels of iron, aconitase dissociates from both mRNAs. This results in increased synthesis of ferritin and iron storage, while the synthesis of the transferrin receptor and the cellular uptake of iron are decreased. Recently, IRE has been identified in the 5'-UTR of the Alzheimer's β-amyloid precursor protein (APP) (Rogers *et al.*, 2002a). APP is the precursor protein of the main component of the neurological amyloid plaques that are formed during the progression of Alzheimer's disease. Reduction of the intracellular concentration of iron using chelators resulted in a decrease of APP 5'-UTR controlled translation (Rogers *et al.*, 2002a). A cell-based assay consisting of a luciferase reporter whose translation is driven by the 5'-UTR of APP mRNA, has been used to identify APP inhibitors by screening (Rogers *et al.*, 2002b). Several compounds with activities in the (sub)millimolar range have been identified (Rogers *et al.*, 2002b).

Tumor necrosis factor-α (TNFα) is a pro-inflammatory cytokine involved in various immune diseases. Like other cytokines and proto-oncogene, TNFα is regulated at the post-transcriptional level through adenylate/uridylate (AU)-rich elements (AREs) in the 3'-UTR of its mRNA (Kruys *et al.*, 1989). Several proteins have been identified that bind AREs to mediate RNA decay, and hence translational efficiency (Kruys and Huez, 1994; Chen and Shyu, 1995). It has been shown that the synthesis of TNFα can be inhibited by thalidomide (Sampaio *et al.*, 1991). Thalidomide is a glutamic acid derivative that was introduced as a sedative hypnotic in 1956, but was withdrawn in 1961, due to the development of grove congenital abnormalities in babies born to mothers using it for treating morning sickness. The compound was re-introduced as a therapeutic for leprosy and more recently has demonstrated potency in the treatment of a variety of cancers (Zhu *et al.*, 2003). The initial mechanism underpinning the action of this compound was shown to be inhibition of TNFα protein expression and it was further demonstrated to act at the post-transcriptional level to facilitate turn-over of the mRNA (Moreira *et al.*, 1993). The action of thalidomide to lower TNFα levels is not particularly potent, but analogues have been synthesized that inhibit TNFα production in peripheral blood monocytes in the low μM range (Zhu *et al.*, 2003). By using luciferase reporter assays, it was shown that the thalidomide analogs act through the ARE-containing sequences in the 5'-UTR of the TNFα mRNA (Zhu *et al.*, 2003).

While targeting RNA creates unique opportunities for drug discovery in many therapeutic areas, it may be particularly attractive for the development of new male contraceptives. Spermatogenesis is a strictly regulated cellular development process, in which several RNA binding proteins play a critical role (Venables and Eperon, 1999). This is evidenced from genetic deficiencies that cause male sterility. For instance, deletions in the Y chromosome encompassing the genes coding for the RNA binding proteins RBM (Elliot *et al.*, 1997) or DAZ (Reijo *et al.*, 1995; 1996) correlate with severe impairment of spermatogenesis. Another RNA binding protein that is involved in post-transcriptional control, and that is expressed in elongating spermatides is TRBP (Siffroi *et al.*, 2001). The abbreviation TRBP stands for *T*AR *R*NA *b*inding *p*rotein, as it was originally identified by its ability to activate HIV TAR-dependent translation (Gatignol *et al.,* 1991). The murine homologue of TRBP is the *P*rotamine *R*NA *b*inding *p*rotein (Prbp), which binds to a structured element in the 3'-UTR of *protamine-1* mRNA (Lee *et al.,* 1996) (Fig. 11). Protamines mediate the compaction of nuclear DNA during spermatogenesis. After synthesis, the *protamine-1* mRNAs are packaged as cytoplasmic ribonucleoprotein particles (RNPs). By means of this protein-RNA complex, *protamine-1* mRNA is kept translationally silent for several days, which is essential for normal completion of spermatogenesis. When in transgenic mice the 3'-UTR of *protamine-1* mRNA is deleted, premature translation of the protamine is observed, which leads to spermatogenic arrest (Lee *et al.*, 1995). Deletion studies have shown that a repeated sequence in the 3'-UTR is essential in the proper timing of protamine-1 mRNA translation (Fajardo et al., 1997; Zhong et al., 2001). The interaction of Prbp/TRBP with the 3'-UTR of *protamine-1* mRNA is an interesting target for the discovery of new male contraceptives (Zaman *et al.*, 2003b).

Outlook

The targeting of human mRNAs with small synthetic chemical molecules is a relatively new approach, but is expected to create new and unique opportunities for drug development.

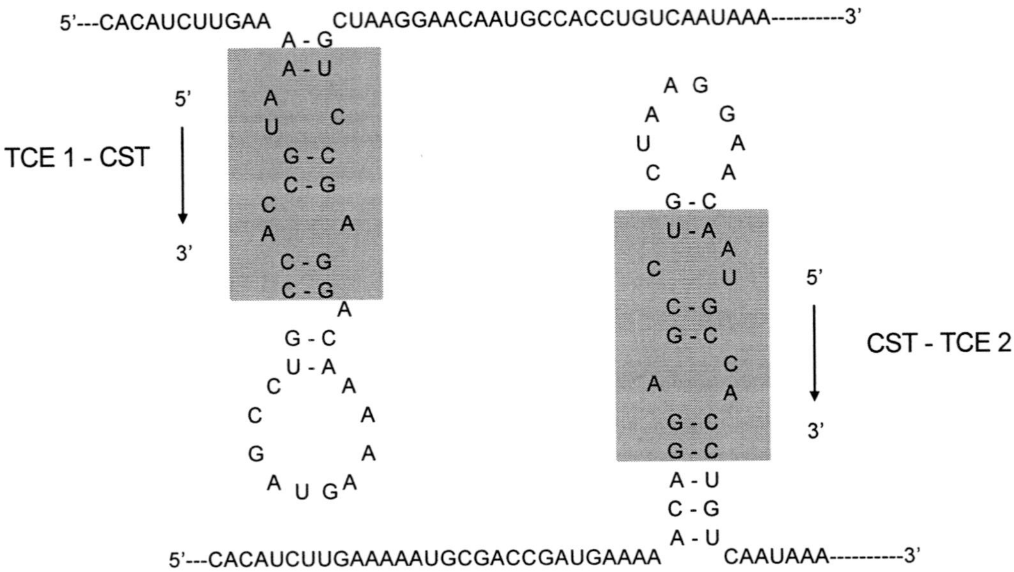

Fig. 11. Target site of Prbp in *protamine-1* mRNA. The 3'-UTR of *protamine-1* mRNA contains a repeated sequence that can fold into two different secondary structures (shaded area) (from Lee *et al.*, 1996).

By using mRNA as a target, it will be possible to target any gene, including those that encode proteins that are not amenable to HTS, but for which clear associations with disease processes follow from (molecular) genetic or pharmacological investigations. Studies on the mechanism of binding of aminoglycosides to rRNA have outlined the basic principles RNA – drug interactions. Binding of aminoglycosides to RNA is relatively non-specific. We believe that in many reports the specificity of binding of aminoglycosides to 16S rRNA is overemphasized. The ribosomal context and the occurrence of selective uptake mechanisms in bacterial cells, however, also account for the antibiotic action of aminoglycosides. The number of new potential targets for drug discovery at the RNA level increases gradually as more functional information on the human genome becomes available. Methodologies to discover small molecules that bind to RNA targets by HTS are based on the same principles and read-out technologies that are applied for drug discovery on protein targets (Zaman *et al.*, 2003b). Usually, the RNA molecules are radiolabeled or labeled with fluorescent probes. RNA binding proteins (Zaman *et al.*, 2003b), or allosteric ribozymes (Soukup and Breaker, 1999) can be used to monitor the binding of small molecules. The screening of large chemical libraries for compounds that fortuitously bind to RNA has, however, not delivered new drug leads that have entered clinical trials. It may be that this approach is not the most optimal. Metabolites on riboswitches and the aminoglycoside antibiotics acting on the ribosome demonstrate that potential drug acting at the RNA level, not necessarily has to bind to an "active" site, or compete with a protein binding site. It may as well bind elsewhere and induce a conformational change, thereby altering the expression of a target protein and having a therapeutic effect. In the search for novel antibiotics it may be more important to identify compounds that are able to shift the equilibrium in the two states of the rRNA towards the structure in which the two adenines are flipped out, then to select compounds on affinity of binding. The same applies to the search of small molecule drugs that target human mRNA. Recently, some successes have been reported in the identification of compounds that inhibit the synthesis of APP or TNFα by applying cell-based reporter assays (Rogers *et al.*, 2002b; Zhu *et al.*, 2003). Although the approach of using more functional and cell-based assays requires extensive knowledge about the mechanism of action, it is of major importance in the strategy for developing new drugs that target RNA.

References

Alper, P.B., Hendrix, M., Sears, P., and Wong, C.-H. (1998) Probing the specificity of aminoglycoside – ribosomal RNA interactions with designed synthetic analogs. *J. Am. Chem. Soc.* 120, 1965-1978.

Barton-Davis, E.R. Cordier, L., Shoturma, D.I., Leland, S.E., and Sweeney, H.L. (1999) Aminoglycoside antibiotics restore dystrophin function to skeletal muscles of mdx mice. *J. Clin. Invest.* 104, 375-381.

Bedwell, D.M., Kaenjak, A., Benos, D.J., Bebok, Z., Bubien, J.K., Hong, J., Tousson, A., Clancy, J.P., and Sorscher, E.J. (1997) Suppression of a CFTR premature stop mutation in a bronchial epithelial cell line. *Nature Med.* 2, 467-469.

Brodersen, D.E., Clemons, Jr., W.M., Carter, A.P., Morgan-Warren, R.J., Wimberly, B.T., and Ramakrishnan (2000) The structural basis of the action of the antibiotics tetracycline, pactamycin, and hygromycin B on the 30S ribosomal subunit. *Cell* 103, 1143-1154.

Carter, A.P., Clemons, W.M., Brodersen, D.E., Morgan-Warren, R.J., Wimberly, B.T., and Ramakrishnan, V. (2000) Functional insights from the structure of the 30S ribosomal subunit and its interaction with antibiotics. *Nature* 407, 340-348.

Carter, R.J., Dubchak, I., and Holbrook, S.R. (2001) A computational approach to identify genes for functional RNAs in genomic sequences. *Nucl. Acids Res.* 29, 3928-3938.

Chen, C.-Y., and Shyu, A.-B. (1995) AU-rich elements: Characterization and importance in mRNA degradation. *Trends Biol. Sci.* 20, 465-470.

Cundliffe, E. (1989) How antibiotic-producing organisms avoid suicide. *Annu. Rev. Microbiol.* 43, 207-233.

Davis, B.D., Chen, L., and Tai, P.C. (1986) Misread protein creates membrane channels: An essential step in the bactericidal action of aminoglycosides. *Proc. Natl. Acad. Sci. USA* 96, 10129-10133.

De Stasio, E.A., Moazed, D., Noller, H.F., and Dahlberg, A.E. (1989) Mutations in 16S ribosomal RNA disrupt antibiotic – RNA interactions. *EMBO J.* 8, 1213-1216.

Du, Z., Lind, K.E., and James, T.L. (2002) Structure of TAR RNA complexed with a Tat-TAR interaction nanomolar inhibitor that was identified by computational screening. *Chem. and Biol.* 9, 707-712.

Dutton, G. (2000) Biotech counters bioterrorism. *Gen. Engineering* 20, No. 21, December 2000.

Elliott, D.J., Millar, M.R., Oghene, K., Ross, A., Kiesewetter, F., Pryor, J., McIntyre, M., Hargreave, T.B., Saunders, P., Vogt, P.H., Chandley A.C., and Cooke H. (1997) Expression of RBM in the nuclei of human germ cells is dependent on a critical region of the Y chromosome arm. *Proc. Natl. Acad. Sci. USA* 94, 3848-3853.

Fajardo, M.A., Haugen, H.S., Clegg, C.H., and Braun, R.E. (1997) Separate elements in the 3' untranslated region of the mouse Protamine 1 mRNA regulate translational repression and activation during murine spermatogenesis. *Dev. Biol.* 191, 42-52.

Fourmy, D., Recht, M.I., Blanchard, S.C., and Puglisi, J.D. (1996) Structure of the A-site of *E. coli* 16S ribosomal RNA complexed with an aminoglycoside antibiotic. *Science* 274, 1367-1371.

Fox, S., Wang, H.S.L., Sopchak, L., and Farr-Jones, S.J. (2002) High-throughput screening 2002: moving toward increased success rates. *J. Biomol. Screen.* 7, 313-316.

Gatignol, A., Buckler-White, B., Berkhout, B., and Jeang, K.-T. (1991) Characterization of a human TAR RNA-binding protein that activates the HIV-1 LTR. *Science* 251, 1597-1600.

Griffey, R.H., Hofstadler, S.A., Sannes-Lowery, K.A., Ecker, D.J., and Crooke, S.T. (1999) Determinants of aminoglycoside-binding specificity for rRNA by using mass spectrometry. *Proc. Natl. Acad. Sci. USA* 96, 10129-10133.

Helip-Wooley, A., Park, M.A., Lemons, R.M., and Thoene, J.G. (2002) Expression of CTNS alleles: subcellular localization and aminoglycoside correction in vitro. *Mol. Genet. Metab.* 75, 128-133.

Hopkins, A.L., and Groom, C.R. (2002) The druggable genome. *Nature Reviews Drug Discovery* 1, 727-730.

Howard, M., Frizzell, R.A., and Bedwell, D.M. (1996) Aminoglycoside antibiotics restore CFTR function by overcoming premature stop mutations. *Nature Med.* 2, 467-469.

Hutchin, T., and Cortopassi, G. (1994) Proposed molecular and cellular mechanism for aminoglycoside ototoxicity. *Antimicrob. Agents Chemother.* 38, 2517-2520.

Kaul, M., Barbieri, C.M., Kerrigan, J.E., and Pilch, D.S. (2003) Coupling of drug protonation to the specific binding of aminoglycosides to the A site of 16 S rRNA: Elucidation of the number of drug amino groups involved and their identities. *J. Mol. Biol.* 326, 1373-1387.

Keeling, K.M., Brooks, D.A., Hopwood, J.J., Li, P., Thompson, J.N., and Bedwell, D.M. (2001) Gentamicin-mediated suppression of Hurler syndrome stop mutations restores a low level of α-L-uronidase activity and reduces lysosomal glycosaminoglycan accumulation. *Hum. Mol. Genet.* 10, 291-299.

Keeling, K.M., and Bedwell, D.M. (2002) Clinically relevant aminoglycosides can suppress disease-associated premature stop mutations in the IDUA and P53 cDNAs in a mammalian translation system. *J. Mol. Med.* 80, 367-376.

Koradi, R. Billeter M., and Wuthrich, K. (1996) MOLMOL: a program for display and analysis of macromolecular structures. *J. Mol. Graph.* 14, 51-55.

Krebs, A., Ludwig, V., Boden, O., and Göbel, M.W. (2003) Targeting the HIV trans-activation responsive region: Approaches towards RNA-binding drugs. *ChemBioChem* 4, 972-978.

Kruys, V., Marinx, O., Shaw, G., Deschamps, J., and Huez, G. (1989) Translational blockade imposed by cytokine-derived UA-rich sequences. *Science* 245, 852-855.

Kruys, V., and Huez, G. (1994) Translational control of cytokine expression by 3' AU-rich sequences. *Biochemie* 76, 862-866.

Lee, K., Haugen, H.S., Clegg, C.H., and Braun, R.E. (1995) Premature translation of protamine 1 mRNA causes precocious nuclear condensation and arrests spermatid differentiation in mice. *Proc. Natl. Acad. Sci. USA* 92, 12451-12455.

Lee, K., Fajardo, M.A., and Braun, R.E. (1996) A testis cytoplasmic RNA-binding protein that has the properties of a translational repressor. *Mol. Cell. Biol.* 16, 3023-3034.

Mandal, M., Boese, B., Barrick, J.E., Winkler, W.C., and Breaker, R.R. (2003) Riboswitches control fundamental biochemical pathways in *Bacillus subtilus* and other bacteria. *Cell* 113, 577-586.

Mei, H.-Y., Cui, M., Heldsinger, A. , Lemrow, S.M., Loo, J.A., Sannes-Lowery, K.A., Sharmeen, L., and Czarnik, A.W. (1998) Inhibitors of protein-RNA complexation that target the RNA: Specific recognition of human immunodeficiency virus type 1 TAR RNA by small organic molecules. *Biochemistry* 37, 14204-14212.

Mironov, A.S., Gusarov, I., Rafikov, R., Lopez, L.E., Shatalin, K., Kreneva, R.A., Perumov, D.A., and Nudler, E. (2002) Sensing small molecules by nascent RNA: A mechanism to control transcription in bacteria. *Cell* 111, 747-756.

Moazed, D., and Noller, H.F. (1987) Interaction of antibiotics with functional sites in 16S ribosomal RNA. *Nature* 327, 389-394.

Moreira, A.L., Sampaio, E.P., Zmuidzinas, A., Frindt, P., Smith, K.A., Kaplan, G. (1993) Thalidomide exerts its inhibitory action on tumor necrosis factor α by enhancing mRNA degradation. *J. Exp. Med.* 177, 1675-1680.

Nahvi, A., Sudarsan, N., Ebert, M.S., Zou, X., Brown, K.L., and Breaker, R.R. (2002) Genetic control by a metabolite binding mRNA. *Chem. Biol.* 9, 1043-1049.

Purohit, P., and Stern, S. (1994) Interactions of a small RNA with antibiotic and RNA ligands of the 30S subunit. *Nature* 370, 659-662.

Recht, M.I., Douthwaite, S., and Puglisi, J.D. (1999) Basis for prokaryotic specificity of action of aminoglycoside antibiotics. *EMBO J.* 18, 3133-3138.

Reijo, R., Lee, T.Y., Salo, P., Alagappan, R., Brown, L.G., Rosenberg, M., Rozen, S., Jaffe, T., Straus, D., Hovatta, O. *et al.* (1995) Diverse spermatogenic defects in humans caused by Y-chromosome deletions encompassing a novel RNA-binding protein gene. *Nat. Genet.* 10, 383-393.

Reijo, R., Alagappan, R.K., Patrizio, P., and Page, D.C. (1996) Severe oligozoospermia resulting from deletions of azoospermia factor gene on Y chromosome. *Lancet* 347, 1290-1293.

Rodriguez-Fonseca, C., Amils, R., and Garrett, R.A. (1995) Fine structure of the peptidyl transferase centre on 23S-like rRNAs deduced from chemical probing of antibiotic-ribosome complexes. *J. Mol. Biol.* 247, 224-235.

Rogers, J.T., Randall, J.D., Cahill, C.M., Eder, P.S., Huang, X., Gunshin, H., Leiter, L., McPhee, J., Sarang, S.S., Utsuki, T., Greig, H.H., Lahiri, D.K., Tanzi, R.E., Bush, A.I., Giordano, T., and Gullans, S.R. (2002a) An iron-responsive element type II in the 5'-

untranslated regions of the Alzheimer's amyloid precursor protein transcript. *J. Biol. Chem.* 277, 45518-45528.

Rogers, J.T., Randall, J.D., Eder, P.S., Huang, X., Bush, A.I., Tanzi, R.E., Venti, A., Payton, S.M., Giordano, T., Nagano, S., Cahill, C.M., Moir, R., Lahiri, D.K., Greig, N., Sarang, S.S., and Gullans, S.R. (2002b) Alzheimer's disease drug discovery targeted to the APP mRNA 5' untranslated region. *J. Mol. Neurosci.* 19, 77-82.

Ryu, D.H., and Rando, R.R. (2002) Decoding region bubble size and aminoglycoside antibiotic binding. *Bioorg. Med. Chem. Lett.* 12, 2241-2244.

Sampaio, E.P., Sarno, E.N., Gallily, R., Cohn, Z.A., and Kaplan, G. (1991) Thalidomide selectively inhibits tumor necrosis factor α production by stimulated human monocytes. *J. Exp. Med.* 173, 699-703.

Schmitz, C., Hilpert, J., Jacobsen, C., Boensch, C., Christensen, E.I., Luft, F.C., and Willnow, T.E. (2002) Megalin deficiency offers protection from renal aminoglycoside accumulation. *J. Biol. Chem.* 277, 618-622.

Siffroi, J.P., Pawlak, A., Alfonsi, M.F., Troalen, F., Guellaen, G., and Dadoune, J.P. (2001) Expression of the TAR RNA binding protein in human testis. *Mol. Hum. Reprod.* 7, 219-225.

Sleat, D.E., Sohar, I., Gin, R.M., and Lobel, P. (2001) Aminoglycoside-mediated suppression of nonsense mutations in late infantile neuronal ceroid lipofuscinosis. *Eur. J. Paediatr. Neurol.* 5: Suppl. A, 57-62.

Soukup, G.A., and Breaker, R.R. (1999) Engineering precision RNA molecular switches. *Proc. Natl. Acad. Sci. USA* 6, 3584-3589.

Sudarsan, N., Barrick, J.E., and Breaker, R.R. (2003) Metabolite-binding RNA domains are present in the genes of eukaryotes. *RNA* 9, 644-647.

Theil, E.C. (2000) Targeting mRNA to regulate iron and oxygen metabolism. *Biochem. Pharmacol.* 59, 87-93.

Tok, J.B.-H., Cho, J., and Rando, R.R. (1999) Aminoglycoside antibiotics are able to specifically bind the 5'-untranslated region of thymidylate synthase messenger RNA. *Biochemistry* 38, 199-206.

Venables, J.P. and Eperon, I.C. (1999) The roles of RNA-binding proteins in spermatogenesis and male infertility. *Curr. Opinion Genet. and Dev.* 9, 346-354.

Venter, J.C. *et al.* (2001) The sequence of the human genome. *Science* 291, 1304-1351.

Verhelst, S.H., Michiels, P.J.A., van der Marel, G.A., van Boom, J.H., and van Boeckel, C.A.A. (2004) Surface plasmon resonance evaluations of various aminoglycoside – RNA hairpin interactions reveals a low degree of selectivity. *ChemBioChem* 5, 937-942.

Vicens, Q., and Westhof, E. (2003) RNA as a drug target: The case of aminoglycosides. *ChemBioChem* 4, 1018-1023.

Wagner, K.R., Hamed, S., Hadley, D.W., Gropman, A.L., Burstein A.H., Escolar, D.M., Hoffman, E.P., and Fischbeck KH. (2001) Gentamicin treatment of Duchenne and Becker muscular dystrophy due to nonsense mutations. *Ann. Neurol.* 49, 706-711.

Wilschanski, M., Yahav, Y., Yaacov, Y., Blau, H., Bentur, L., Rivlin, J., Aviram, M., Bdolah-Abram, T., Bebok, Z., Shushi, L., Kerem, B., and Kerem, E. (2003) Gentamicin-induced correction of CFTR functions in patients with cystic fibrosis and *CFTR* stop mutations. *New Engl. J. Med.* 349, 1433-1441.

Wilson, W.D., and Li, K. (2000) Targeting RNA with small molecules. *Curr. Med. Chem.* 7, 73-98.

Winkler, W., Nahvi, A., and Breaker, R.R. (2002) Thiamine derivatives bind messenger RNAs directly to regulate bacterial gene expression. *Nature* 419, 952-956.

Winkler, W.C., and Breaker, R.R. (2003) Genetic control by metabolite-binding riboswitches. *ChemBioChem* 4, 1024-1032.

Wong, C.-H., Hendrix, M., Manning, D.D., Rosenbohm, C., and Greenberg, W.A. (1998) A library approach to the discovery of small molecules that recognize RNA: use of a 1,3-hydroxyamine motif as core. *J. Am. Chem. Soc.* 120, 8319-8327.

Zaman, G.J.R., Garritsen, A, de Boer, Th, and van Boeckel, C.A.A. (2003a) Fluorescence assays for high-throughput screening of protein kinases. *Comb. Chem. and HTS* 6, 313-320.

Zaman, G.J.R., Michiels, P.J.A., and van Boeckel, C.A.A. (2003b) Targeting RNA: new opportunities to address drugless targets. *Drug Discov. Today* 8, 297-306.

Zhong, J., Peters, A.H.F.M., Lee, K., and Braun, R.E. (1999) A double-stranded RNA binding protein required for activation of repressed messages in mammalian germ cells. *Nature Genet.* 22, 171-174.

Zhu, X., Giordano, T., Yu, Q.-S., Holloway, H.W., Perry, T.A., Lahiri, D.K., Brossi, A., and Greig, N.H. (2003) Thiothalidomides: Novel isosteric analogues of thalidomide with enhanced TNF-α inhibitory activity. *J. Med. Chem.* 46, 5222-5229.

In: Trends in RNA Research
Editor: Peter A. McNamara, pp. 23-37

ISBN: 1-59454-506-5

Chapter II

RNA Interference: Applications in Research in Amyloidogenesis in Alzheimer's Disease

Henrik Zetterberg[1,2,3], Kaj Blennow[1,2] and Weiming Xia[3]
[1]Institute of Clinical Neuroscience, Department of Experimental Neuroscience, Sahlgrenska University Hospital, Göteborg University, Sweden;
[2]Institute of Laboratory Medicine, Department of Clinical Chemistry and Transfusion Medicine, Sahlgrenska University Hospital, Göteborg University; Sweden
[3]Center for Neurologic Diseases, Brigham and Women's Hospital and Harvard Medical School, Boston, MA 02115, USA

Abstract

Alzheimer's disease (AD) is a complex neurodegenerative disorder characterized by the progressive accumulation of β-amyloid (Aβ) in specific brain regions. Aβ is derived from the amyloid precursor protein (APP) after β- and γ-secretase cleavages. β-secretase was recently identified as a membrane-anchored aspartyl protease, but the identity of γ-secretase has remained elusive. Recent studies have demonstrated that γ-secretase is a multiprotein complex consisting of at least four proteins: presenilin (PS), which may harbor the catalytic site, nicastrin (NCT), Aph-1 and Pen-2. γ-Secretase cleaves not only APP but also Notch and several other transmembrane receptors indicating that it serves as a mediator of diverse signaling pathways with multiple biological functions. The discovery that double-stranded RNA is able to silence gene expression was made only a few years ago but methods for experimentally silencing genes have already been extended to a broad diversity of organisms and cell types, including neurons. Here, we review recent studies that have employed RNA-mediated gene silencing techniques to

[1] Corresponding author : Henrik Zetterberg, MD, PhD, Department of Clinical Chemistry and Transfusion Medicine, Sahlgrenska University Hospital Göteborg University, S-413 45 Gothenburg, SWEDEN, Telephone: +46 31 3423054, Fax: +46 31 828458, e-mail: henrik.zetterberg@clinchem.gu.se

dissect the roles of the different factors involved in β- and γ-secretase-mediated amyloidogenic cleavage of APP in AD.

Introduction

Alzheimer's disease (AD) is one of the major neuropsychiatric disorders of our times, affecting more than 20 million individuals worldwide. Senile plaques and neurofibrillary tangles throughout cortical and limbic regions of the brain are the two neuropathological hallmarks of AD [1]. The major component of senile plaques is β-amyloid (Aβ) that is generated from amyloid precursor protein (APP), a type I integral membrane protein with one transmembrane domain, by enzymatic digestion involving β- and γ-secretase activities [2] (Fig. 1). Cleavage of APP by β-secretase generates an approximately 100 kDa soluble N-terminal fragment and a 12 kDa C-terminal fragment (C99), which can be further cleaved by γ-secretase to yield 40 and 42 amino acid long Aβ peptides (Aβ40 and Aβ42), the latter of which seems to be the most fibrillogenic (the amyloidogenic pathway) [3], and the APP intracellular domain (AICD). Another enzymatic activity, α-secretase, cuts APP approximately in the middle of the Aβ region, thereby precluding production of Aβ peptides (the non-amyloidogenic pathway). The candidate α-secretases identified so far are ADAM9 [4, 5], ADAM10 [6, 7] and ADAM17 [8, 9], which all are membrane-anchored cell surface proteins belonging to the ADAM (a disintegrin and metalloprotease) family. Besides its role in APP processing, γ-secretase can also modulate the proteolytic cleavage of the Notch receptor. The Notch signaling pathway mediates a wide range of developmental cell fate decisions in multicellular organisms. In response to ligand-binding, Notch receptors undergo an extracellular cleavage at a site near the membrane, generating a membrane-tethered C-terminal domain. Subsequent cleavage by γ-secretase at a site within the transmembrane domain releases the Notch intracellular domain (NICD) that translocates to the nucleus where it associates with RBP-Jκ and other transcription factors to regulate the promoters of specific target genes of the pathway [10-12]. In addition to APP and Notch, γ-secretase also catalyzes the proteolytic release of the intracellular domains of numerous other type I transmembrane receptors, underscoring its complex role in cell signaling [13]. No specific target genes have been identified so far that are regulated by physiological levels of the AICD fragment of APP but the finding that AICD interacts with the transcriptional co-activator Fe65 [14, 15] and the histone acetyltransferase Tip60 [14, 16] provides a strong case for a nuclear transcriptional activity associated with APP, thereby possibly closing the circle to Notch.

On the basis of data generated from epidemiological, genetic and biochemical studies several hypotheses on the pathogenesis of AD have been erected. The amyloid hypothesis is at present the most prevailing and predicts that a gradual elevation of Aβ42 levels in brain interstitial fluid leads to the oligomerization of the peptide and eventually to its deposition in senile plaques, and that this is the primary pathogenic process in AD [17]. A large body of evidence supports this hypothesis. Aβ42 accumulation can clearly be initiated by rare mutations in the genes that encode APP and presenilin (PS), the latter of which may harbor the active site of γ-secretase, and these mutations cause autosomal dominant forms of familial AD. Furthermore, trisomi 21 (Down syndrome) leads to APP overexpression due to increased dosage of the APP gene on chromosome 21 and early onset of AD with classical AD neuropathology in the form of Aβ42 deposition. Finally, the strongest susceptibility gene for sporadic AD is the apolipoprotein E (*APOE*) ε4 gene variant that encodes the E4 isoform of the protein. The AD-promoting effect of inheriting one or two ε4 alleles seems to involve an enhanced aggregation or decreased clearance of Aβ42 [18-20]. Transgenic mice that express E4 display higher density of Aβ plaques and vascular deposits than E3-expressing mice [21, 22], which provides an additional linchpin for the amyloid hypothesis. In conclusion, the genetics of AD points to a central role for Aβ accumulation in the pathogenesis of the disease and this view is supported by biochemical and neuropathological studies. Hence, the two amyloidogenic secretase activities in APP processing (β- and γ-secretase) have become desirable therapeutic targets.

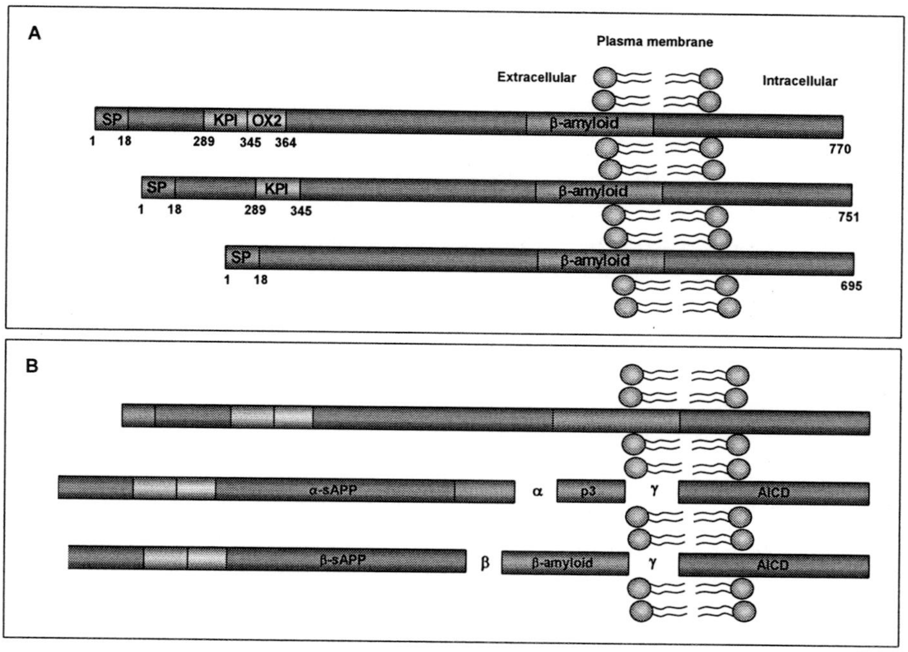

Fig. 1. Schematic representation of the amyloid precursor protein (APP) and its principal proteolytic cleavage products. (A) Three major isoforms (APP_{770}, APP_{751} and APP_{695}) exist due to differential splicing of exon 7 that contains the Kunitz protease inhibitor domain (KPI) and exon 8 that contains the OX2 antigen domain. A 17 amino acid signal peptide (SP) is indicated at the N-terminus and a single membrane-spanning domain occurs at amino acids 700-723. The coordinates are codon numbers in the APP_{770} isoform. The β-amyloid (Aβ) fragment includes 28 residues just outside the membrane plus the first 12 to 14 residues of the transmembrane domain. (B) Proteolytic cleavages of the largest isoform of APP (APP_{770}). In the non-amyloidogenic pathway, a cleavage by α-secretase occurs after residue 687, which enables the secretion of the large, soluble ectodomain of APP (α-sAPP) into the medium and retention of the 83-residue C-terminal fragment (C83) in the membrane. The C83 fragment can undergo cleavage by γ-secretase at residue 711 or 713 to release the p3 peptides. In the amyloidogenic pathway, β-secretase cleaves after residue 671, which causes the secretion of the slightly truncated β-sAPP molecule and the retention of a 99 residue C-terminal fragment (C99). This fragment can undergo further cleavage by γ-secretase to release 40 or 42 amino acid long Aβ fragments. Cleavage of both C83 and C99 by γ-secretase releases the APP intracellular domain (AICP) into the cytoplasm.

A prerequisite for finding a specific and non-toxic inhibitor of Aβ production is the identification and functional characterization of the individual genes that encode the catalytic and regulatory components of the β- and γ-secretases. Classical approaches to this type of complex problem are targeting of candidate genes by homologous recombination, ribozymes, antisense technologies and/or creation of knock-out animals [23]. All these approaches, however, have their limitations and none can be applied universally. The discovery of RNA-mediated gene silencing (RNA interference; RNAi) has facilitated the study of gene function in complex systems in a rapid and inexpensive way. As an example, large-scale functional analyses of almost all the 19,000 genes of *Caenorhabditis elegans* have been carried out using RNAi technology [24], and the technique is proving to be useful to identify the function of novel genes predicted from genome sequencing data in a wide variety of organisms and in complex biological processes. Here, we review the mechanism of RNAi and recent studies that have employed RNAi techniques to dissect the roles of the different factors involved in β- and γ-secretase-mediated amyloidogenic cleavage of APP in AD.

The Mechanism of RNAi

It has been known for several years that double-stranded RNA (dsRNA) has the ability to silence gene expression in different plant systems [25]. However, until recently the biochemical nature of this phenomenon was unknown and it took a few years before dsRNA itself was shown to be the trigger of gene silencing and before similar responses were noted in animals. In fact, the discovery was made in the roundworm *Caenorhabditis elegans* somewhat through experimental serendipity. In a study of the efficacy of certain antisense RNA molecules, sense RNA was used as negative control [26]. Surprisingly, the combination of sense and antisense RNA (dsRNA) caused a very efficient gene silencing. A subsequent study in the same model eventually demonstrated that dsRNA was a much more potent silencing trigger than either strain alone [27]. This phenomenon was named RNA interference (RNAi), distinguishing it mechanistically from classical antisense-mediated gene suppression. RNAi has turned out to be a general occurrence and seems to offer a protection mechanism against foreign nucleic acids, e.g. RNA viruses, in eukaryotic cells.

Through a combination of genetic and biochemical techniques many of the molecular mechanisms mediating the RNAi effect have been unraveled. In a simplified model, cells activate a two-step, enzymatically catalyzed reaction in response to dsRNA (Fig. 2). Firstly, long dsRNA is processed by a ribonuclease (RNase) III enzyme dimer called Dicer into small interfering RNAs (siRNAs) with a size of approximately 22 bp. Secondly, while still attached to Dicer-containing complexes, these siRNAs are incorporated in an enzyme complex termed the RNA-induced silencing complex (RISC) together with ATP [28, 29]. ATP-dependent unwinding of siRNAs activates RISC that is guided through sequence homology by the single-stranded siRNAs to specifically degrade homologous mRNAs causing gene silencing. The penetrance of this homology-driven RNA degradation approaches 100% in many cases [30]. This high efficacy leads to the immediate question whether some aspect of the silencing signal might be amplified. Some data derived from studies in plants and in *Drosophila*, *Neurospora* and *Caenorhabditis elegans* suggest that RNA-dependent RNA polymerases

(RdRP) might serve to amplify the silencing signal initiated by very small amounts of dsRNA by converting degraded mRNAs into novel dsRNAs [31-36], but the generality of these results remains to be established. For instance, the human genome lacks a recognizable RdRP homologue [37].

Similar small RNA molecules that can silence gene activity are micro-RNAs (miRNAs). These single-stranded RNA molecules, which have wide-spread roles in growth and development, are also the result of Dicer activity, but in this case, the stem and loop precursor molecules are encoded within the animal or plant genomes, and silencing can occur either via destruction of the mRNA or by blocking its translation [38]. Recently, it was shown that certain viruses, e.g. the Epstein-Barr virus, can synthesize miRNAs and suppress cellular gene expression as a means for conquering the infected cell [39].

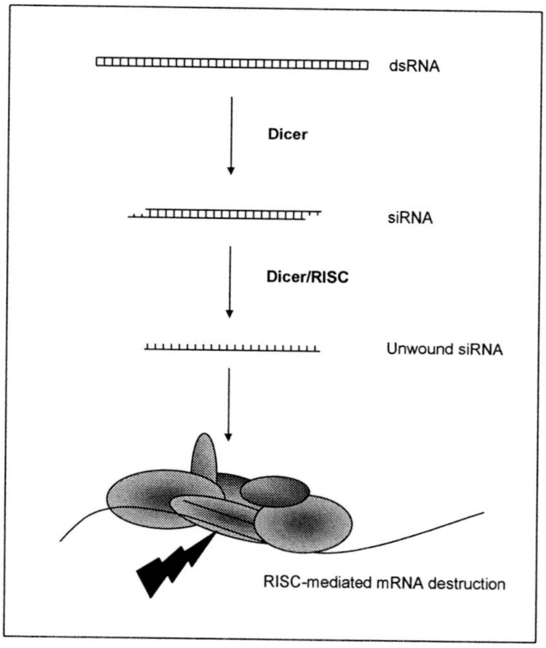

Fig. 2. Model for RNA-mediated gene silencing. In an ordered biochemical pathway, siRNAs are processed from double-stranded precursor molecules by an RNase III enzyme dimer designated Dicer into small interfering RNAs (siRNAs) with a size of approximately 22 bp. While still attached to Dicer-containing complexes, these siRNAs are incorporated in an enzyme complex termed the RNA-induced silencing complex (RISC) together with ATP. ATP-dependent unwinding of siRNAs activates the multiprotein complex RISC that is guided through sequence homology by the single-stranded siRNAs to specifically degrade homologous mRNAs causing gene silencing.

Amyloidogenesis in Alzheimer's Disease and RNAi

Following the discovery that Aβ is the principal component of the senile plaques in AD brains, an intense search was launched to identify the enzymes responsible for generating this

40-42 amino acid fragment of APP. With no knowledge of their likely identity, these enzymes were called β- and γ-secretases in reference to their sites of cleavage at the N- and C-terminal sites, respectively, in the Aβ region of APP (Fig. 3).

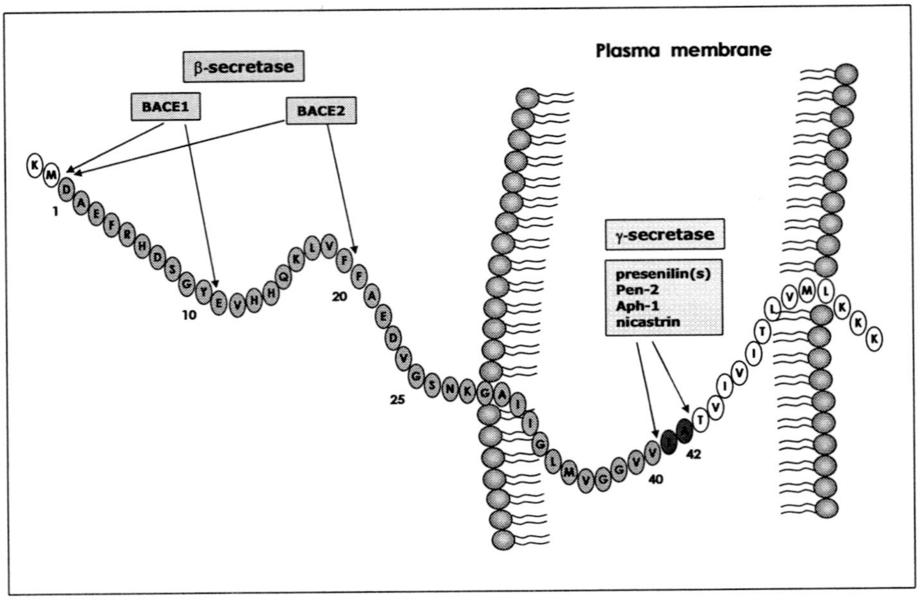

Fig. 3. Detailed map of cleavage sites within APP by β- and γ-secretase. The sequence that contains the Aβ and transmembrane regions is expanded and shown by the single-letter amino acid code. BACE1 cleaves at the N-terminus of Aβ (after residue 671 according to the APP_{770} isoform) and at a second site between Aβ residues 10 and 11. BACE2 also cleaves at the N-terminus of Aβ and at a second site between Aβ residues 19 and 20. Cleavage of APP at either site by BACE1 is amyloidogenic, whereas the second cleavage site for BACE2 precludes formation of Aβ. γ-Secretase cleaves either between Aβ residues 40 and 41 or between residues 42 and 43 (after residues 711 and 713, respectively, according to the APP_{770} isoform). Presenilin(s), Pen-2, Aph-1 and nicastrin constitute the minimal core of γ-secretase and are essential for its assembly and activity. Both APP and presenilin mutations that cause the rare autosomal dominant forms of AD selectively increase the γ-secretase cleavage of C99 to yield more Aβ42 peptides compared with Aβ40 (the less hydrophobic fragment of Aβ).

BACE1 and -2

In 1999 several independent research groups published evidence demonstrating that a significant part of β-secretase activity originates from a novel integral membrane aspartyl protease encoded by the β-site APP-cleaving enzyme 1 (BACE1) gene [40-43]. Cellular antisense [42, 43] and knock-out mouse models [44-46] unambiguously confirmed the role of BACE1 in promoting Aβ production. A second integral membrane aspartyl protease related to BACE1, referred to as BACE2, was also demonstrated to cleave APP at the β-cleavage site in transfected cells [43, 47-51]. Both enzymes have the ability to cleave APP at a second site within the Aβ region. This second site for BACE1 is between amino acids 10 and 11 and between amino acids 19 and 20 for BACE2 (Fig. 3). Cleavage of APP by BACE1 at either site is amyloidogenic, whereas the second cleavage site for BACE2 precludes formation of Aβ. Many tissues, including the brain, co-express BACE1 and BACE2 [47-49, 51], but the individual contribution of these two enzymes in the generation of Aβ in the brain has been elusive. Recently, Basi *et al.* reported results from selective RNAi-mediated knock-down of either enzyme in HEK293 cells expressing wild-type or the Swedish mutant form of APP [52], demonstrating that selective BACE1 inactivation decreases Aβ secretion from cells, while selective BACE2 inactivation results in increased production of amyloidogenic Aβ. The results confirm the essential role of β-secretase in Aβ production and suggest that BACE1 and BACE2 may act as antagonists in amyloidogenesis, which may be of significance in the search for amyloid-based therapeutics targeting these enzymes. The BACE1 results were recently confirmed in a study of primary cortical neurons where selective RNAi-mediated BACE1 suppression specifically reduced Aβ production in wild-type and transgenic mice harboring the Swedish APP mutant [53]. Two possible reservations may, however, be considered when inferring these data to the pathogenesis of sporadic AD. Firstly, BACE1 seems to have preference for cleaving the Swedish APP mutant at the β-secretase site [40-43, 47]. This may lead to an overestimation of the importance of BACE1 for amyloidogenic APP processing in sporadic AD. Secondly, a recent study suggests that only a minor portion of Aβ is secreted via the constitutive secretory pathway [54]. Instead, most of the Aβ is claimed to be produced upon activation of neurons by neurotransmitters ("the regulated secretory pathway") in a proteolytic process that involves multiple cysteine proteases in addition to BACE1 [54].

The γ-Secretase Complex

Accumulating data have shown that γ-secretase is an unusual high molecular weight intramembrane-cleaving protease complex that contains presenilins (PS) [55]. The presenilins, encoded by two homologous genes (PS1 and PS2), are multipass proteins with eight TM domains that localize predominantly in the endoplasmic reticulum (ER) and Golgi compartments, although small amounts are also present on the cell surface [56-59]. Presenilins are synthesized as a holo-protein, which undergoes an internal proteolytic cleavage, yielding N- and C-terminal fragments (NTF and CTF) of 27 and 17 kDa, respectively [60]. There is now substantial evidence that the PS NTF/CTF heterodimers constitute the active site of γ-secretase [61-63].

Recent work has identified three additional components of the γ-secretase complex: nicastrin (NCT), Pen-2 and Aph-1 (Fig. 3). NCT is a type I integral membrane protein that associates with PS and C-terminal fragments of APP [64, 65]. Aph-1 and Pen-2 are polytopic membrane proteins containing at least seven and two transmembrane segments, respectively. The genes encoding these proteins were identified through genetic screens in *Caenorhabditis elegans* and are phylogenetically conserved [66, 67]. The identification of NCT, Aph-1 and Pen-2 as possible γ-secretase components gave rise to a host of questions about their possible roles in γ-secretase assembly, trafficking and enzymatic activation.

The first evidence that NCT is indeed an essential component of γ-secretase came from RNAi experiments in *Caenorhabditis elegans*, where silencing of NCT expression produced an embryonic lethal phenotype highly reminiscent of those produced by reduction of the activity of genes involved in the Notch signaling pathway or the simultaneous deletion of both PS genes [65]. Francis *et al.* demonstrated that RNAi-mediated inactivation of Aph-1, Pen-2 or NCT in cultured *Drosophila* cells reduces γ-secretase cleavage of APP and Notch substrates as well as the levels of processed PS [66]. These data confirm that both Aph-1 and Pen-2, like NCT, are required for the assembly and activity of γ-secretase. Takasugi *et al.* extended the results by showing that *Drosophila* Aph-1 increases the stability of PS holoprotein in the complex [68]. Pen-2 interacts with PS [69] and depletion of Pen-2 by RNAi prevents endoproteolysis of presenilin and promotes stabilization of the holoprotein in both *Drosophila* and mammalian cells, including primary neurons [68, 69]. Co-expression of *Drosophila* Pen-2 with Aph-1 and NCT increases the formation of PS fragments as well as γ-secretase activity, supporting the view that these genes must be provided together to rescue the phenotype of knock-down cells. Additionally, overexpression of PS, NCT, Aph-1 and Pen-2 results in increased PS endoproteolysis and γ-secretase activity, both in mammalian cells [13, 68-71] and in yeast [72]. Altogether, RNAi-mediated inactivation of NCT, Pen-2 and Aph-1, in conjunction with overexpression experiments, has confirmed the essential roles of these proteins in the assembly of an active γ-secretase and further corroborated the view that these proteins together with PS constitute the minimal core of the γ-secretase complex.

Some Practical Aspects on the siRNA Technology

The success of RNAi-mediated gene silencing depends on the careful consideration of a few variables. First of all, siRNAs synthesized against different regions of the same target mRNA show different silencing efficiencies [73]. Still, no consensus on selecting an efficient siRNA sequence has evolved, although some general advice can be found in the literature [74]. As a general rule the sequence should be selected in the region 50 to 100 bp downstream of the start codon. The 5' or 3' untranslated regions (UTR) should be avoided, assuming that UTR-binding protein complexes may interfere with the binding of siRNA to the RISC complex. The GC content of the siRNA should be kept between 30 and 70%. Regarding the chemical properties of siRNA some general guidelines are that siRNAs should have 2-nucleotide 3' overhang of uridine residues (A or T), a length of 21 nucleotides and phosphate and hydroxyl groups at the 5' and 3' end, respectively. Several computer programs have been developed that offer helpful strategies in the selection of appropriate siRNA sequence (e.g. www.Ic.sunysb.edu/stu/shiklin/rnai.html; www.ambion.com; www.invitrogen.com). The siRNAs can be synthesized either chemically or from plasmids. One drawback with the use of chemically synthesized siRNA is the high synthesis cost. To circumvent this problem, two approaches have been developed for expressing siRNAs from plasmids. In the first approach, sense and antisense strands are transcribed by individual promoters [75], and, in the second, siRNAs are expressed as fold-back stem-loop structures that give rise to short hairpin RNAs (shRNAs). Many of these plasmid-based vectors, such as pSilencer 1.0 (Ambion), pSuper (DNA Engine) and the Block-IT system (Invitrogen), are now commercially available. shRNAs activate the miRNA pathway described above to silence gene expression and can be produced within cells following transfection from standard mammalian expression vectors or stably replicating plasmids. In this way, shRNA expression makes possible the creation of continuous cell lines and even transgenic animals in which suppression of a target gene is stably maintained by RNAi. As a result, the types of RNAi-based gene function analysis that can be carried out in mammals have been greatly expanded [76]. Given the gene-specific features of RNAi, it is conceivable that this method also may play a role in therapeutic applications in neurodegenerative and other disorders, although a lot of obstacles regarding the delivery of siRNAs need to be overcome.

Concluding Remarks

The usefulness of RNAi technology to delineate the contribution of individual factors in complex biochemical processes is becoming increasingly clear. RNAi-mediated silencing of different genes involved in amyloidogenic cleavage of APP has contributed significantly to the understanding of the function of the main molecular partners constituting the minimal core of the γ-secretase complex (PS, NCT, Pen-2 and Aph-1) and the differential effects of BACE1 and BACE2 in Aβ production. These data are highly relevant in the search for specific inhibitors of Aβ production for the treatment of AD. The restriction of RNAi technology with regard to the duration of the effect of microinjected or transfected siRNA molecules in cells can now be circumvented by the employment shRNA-expressing standard or self-replicating plasmids that activate the miRNA pathway for RNA degradation. This opens up a whole new field of RNAi studies with possible therapeutic implications, where siRNA can be stably expressed in cells and animals. It must, however, be borne in mind that the RNAi technique should not be used in isolation, but rather in conjunction with other techniques. The most information is likely to be derived from studies in which RNAi-induced loss-of-function experiments are combined with overexpression or gain-of-function experiments in order to address the function of a specific gene in a proper way.

Acknowledgments

This study was supported by the Fulbright Commission, the Swedish Society for Medical Research, the Göteborg Medical Society, the Wenner-Gren Foundations and the Swedish Society of Medicine.

References

[1] Xia, W. Amyloid metabolism and secretases in Alzheimer's disease. *Curr Neurol Neurosci Rep*, 2001;1:422-427.

[2] Selkoe, DJ. Alzheimer's disease: genes, proteins, and therapy. *Physiol Rev*, 2001;81:741-766.

[3] Lorenzo, A; Yankner, BA. Beta-amyloid neurotoxicity requires fibril formation and is inhibited by congo red. *Proc Natl Acad Sci U S A*, 1994;91:12243-12247.

[4] Koike, H; Tomioka, S; Sorimachi, H; Saido, TC; Maruyama, K; Okuyama, A; Fujisawa-Sehara, A; Ohno, S; Suzuki, K; Ishiura, S. Membrane-anchored metalloprotease MDC9 has an alpha-secretase activity responsible for processing the amyloid precursor protein. *Biochem J*, 1999;343 Pt 2:371-375.

[5] Nelson, KK; Schlondorff, J; Blobel, CP. Evidence for an interaction of the metalloprotease-disintegrin tumour necrosis factor alpha convertase (TACE) with mitotic arrest deficient 2 (MAD2), and of the metalloprotease-disintegrin MDC9 with a novel MAD2-related protein, MAD2beta. *Biochem J*, 1999;343 Pt 3:673-680.

[6] Kojro, E; Gimpl, G; Lammich, S; Marz, W; Fahrenholz, F. Low cholesterol stimulates the nonamyloidogenic pathway by its effect on the alpha -secretase ADAM 10. *Proc Natl Acad Sci U S A*, 2001;98:5815-5820.

[7] Lammich, S; Kojro, E; Postina, R; Gilbert, S; Pfeiffer, R; Jasionowski, M; Haass, C; Fahrenholz, F. Constitutive and regulated alpha-secretase cleavage of Alzheimer's amyloid precursor protein by a disintegrin metalloprotease. *Proc Natl Acad Sci U S A*, 1999;96:3922-3927.

[8] Buxbaum, JD; Liu, KN; Luo, Y; Slack, JL; Stocking, KL; Peschon, JJ; Johnson, RS; Castner, BJ; Cerretti, DP; Black, RA. Evidence that tumor necrosis factor alpha converting enzyme is involved in regulated alpha-secretase cleavage of the Alzheimer amyloid protein precursor. *J Biol Chem*, 1998;273:27765-27767.

[9] Slack, BE; Ma, LK; Seah, CC. Constitutive shedding of the amyloid precursor protein ectodomain is up-regulated by tumour necrosis factor-alpha converting enzyme. *Biochem J*, 2001;357:787-794.

[10] Artavanis-Tsakonas, S; Rand, MD; Lake, RJ. Notch signaling: cell fate control and signal integration in development. *Science*, 1999;284:770-776.

[11] Fortini, ME. Notch and presenilin: a proteolytic mechanism emerges. *Curr Opin Cell Biol*, 2001;13:627-634.

[12] Kopan, R; Goate, A. A common enzyme connects notch signaling and Alzheimer's disease. *Genes Dev*, 2000;14:2799-2806.

[13] De Strooper, B. Aph-1, Pen-2, and Nicastrin with Presenilin generate an active gamma-Secretase complex. *Neuron*, 2003;38:9-12.

[14] Cao, X; Sudhof, TC. A transcriptionally [correction of transcriptively] active complex of APP with Fe65 and histone acetyltransferase Tip60. *Science*, 2001;293:115-120.

[15] Kimberly, WT; Zheng, JB; Guenette, SY; Selkoe, DJ. The intracellular domain of the beta-amyloid precursor protein is stabilized by Fe65 and translocates to the nucleus in a notch-like manner. *J Biol Chem*, 2001;276:40288-40292.

[16] Scheinfeld, MH; Matsuda, S; D'Adamio, L. JNK-interacting protein-1 promotes transcription of A beta protein precursor but not A beta precursor-like proteins, mechanistically different than Fe65. *Proc Natl Acad Sci U S A*, 2003;100:1729-1734.

[17] Selkoe, DJ. Alzheimer's disease is a synaptic failure. *Science*, 2002;298:789-791.

[18] Evans, KC; Berger, EP; Cho, CG; Weisgraber, KH; Lansbury, PT, Jr. Apolipoprotein E is a kinetic but not a thermodynamic inhibitor of amyloid formation: implications for the pathogenesis and treatment of Alzheimer disease. *Proc Natl Acad Sci U S A*, 1995;92:763-767.

[19] Ma, J; Yee, A; Brewer, HB, Jr.; Das, S; Potter, H. Amyloid-associated proteins alpha 1-antichymotrypsin and apolipoprotein E promote assembly of Alzheimer beta-protein into filaments. *Nature*, 1994;372:92-94.

[20] Schmechel, DE; Saunders, AM; Strittmatter, WJ; Crain, BJ; Hulette, CM; Joo, SH; Pericak-Vance, MA; Goldgaber, D; Roses, AD. Increased amyloid beta-peptide deposition in cerebral cortex as a consequence of apolipoprotein E genotype in late-onset Alzheimer disease. *Proc Natl Acad Sci U S A*, 1993;90:9649-9653.

[21] Fagan, AM; Watson, M; Parsadanian, M; Bales, KR; Paul, SM; Holtzman, DM. Human and murine ApoE markedly alters A beta metabolism before and after plaque formation in a mouse model of Alzheimer's disease. *Neurobiol Dis*, 2002;9:305-318.

[22] Holtzman, DM; Bales, KR; Tenkova, T; Fagan, AM; Parsadanian, M; Sartorius, LJ; Mackey, B; Olney, J; McKeel, D; Wozniak, D; Paul, SM. Apolipoprotein E isoform-dependent amyloid deposition and neuritic degeneration in a mouse model of Alzheimer's disease. *Proc Natl Acad Sci U S A*, 2000;97:2892-2897.

[23] Sullenger, BA; Gilboa, E. Emerging clinical applications of RNA. *Nature*, 2002;418:252-258.

[24] Kamath, RS; Fraser, AG; Dong, Y; Poulin, G; Durbin, R; Gotta, M; Kanapin, A; Le Bot, N; Moreno, S; Sohrmann, M; Welchman, DP; Zipperlen, P; Ahringer, J. Systematic functional analysis of the Caenorhabditis elegans genome using RNAi. *Nature*, 2003;421:231-237.

[25] Matzke, MA; Aufsatz, W; Kanno, T; Mette, MF; Matzke, AJ. Homology-dependent gene silencing and host defense in plants. *Adv Genet*, 2002;46:235-275.

[26] Guo, S; Kemphues, KJ. par-1, a gene required for establishing polarity in C. elegans embryos, encodes a putative Ser/Thr kinase that is asymmetrically distributed. *Cell*, 1995;81:611-620.

[27] Fire, A; Xu, S; Montgomery, MK; Kostas, SA; Driver, SE; Mello, CC. Potent and specific genetic interference by double-stranded RNA in Caenorhabditis elegans. *Nature*, 1998;391:806-811.

[28] Zhang, H; Kolb, FA; Brondani, V; Billy, E; Filipowicz, W. Human Dicer preferentially cleaves dsRNAs at their termini without a requirement for ATP. *Embo J*, 2002;21:5875-5885.

[29] Tijsterman, M; Plasterk, RH. Dicers at RISC; the mechanism of RNAi. *Cell*, 2004;117:1-3.

[30] Nykanen, A; Haley, B; Zamore, PD. ATP requirements and small interfering RNA structure in the RNA interference pathway. *Cell*, 2001;107:309-321.

[31] Cogoni, C; Macino, G. Gene silencing in Neurospora crassa requires a protein homologous to RNA-dependent RNA polymerase. *Nature*, 1999;399:166-169.

[32] Dalmay, T; Hamilton, A; Rudd, S; Angell, S; Baulcombe, DC. An RNA-dependent RNA polymerase gene in Arabidopsis is required for posttranscriptional gene silencing mediated by a transgene but not by a virus. *Cell*, 2000;101:543-553.

[33] Lipardi, C; Wei, Q; Paterson, BM. RNAi as random degradative PCR: siRNA primers convert mRNA into dsRNAs that are degraded to generate new siRNAs. *Cell*, 2001;107:297-307.

[34] Sijen, T; Fleenor, J; Simmer, F; Thijssen, KL; Parrish, S; Timmons, L; Plasterk, RH; Fire, A. On the role of RNA amplification in dsRNA-triggered gene silencing. *Cell*, 2001;107:465-476.

[35] Smardon, A; Spoerke, JM; Stacey, SC; Klein, ME; Mackin, N; Maine, EM. EGO-1 is related to RNA-directed RNA polymerase and functions in germ-line development and RNA interference in C. elegans. *Curr Biol*, 2000;10:169-178.

[36] Vaistij, FE; Jones, L; Baulcombe, DC. Spreading of RNA targeting and DNA methylation in RNA silencing requires transcription of the target gene and a putative RNA-dependent RNA polymerase. *Plant Cell*, 2002;14:857-867.

[37] Denli, AM; Hannon, GJ. RNAi: an ever-growing puzzle. *Trends Biochem Sci*, 2003;28:196-201.

[38] Carrington, JC; Ambros, V. Role of microRNAs in plant and animal development. *Science*, 2003;301:336-338.

[39] Pfeffer, S; Zavolan, M; Grasser, FA; Chien, M; Russo, JJ; Ju, J; John, B; Enright, AJ; Marks, D; Sander, C; Tuschl, T. Identification of virus-encoded microRNAs. *Science*, 2004;304:734-736.

[40] Hussain, I; Powell, D; Howlett, DR; Tew, DG; Meek, TD; Chapman, C; Gloger, IS; Murphy, KE; Southan, CD; Ryan, DM; Smith, TS; Simmons, DL; Walsh, FS; Dingwall, C; Christie, G. Identification of a novel aspartic protease (Asp 2) as beta-secretase. *Mol Cell Neurosci*, 1999;14:419-427.

[41] Sinha, S; Anderson, JP; Barbour, R; Basi, GS; Caccavello, R; Davis, D; Doan, M; Dovey, HF; Frigon, N; Hong, J; Jacobson-Croak, K; Jewett, N; Keim, P; Knops, J; Lieberburg, I; Power, M; Tan, H; Tatsuno, G; Tung, J; Schenk, D; Seubert, P; Suomensaari, SM; Wang, S; Walker, D; John, V; et al. Purification and cloning of amyloid precursor protein beta-secretase from human brain. *Nature*, 1999;402:537-540.

[42] Vassar, R; Bennett, BD; Babu-Khan, S; Kahn, S; Mendiaz, EA; Denis, P; Teplow, DB; Ross, S; Amarante, P; Loeloff, R; Luo, Y; Fisher, S; Fuller, J; Edenson, S; Lile, J; Jarosinski, MA; Biere, AL; Curran, E; Burgess, T; Louis, JC; Collins, F; Treanor, J; Rogers, G; Citron, M. Beta-secretase cleavage of Alzheimer's amyloid precursor protein by the transmembrane aspartic protease BACE. *Science*, 1999;286:735-741.

[43] Yan, R; Bienkowski, MJ; Shuck, ME; Miao, H; Tory, MC; Pauley, AM; Brashier, JR; Stratman, NC; Mathews, WR; Buhl, AE; Carter, DB; Tomasselli, AG; Parodi, LA; Heinrikson, RL; Gurney, ME. Membrane-anchored aspartyl protease with Alzheimer's disease beta-secretase activity. *Nature*, 1999;402:533-537.

[44] Cai, H; Wang, Y; McCarthy, D; Wen, H; Borchelt, DR; Price, DL; Wong, PC. BACE1 is the major beta-secretase for generation of Abeta peptides by neurons. *Nat Neurosci*, 2001;4:233-234.

[45] Luo, Y; Bolon, B; Kahn, S; Bennett, BD; Babu-Khan, S; Denis, P; Fan, W; Kha, H; Zhang, J; Gong, Y; Martin, L; Louis, JC; Yan, Q; Richards, WG; Citron, M; Vassar, R. Mice deficient in BACE1, the Alzheimer's beta-secretase, have normal phenotype and abolished beta-amyloid generation. *Nat Neurosci*, 2001;4:231-232.

[46] Roberds, SL; Anderson, J; Basi, G; Bienkowski, MJ; Branstetter, DG; Chen, KS; Freedman, SB; Frigon, NL; Games, D; Hu, K; Johnson-Wood, K; Kappenman, KE; Kawabe, TT; Kola, I; Kuehn, R; Lee, M; Liu, W; Motter, R; Nichols, NF; Power, M; Robertson, DW; Schenk, D; Schoor, M; Shopp, GM; Shuck, ME; Sinha, S; Svensson, KA; Tatsuno, G; Tintrup, H; Wijsman, J; Wright, S; McConlogue, L. BACE knockout mice are healthy despite lacking the primary beta-secretase activity in brain: implications for Alzheimer's disease therapeutics. *Hum Mol Genet*, 2001;10:1317-1324.

[47] Lin, X; Koelsch, G; Wu, S; Downs, D; Dashti, A; Tang, J. Human aspartic protease memapsin 2 cleaves the beta-secretase site of beta-amyloid precursor protein. *Proc Natl Acad Sci U S A*, 2000;97:1456-1460.

[48] Bennett, BD; Babu-Khan, S; Loeloff, R; Louis, JC; Curran, E; Citron, M; Vassar, R. Expression analysis of BACE2 in brain and peripheral tissues. *J Biol Chem*, 2000;275:20647-20651.

[49] Hussain, I; Powell, DJ; Howlett, DR; Chapman, GA; Gilmour, L; Murdock, PR; Tew, DG; Meek, TD; Chapman, C; Schneider, K; Ratcliffe, SJ; Tattersall, D; Testa, TT; Southan, C; Ryan, DM; Simmons, DL; Walsh, FS; Dingwall, C; Christie, G. ASP1 (BACE2) cleaves the amyloid precursor protein at the beta-secretase site. *Mol Cell Neurosci*, 2000;16:609-619.

[50] Acquati, F; Accarino, M; Nucci, C; Fumagalli, P; Jovine, L; Ottolenghi, S; Taramelli, R. The gene encoding DRAP (BACE2), a glycosylated transmembrane protein of the aspartic protease family, maps to the down critical region. *FEBS Lett*, 2000;468:59-64.

[51] Farzan, M; Schnitzler, CE; Vasilieva, N; Leung, D; Choe, H. BACE2, a beta -secretase homolog, cleaves at the beta site and within the amyloid-beta region of the amyloid-beta precursor protein. *Proc Natl Acad Sci U S A*, 2000;97:9712-9717.

[52] Basi, G; Frigon, N; Barbour, R; Doan, T; Gordon, G; McConlogue, L; Sinha, S; Zeller, M. Antagonistic effects of beta-site amyloid precursor protein-cleaving enzymes 1 and 2 on beta-amyloid peptide production in cells. *J Biol Chem*, 2003;278:31512-31520.

[53] Kao, SC; Krichevsky, AM; Kosik, KS; Tsai, LH. BACE1 suppression by RNA interference in primary cortical neurons. *J Biol Chem*, 2004;279:1942-1949.

[54] Hook, VY; Toneff, T; Aaron, W; Yasothornsrikul, S; Bundey, R; Reisine, T. Beta-amyloid peptide in regulated secretory vesicles of chromaffin cells: evidence for multiple cysteine proteolytic activities in distinct pathways for beta-secretase activity in chromaffin vesicles. *J Neurochem*, 2002;81:237-256.

[55] Esler, WP; Kimberly, WT; Ostaszewski, BL; Ye, W; Diehl, TS; Selkoe, DJ; Wolfe, MS. Activity-dependent isolation of the presenilin- gamma -secretase complex reveals nicastrin and a gamma substrate. *Proc Natl Acad Sci U S A*, 2002;99:2720-2725.

[56] Annaert, WG; Levesque, L; Craessaerts, K; Dierinck, I; Snellings, G; Westaway, D; George-Hyslop, PS; Cordell, B; Fraser, P; De Strooper, B. Presenilin 1 controls gamma-secretase processing of amyloid precursor protein in pre-golgi compartments of hippocampal neurons. *J Cell Biol*, 1999;147:277-294.

[57] Cupers, P; Bentahir, M; Craessaerts, K; Orlans, I; Vanderstichele, H; Saftig, P; De Strooper, B; Annaert, W. The discrepancy between presenilin subcellular localization and gamma-secretase processing of amyloid precursor protein. *J Cell Biol*, 2001;154:731-740.

[58] Kaether, C; Lammich, S; Edbauer, D; Ertl, M; Rietdorf, J; Capell, A; Steiner, H; Haass, C. Presenilin-1 affects trafficking and processing of betaAPP and is targeted in a complex with nicastrin to the plasma membrane. *J Cell Biol*, 2002;158:551-561.

[59] Walter, J; Capell, A; Grunberg, J; Pesold, B; Schindzielorz, A; Prior, R; Podlisny, MB; Fraser, P; Hyslop, PS; Selkoe, DJ; Haass, C. The Alzheimer's disease-associated presenilins are differentially phosphorylated proteins located predominantly within the endoplasmic reticulum. *Mol Med*, 1996;2:673-691.

[60] Thinakaran, G; Borchelt, DR; Lee, MK; Slunt, HH; Spitzer, L; Kim, G; Ratovitsky, T; Davenport, F; Nordstedt, C; Seeger, M; Hardy, J; Levey, AI; Gandy, SE; Jenkins, NA; Copeland, NG; Price, DL; Sisodia, SS. Endoproteolysis of presenilin 1 and accumulation of processed derivatives in vivo. *Neuron*, 1996;17:181-190.

[61] Esler, WP; Kimberly, WT; Ostaszewski, BL; Diehl, TS; Moore, CL; Tsai, JY; Rahmati, T; Xia, W; Selkoe, DJ; Wolfe, MS. Transition-state analogue inhibitors of gamma-secretase bind directly to presenilin-1. *Nat Cell Biol*, 2000;2:428-434.

[62] Li, YM; Xu, M; Lai, MT; Huang, Q; Castro, JL; DiMuzio-Mower, J; Harrison, T; Lellis, C; Nadin, A; Neduvelil, JG; Register, RB; Sardana, MK; Shearman, MS; Smith, AL; Shi, XP; Yin, KC; Shafer, JA; Gardell, SJ. Photoactivated gamma-secretase inhibitors directed to the active site covalently label presenilin 1. *Nature*, 2000;405:689-694.

[63] Wolfe, MS; Xia, W; Ostaszewski, BL; Diehl, TS; Kimberly, WT; Selkoe, DJ. Two transmembrane aspartates in presenilin-1 required for presenilin endoproteolysis and gamma-secretase activity. *Nature*, 1999;398:513-517.

[64] Chen, F; Yu, G; Arawaka, S; Nishimura, M; Kawarai, T; Yu, H; Tandon, A; Supala, A; Song, YQ; Rogaeva, E; Milman, P; Sato, C; Yu, C; Janus, C; Lee, J; Song, L; Zhang, L; Fraser, PE; St George-Hyslop, PH. Nicastrin binds to membrane-tethered Notch. *Nat Cell Biol*, 2001;3:751-754.

[65] Yu, G; Nishimura, M; Arawaka, S; Levitan, D; Zhang, L; Tandon, A; Song, YQ; Rogaeva, E; Chen, F; Kawarai, T; Supala, A; Levesque, L; Yu, H; Yang, DS; Holmes, E; Milman, P; Liang, Y; Zhang, DM; Xu, DH; Sato, C; Rogaev, E; Smith, M; Janus, C; Zhang, Y; Aebersold, R; Farrer, LS; Sorbi, S; Bruni, A; Fraser, P; St George-Hyslop, P. Nicastrin modulates presenilin-mediated notch/glp-1 signal transduction and betaAPP processing. *Nature*, 2000;407:48-54.

[66] Francis, R; McGrath, G; Zhang, J; Ruddy, DA; Sym, M; Apfeld, J; Nicoll, M; Maxwell, M; Hai, B; Ellis, MC; Parks, AL; Xu, W; Li, J; Gurney, M; Myers, RL; Himes, CS; Hiebsch, R; Ruble, C; Nye, JS; Curtis, D. aph-1 and pen-2 are required for Notch pathway signaling, gamma-secretase cleavage of betaAPP, and presenilin protein accumulation. *Dev Cell*, 2002;3:85-97.

[67] Goutte, C; Tsunozaki, M; Hale, VA; Priess, JR. APH-1 is a multipass membrane protein essential for the Notch signaling pathway in Caenorhabditis elegans embryos. *Proc Natl Acad Sci U S A*, 2002;99:775-779.

[68] Takasugi, N; Tomita, T; Hayashi, I; Tsuruoka, M; Niimura, M; Takahashi, Y; Thinakaran, G; Iwatsubo, T. The role of presenilin cofactors in the gamma-secretase complex. *Nature*, 2003;422:438-441.

[69] Luo, WJ; Wang, H; Li, H; Kim, BS; Shah, S; Lee, HJ; Thinakaran, G; Kim, TW; Yu, G; Xu, H. PEN-2 and APH-1 coordinately regulate proteolytic processing of presenilin 1. *J Biol Chem*, 2003;278:7850-7854.

[70] Baulac, S; LaVoie, MJ; Kimberly, WT; Strahle, J; Wolfe, MS; Selkoe, DJ; Xia, W. Functional gamma-secretase complex assembly in Golgi/trans-Golgi network: interactions among presenilin, nicastrin, Aph1, Pen-2, and gamma-secretase substrates. *Neurobiol Dis*, 2003;14:194-204.

[71] Kimberly, WT; LaVoie, MJ; Ostaszewski, BL; Ye, W; Wolfe, MS; Selkoe, DJ. Gamma-secretase is a membrane protein complex comprised of presenilin, nicastrin, Aph-1, and Pen-2. *Proc Natl Acad Sci U S A*, 2003;100:6382-6387.

[72] Edbauer, D; Winkler, E; Regula, JT; Pesold, B; Steiner, H; Haass, C. Reconstitution of gamma-secretase activity. *Nat Cell Biol*, 2003;5:486-488.

[73] Holen, T; Amarzguioui, M; Wiiger, MT; Babaie, E; Prydz, H. Positional effects of short interfering RNAs targeting the human coagulation trigger Tissue Factor. *Nucleic Acids Res*, 2002;30:1757-1766.

[74] Agrawal, N; Dasaradhi, PV; Mohmmed, A; Malhotra, P; Bhatnagar, RK; Mukherjee, SK. RNA interference: biology, mechanism, and applications. *Microbiol Mol Biol Rev*, 2003;67:657-685.

[75] Donze, O; Picard, D. RNA interference in mammalian cells using siRNAs synthesized with T7 RNA polymerase. *Nucleic Acids Res*, 2002;30:e46.

[76] Hannon, GJ; Conklin, DS. RNA interference by short hairpin RNAs expressed in vertebrate cells. *Methods Mol Biol*, 2004;257:255-266.

In: Trends in RNA Research
Editor: Peter A. McNamara, pp. 39-71

ISBN: 1-59454-506-5

Chapter III

CRD-BP/IMP1: An RNA Binding Protein with Tumorigenic Characteristics

***Panayotis Ioannidis*[1, 2] *and Theoni Trangas*[2]**
[1]Dept. of Cytology, "Sotiria"Chest Dis. Hospital, 152 Mesogeion Ave., Athens, 15161 Greece
[2]Dept. of Genetics, "G. Papanikolaou" Research Center of Oncology, "St. Savvas" Hospital 171 Alexandras Ave., Athens, 11522, Greece,

Abstract

CRD-BP/IMP1 belongs to a family of highly conserved RNA binding proteins which contain four hnRNP K homology (KH) domains and two RNA recognition motifs (RRM) and whose members appear to be involved in the localization, stability and/or translational control of their RNA targets. In humans, besides CRD-BP/IMP1 this family is represented by two other members, KOC/IMP3 and IMP-2 isolated in screens designed to identify genes or tumor-associated antigens specifically expressed in pancreatic and hepatocellular carcinomas respectively. CRD-BP/IMP-1 was initially isolated by means of its high specific affinity to the *c-myc* mRNA Coding Region Determinant (CRD) instability element. In a cell-free mRNA decay system CRD-BP/IMP1 binding protects the message of this proto-oncogene from rapid destruction *via* endonucleolytic cleavage. CRD-BP/IMP-1, IMP-2 and KOC/IMP-3 were also isolated independently, as members of the protein family that binds to the 5' UnTranslated Region (UTR) of the Insulin-like Growth Factor II (*IGF-II*) leader 3' mRNA species and postulated to play a role in its translational regulation. This mRNA species represents the major embryonic form of IGF-II transcripts. Besides leader 3' *IGF-II* mRNA, IMP-1, 2 and 3, bind multiple sites of the *H19* RNA as well, a transcript with no open reading frame that exhibits an oncofetal pattern of expression and whose role remains enigmatic. Recently was shown that CRD-BP/IMP1 co-associates with the G3BP and HuD proteins to the *Tau* mRNA that is localized primarily in the cell body and axon of neurons. CRD-BP/IMP1 is able to

[1] Tel:+30 210-6409168, e-mail: biochem@naf.forthnet.gr; pioannidis@ath.forthnet.gr

translocate to the nucleus and also contains two nuclear export signals. In the cytoplasm CRD-BP/IMP1 is distributed along with microtubules and transports towards the leading edge in migrating cells. It has been postulated that CRD-BP/IMP-1 initially binds to its targets in the nucleus and subsequently plays a role in the polarization of genetic information *via* specific cytoplasmic RNA localization, as it has been shown for two other members of this protein family, the Xenopus Vg1RBP/Vera and chicken ZBP-1 that promote the localization of *Vg1* mRNA to the vegetal pole of the oocyte and beta-actin mRNA to the leading edge of motile cells, respectively. Normally, CRD-BP/IMP-1 is expressed only during fetal life and its expression becomes undetectable shortly after birth. CRD-BP/IMP1 null mice exhibit high perinatal lethality, dwarfism and impaired gut development. Growing evidence demonstrates the *de novo* activation of this gene in high percentages of tumor specimens from different types of human malignancies, including some of the most common, such as breast and colon carcinomas. Furthermore, transgenic mice exhibiting targeted expression of CRD-BP/IMP-1 in adult mammary tissue develop tumors. These findings imply that the *de novo* activation of CRD-BP/IMP-1 in adult tissue, whatever its precise role in neoplasia, may be causally related to the onset or progression of tumorigenesis.

Post-Transcriptional Regulatory Mechanisms and Their Role in the Control of Gene Expression

Mature mRNAs are not passive intermediates between genes and proteins since, besides carrying the genetic information in a translatable form, they may also contain specific information regarding the regulation of their translation, localisation and decay. This information is embedded in specific sequences and/or structures, within discrete areas in the 5’, 3’ UTRs and also in the coding region of the message, usually serving as recognition platforms for RNA-binding proteins. In addition to proteins that recognise specific regions or structures of certain mRNAs, other proteins exhibiting non-specific RNA binding activity, help to prevent misfolding of the molecule and assist in processes such as mRNA transcription and maturation, nucleocytoplasmic transport, translation or decay. Thus, mRNAs *in vivo,* just like the DNA in the nucleus that is present in the form of a dynamic and functional nucleoprotein complex, are associated with a number of proteins determining their common as well as their individual fates that may vary considerably depending on specific mRNA species, cell type and physiological status. A number of findings demonstrate the importance of these RNA-protein interactions in the post-transcriptional regulation of gene expression both at the nuclear and the cytoplasmic compartment. In the cytoplasm, the regulatory mechanisms affecting mature mRNA fate, control mainly the half-life as well as the site, time and duration of translation of a given transcript and play a pivotal role in development and differentiation and for the proper execution of many and diverse biological processes.

The impact of postranscriptional regulation on gene expression has been established by several extensively studied systems. The utilization of altered mRNA stability and its control is illustrated by histone mRNA half-life modulation for the proper execution of the cell cycle. The wide fluctuation in histone mRNA abundance in the absence of corresponding changes in transcription rates, as the cells transit the S phase, is achieved by posttranscriptional mechanisms involving the stem-loop structure found at the 3' end of all replication-dependent histone mRNAs [Heintz *et al.,* 1983; Plumb *et al.,* 1983; Baumbach *et a.l,* 1987]. These sequences direct a regulated processing step that keeps histone mRNA levels low in resting G1/Go cells and contributes to the rapid rise in histone mRNA levels as cells begin proliferative growth [Luscher and Schumperli, 1987]. A processing factor that regulates histone mRNA 3' processing becomes limiting in Gl-arrested cells [Stauber *et al.,* 1986 a; Stauber *et al.,* 1986 b]. These stem-loop sequences at the 3' end have also been shown to play a role in the regulation of histone mRNA stability since transcripts that contain them are targeted for degradation when DNA synthesis is halted [Levine *et al.*, 1987; Pandey *et al.*, 1987]. A stem-loop binding protein (SLBP) binds the 3' end of histone mRNA and participates in all aspects of histone mRNA metabolism. SLBP expression is restricted to the S phase of the cell cycle and accounts for the increase in histone pre-mRNA processing and stabilization observed as mammalian cells enter S phase [Whitfield *et al.,* 2000; Zheng *et al.,* 2003]

Another well-known example of gene expression control, by posttranscriptional mechanisms, involves the regulation of iron metabolism in mammalian cells serving the rapid adaptation of the cell to intracellular iron changes. Key components of the iron homeostatic mechanism are the transferrin receptor that imports iron into the cell and ferritin that stores it within the cell. When intracellular iron levels are depleted the transferrin receptor mRNA is stabilized while ferritin mRNA translation is abrogated [Klausner *et al.,* 1994]. This is attained by binding of the iron regulatory proteins IRP1 and IRP2 on an RNA 23-27 bp stem and 6 bp loop structure known as the iron responsive elements (IREs) [Mullner *et al.,* 1988; Casey *et al.,* 1989; Koeller *et al.,* 1991]. The affinity of IRPs for IREs is regulated by the availability of intracellular iron. In the absence of iron, binding of IRP 1 on the IREs found within the 3'UTR of the transferrin receptor mRNA results in its protection from endonucleolytic attack and stabilization, whereas binding in the 5'UTR of ferritin mRNA results in attenuation of its translation [Klausner *et al.,* 1993; Hentze *et al.,* 1996]. The IRE/IRP system also controls the expression of additional IRE-containing mRNAs, encoding proteins of iron and energy metabolism. The activities of IRP1 and IRP2 are regulated by distinct postranslational mechanisms in response to cellular iron levels [Pantopoulos, 2004].

The importance of localization of mRNAs within the cytoplasm has been documented in numerous cell types including developing oocytes, neurons, oligodentrocytes, myocytes and fibroblasts as a means of partitioning and distributing cytosolic proteins. RNA binding proteins play a crucial role in the creation of asymmetries in development. Messenger RNA localization involves several steps which include the formation of RNP particles by binding of proteins to specific mRNA sequences mostly contained within the 3'UTR, translational repression of the mRNA, motor driven transport along the cytoskeleton and anchoring to actin rich cortex and initiation of translation.

In development, translational regulation of maternal mRNAs serves to restrict their activities in both time and space. Both types of constraint are critical for normal development as is exemplified by the mechanism determining the anterior posterior polarity in the *Drosophila* embryo. The nurse cells of the ovary deposit mRNAs in the developing oocyte, and these maternal mRNAs are distributed to different regions.

Bicoid and *Hunchback* proteins control head and thorax formation [Driever *at al.*, 1988; Struhl *et al.,* 1992] while *Nanos* and *Caudal* are critical for the formation of the abdominal segments [Lehmann and Nusslein-Volhard, 1991; Wu *et al.*, 1998]. The *bicoid* mRNAs are located in the anterior portion of the unfertilized egg, and are tethered to the anterior microtubules [Driever *et al.,* 1989 a]. Upon fertilization the *bicoid* mRNA is translated, forming a protein gradient highest at the anterior [Driever *et al.,* 1988; Ferrandon *et al.,* 1997]. *Bicoid* protein inhibits the translation of the caudal mRNA, permitting *Caudal* protein to be synthesized only in the posterior of the cell [Dubnau and Struhl, 1996; Rivera-Pomar *et al.,* 1996]. *Bicoid* at the same time elevates the level of *Hunchback* protein in the anterior of the embryo by enhancing the transcription of the *hunchback* gene [Struhl *et al.,* 1989, Driever *et al.,* 1989 b]. *Nanos* mRNA is produced by the ovarian nurse cells and is transported into the posterior region of the egg (farthest away from the nurse cells). The *nanos* message is bound to the cytoskeleton in the posterior region of the egg through its 3′ UTR and its association with the products of several other genes such as *oskar* whose activity has to be localized posterior as well, and this is achieved by the binding of the ovarian protein *bruno* to the 3’ UTR of *oskar* mRNA [Castagnetti *et al.*, 2000]. The *nanos* message is dormant in the unfertilized egg, as it is repressed by the binding of the *Smaug* protein to its 3′ UTR [Smibert *et al.*, 1996]. At fertilization, this repression is removed, and *Nanos* protein can be synthesized. Translation of *nanos* mRNA is confined at the posterior pole. *Nanos* in conjunction with *Pumilio* protein, represses translation of maternal *hunchback* mRNA by binding specific sequences in its 3’untranslated region called Nanos-response elements (NREs) [Barker *et al.,* 1992]. In embryos, the length of the poly(A) tail increases immediately prior to translation of *hunchback* mRNA whereas *Nanos* appears to decrease translation by promoting deadenylation of *hunchback* mRNA [Wreden *et al.,* 1997]. Therefore it is apparent that an orchestrated interplay between RNA binding proteins and their target RNAs dictates the formation of morphogenetic gradients.

Given the importance of these post-transcriptional regulatory mechanisms in the regulation of gene expression, it is not unexpected that the inappropriate execution of these processes could lead to aberrant gene expression, and as a result may open the road to cell malfunction and disease. In fact, perturbation of the normal functions of RNA binding proteins that play a role mainly in cytoplasmic processes, has been implicated in clinical disorders.

A disease involving aberrant function of an RNA binding protein is Fragile-X syndrome that is the most common form of inherited mental retardation. Defects have been observed in dendritic cells and affect the number and function of synapses [Bardoni *et al.,* 2002]. The disease is associated with loss-of-function mutations in the *FMR1* gene, usually by trinucleotide repeat expansion. *FMR1* encodes the RNA binding protein FMRP [Ashley *et al.,* 1993] that is widely expressed and has been proposed to act in the regulation of mRNA trafficking from the nucleus to specific cytoplasmic compartments and/or to have a role in the regulation of translation [Inoue *et al*, 2000]. FMRP is a protein with multiple sites of interaction with RNA. Sequence specificity is most likely achieved by the whole block that comprises two K-homologous (KH) modules within the first approximately 400 residues, whereas the C-terminus featuring an RGG box provides a nonspecific-binding surface [Inoue *et al.*, 2000]. Essential for FMRP binding to its target is a purine quartet motif that is also found within its own mRNA implying autoregulation [Schaeffer *et al,* 2001]. Among the mRNAs that harbour FMRP binding elements are those encoding proteins involved in synaptic or developmental neurobiology. The majority of these mRNAs have an altered polysome association in fragile X patient cells [Darnell *et al.,* 2001]. FMRP associates with mRNAs through the non-translatable RNA *BC*1, a short transcript abundant in brain and in dendrites [Muslimov *et al.,* 1997] and evidence exists that its function is to suppress translation of the mRNAs that bind to the *BC*1-FMRP complex [Laggerbauer *et al.,* 2001; Zafla *et al.,* 2003].

The alpha-CP proteins were originally discovered by virtue of their connection to alpha thalassemia. A naturally occurring α globin mutant, α-Constant Spring (α-CS) has been extensively studied, and the findings clearly demonstrate the importance of mRNA stability in α-globin gene expression. The α-CS mutation was determined to be a base change in the normal stop codon of the α2-globin reading frame promoting read through into the 3'-UTR for an additional 31 codons and resulting in destabilization of the α2-globin mRNA, loss of >95% of α2-globin gene expression from the affected locus, and consequently clinical disease (α-thalassemia) [Clegg *et al.,* 1971; Morales *et al.*, 1997]. Alpha2 -globin mRNA stability is dependent upon the assembly of a sequence-specific 3'-UTR RNP α-Complex [Wang *et al.,* 1995; Holcik *et al.,* 1997]. Two closely related (80% identical) proteins αCP-1 and -2 have been purified by affinity chromatography using α-globin 3'-UTR sequences, and the corresponding genes subsequently cloned [Kiledjian *et al.,* 1995]. Both these proteins, αCP-1 and -2 containing each three copies of KH domains [Gibson *et al.,* 1993], are incorporated into the α-Complex *in vitro.* The α-Complex is a binary complex containing a single molecule of the α-CP protein bound to the pyrimidine-rich *cis* element in the α-globin mRNA 3'UTR [Chkheidze *et al.,* 1999]. The α-Complex appears to stabilize the α globin mRNA by protecting it from both deadenylation and endonucleolytic cleavage [Wang *et al.,* 2000].

The abnormal expression of RNA binding proteins has also been detected in malignancies and may associate with the process of tumorigenesis or certain cancer cell characteristics. The mRNA-stability factor HuR overexpression is associated with aggressive colon carcinomas and has been linked to the elevated expression of COX-2, VEGF and IL-8 [Dixon *et al.,* 2001]. Whereas the role of COX-2 in promoting colon carcinogenesis is established, the mechanism underlying its overexpression is largely unknown. The above finding links the deregulated expression of COX-2 to altered posttranscriptional processing resulting from HuR overexpression. In addition to colon cancer, increased nuclear and cytoplasmic expression of HuR was also observed in ovarian carcinomas as compared to borderline tumors and normal ovaries. Elevated HuR expression correlated with increased COX-2 expression and was found to be an independent prognostic factor for progression free as well as for overall survival of ovarian cancer patients [Denkert *et al.,* 2004]. Overproduction of IL-10 in melanoma cells in comparison to normal melanocytes could help tumors to evade immune surveillance and escape rejection. The overexpression of IL-10 is probably due to the relative stabilization of the IL-10 mRNA which in melanoma cells exhibits a half-life of 75 min compared to a half-life of 7 min in normal melanocytes. It has been postulated that IL-10 mRNA half-life change is likely a result of the altered cellular distribution of the AUF1 protein whose expression in tumor cells is restricted to the nucleus.

AUF1 is an RNA-binding protein recognizing specifically (A+U)-rich elements (ARE) found in the 3' UTR of several unstable messenger RNAs, and probably acts as a destabilizing factor [Brewer *et al.,* 2003].

Another RNA binding protein associated with tumorigenesis is CRD-BP/IMP1 that normally is expressed only during embryogenesis. Following in this chapter we present and discuss the accumulating evidence implicating the *de novo* activation of CRD-BP/IMP1 in the onset and progression of malignancies.

Identification of CRD/IMP1 Protein, Its Known RNA Targets and Their Function

CRD-BP/IMP1 was first identified, by J. Ross and colleagues, as a protein binding to the *C*oding *R*egion *D*eterminant instability region of c-*myc* mRNA (*CRD-B*inding *P*rotein), protecting the message from endonucleolytic cleavage and extending its half-life, up to eight fold, implying a role for this protein in the regulation of c-*myc* expression [Bernstein *et al.,* 1992]. CRD-BP/IMP1 has been cloned independently, and named IMP1, by Nielsen *et al.* [1999] as a member of the IMP protein family binding to the 5' UTR of the leader 3' *IGF-II* mRNA species, that represents the major embryonic form of the *IGF-II* transcripts. The same group has shown that CRD-BP/IMP1 also binds to the *H19* RNA, an RNA whose role remains unknown [Runge *et al.,* 2000]. The aberrant expression of the above CRD-BP/IMP1 RNA targets is frequently encountered in cancer cells. Recently it has been shown that CRD-BP/IMP1 also recognises the axonally localized *tau* mRNA that is expressed in developing neurons [Atlas *et al.,* 2004].

C-*myc* mRNA: C-*myc* encodes for a helix-loop-helix basic leucine transcription factor that upon dimerisation with its obligatory partner Max recognizes the E-box DNA sequence (CACGTG) and acts as an activator, but for some promoters as a repressor, of transcription [Reviewed: Amati and Land, 1994; Grandori *et al.,* 2000]. These same DNA sequences are also recognized by alternative dimers, containing as an obligatory partner Max and as the second one usually one of the Mad family proteins. The two types of dimers are thought to exert opposing effects upon transcription. Since Max protein levels are constant, the activation or repression of the target genes depends upon the relative levels of *Myc* family members (N-*Myc* and L-*Myc;* being the two other members) and those of their competitors. In general Myc:Max heterodimers are characteristic of proliferating cells, while Mad:Max heterodimers are found in differentiated cells. The aberrant expression of c-*myc* (or N-*myc* / L-*myc*, in some tumour types) presumably perturbs the balance between Myc:Max and Mad:Max dimers, altering the genetic programme of the cell. Since its discovery, thousands of studies have made apparent the association of aberrant c-*myc* expression, caused by various mechanisms such as retroviral insertions, chromosomal translocations, amplification, or mutations, with cancer [Reviewed: Spencer and Groudine, 1991; Nesbit *et al.*, 1999]. Experimental data showing that c-*myc* deregulated expression affects cell proliferation, apoptosis, growth, metabolism and differentiation, have linked its aberrant expression with characteristics of cancer phenotype [Karn *et al.,* 1989; Evan and Littlewood, 1998; Iritani and

Eisenman, 1999; Pelengaris *et al.,* 2002]. Furthermore, animal models have established its causal relationship to the onset and/or progression of carcinogenesis [Schoenenberger *et al.,* 1988; Skoda *et al.,* 1995; Felser and Bishop, 1999; Pelengaris *et al.,* 2000]. Despite the widespread presence of E-boxes in the genome, due to the weak c-Myc transactivating action, few unequivocal targets were identified until recently [Reviewed: Dang, 1999]. However, accumulating data, mainly from microarray [Coller *et al.*, 2000; Schuldiner and Benvenisty, 2001; Watson *et al.,* 2002] and SAGE [Menssen and Hermeking, 2002] analysis, now indicate that c-*myc* seems to affect directly the expression of hundreds of genes, encoding proteins involved in practically "every major biochemical and regulatory process in the cell"[Levens, 2002]. The transcription programme orchestrated by c-*myc* could be even more complex as it has been shown that all E-boxes do not exhibit the same affinity for Myc/Max dimers [Fernandez *et al.,* 2002]. The presence of these "weaker" targets suggest that in normal life there could be instances where varying levels of c-*myc* expression may affect varying numbers of target genes and furthermore, imply that the aberrantly elevated expression of c-*myc* may activate "illegitimate" targets, arguing against a simple on-off mode of action.

Being a "senior administrator of the cell" and "as an oncogene a high-ranking member of the gang" of oncogenes [Levens, 2003], it is not surprising that the tight regulation of c-*myc* expression is under the control of multiple and complex mechanisms, acting both at the transcriptional and the post-transcriptional levels. C-*myc* is transcribed to a highly unstable mRNA that in turn is translated to a highly unstable protein [Dani *et al.*, 1984; Ramsay *et al.,* 1984]. The c-*myc* mRNA contains in its 3' UTR an Adenylate and Uridylate rich (A+U-rich) element (ARE) that is a common characteristic of many unstable transcripts encoding cytokines, growth factors and oncogenes. Several approaches such as deletions, extensive mutagenesis or transfer of the 3' UTR of unstable mRNAs to stable reporter messages highlighted the importance of these AREs as mRNA destabilizing elements [Shaw and Kamen, 1986; Zumbiaga *et al.,* 1995; Chen and Shyu, 1995; Peng *et al.,* 1996]. The effect of AREs on mRNA stability is mediated by proteins binding specifically the AU-rich elements (AU-BPs) and may either stabilize or destabilize the cognate mRNAs. Examples of stabilizing AU-BPs are the mammalian homologues of *Drosophila* ELAV, HelNi, HuC, HuD and HuR [Reviewed: Burd and Dreyfuss, 1994; Keene, 2001; Tourriere *et al.,* 2002] whereas a c-*myc* mRNA destabilizing AU-BP is the hnRNP D protein [Brewer, 1991]. While transfer of the c-*myc* ARE to stable reporter genes induced a destabilizing effect, other methodological approaches, such as defined deletions or even the complete removal of this region, resulted in minor effects on the half life of the truncated c-*myc* mRNAs [Jones and Cole, 1987; Bonnieu *et al.,* 1990; Laird-Offringa *et al.,* 1991]. These observations suggested the existence of additional destabilizing elements elsewhere in the message. Analysis of the half-life of hybrid mRNAs, containing portions of the c-*myc* coding region fused to stable reporter genes, identified an additional instability element, comprising the last 180 coding nucleotides, that was named CRD [Wisdom and Lee, 1991; Bernstein *et al.,* 1992]. The fact that the CRD and the ARE of c-*myc* mRNA act independently to regulate c-*myc* half-life comes from the following observations: 1) Removal of CRD from the intact c-*myc* mRNA stabilizes the message 2) Inserting in frame the c-*myc* CRD region in the beta-globin mRNA destabilizes this normally very stable message, and 3) CRD is indispensable for the modulation of c-*myc* mRNA half-life in mouse hepatocytes undergoing liver regeneration and in myoblasts induced to differentiate. In addition to the studies identifying the *cis*-acting destabilizing elements other studies have shown that c-*myc* mRNA half-life may vary in response to external stimuli, indicating that the mechanisms ensuring the short half-life of c-*myc* mRNA do not have a passive role and do not exist merely to prevent the accumulation of elevated MYC levels and to warrant their rapid clearance from the cell upon transcriptional shut down. The c-*myc* mRNA half-life, usually 10-20 min, can be altered when necessary. In serum starved G_0 arrested Chinese hamster fibroblasts the dramatic increase of c-*myc* mRNA levels, following serum stimulation, has been shown to be due to c-*myc* mRNA stabilisation and not to altered transcription rates [Blanchard *et al.*, 1985]. In some cell lines the antiproliferative action of type I Interferons is preceded by c-*myc* mRNA downregulation as a result of a two to three fold reduction of its half-life [Knight *et al.*, 1984; Dani *et al.*, 1985]. Following partial hepatectomy, in proliferating adult hepatocytes the c-*myc* mRNA is stabilized several fold [Lavenu *et al.*, 1995], whereas, in the C2C12 myoblastic cell line the c-*myc* mRNA half-life is reduced upon induction of differentiation to myotubules [Yeilding *et al.*, 1998]. The above-mentioned examples indicate the flexibility of the mechanisms controlling c-*myc* mRNA degradation and their active role in regulating c-*myc* expression *via* alterations of c-*myc* mRNA stability.

Various methodological approaches have been used to study the decay pathways of mRNAs in eukaryotic cells. For most studied mRNAs, both stable and labile, the first step that precedes the decay of the main body of the message is the gradual removal of the poly(A) tail usually leading to the generation of an oligoadenylated intermediate [Reviewed :Sachs, 1993; Decker and Parker, 1994]. However, this routine poly(A) shortening of labile messages exhibits different kinetics from those observed with stable ones. The AREs of labile messages, divided into three classes, on the basis of sequence features and functional properties, promote the rapid decay of the message albeit with distinct kinetics [Xu *et al.*, 1997]. Class II AREs, usually found in cytokine mRNAs, direct asynchronous deadenylation that leads to the generation of a completely deadenylated and highly unstable message. Both class I AREs, mostly found in the mRNAs of early response genes, like c-*fos* and c-*myc,* and class III AREs direct a biphasic pattern of decay. In the fist phase all mRNA molecules are synchronously deadenylated with little or no degradation of the message whereas in the second phase poly(A) shortened messages are rapidly degraded. Experimental data have shown that c-*myc* mRNA decay follows this biphasic process, and indicate that the half-life of these oligoadenylated intermediates may be twice as long as that of their polyadenylated counterparts [Swartwood *et al.,* 1987; Ioannidis *et al.,* 1996]. The enhanced translatability of the oligodenylated c-*myc* mRNA synthesized in the presence of the polyadenylation inhibitor cordycepin, implies that these intermediates may serve as templates as long as they persist in the cell [Ioannidis *et al.,* 1999]. Indications that c-*myc* mRNA may follow an alternative degradation pathway as well, was postulated for the first time by Swartwout and Kinniburgh [1989], who compared the decay rates of polyadenylated and oligoadenylated c-*myc* mRNA species in the HL-60 cell line before and early after the induction for differentiation. The permanent shut down of c-*myc* expression in differentiated HL-60 cells is accomplished by the transcriptional silencing of the gene a few days following the differentiation stimulus. However, early upon the induction for differentiation the c-*myc* mRNA levels decline due to the rapid loss of the polyadenylated message without its prior deadenylation, implying the activation of an alternative degradation mechanism. The *in vivo* generation of truncated 5' and 3'adenylated c-*myc* mRNA species was first demonstrated in a HeLa cell line overexpressing c-*myc* mRNA [Ioannidis *et al.,* 1996]. These fragments, whose production was augmented upon exposure to Interferon-alpha, a factor known to modulate c-*myc* mRNA half-life, were generated by endonucleolytic cleavage in the 3'UTR region of the message by a mechanism not necessarily requiring the prior removal of the poly(A) tail. Subsequently, it was shown that the c-*myc* 3' UTR may serve as target of the G3BP endoribonuclease [Gallouzi *et al.,* 1998; Tourriere *et al.,* 2001]. G3BP is a phosphorylation dependent endoribonuclease that associates with RasGAP (Ras-GTPase activating protein). Ras-GAP protein accelerates the hydrolysis of the Ras- associated GTP converting Ras to its inactive form. RasGAP mediates interactions with additional to Ras proteins, and G3BP is likely one of the effectors of its action [Tocque *et al.,* 1997]. G3BP, whose RNase activity is abolished *in vitro* by alkaline phosphatase treatment, is hyperphosphorylated in quiescent cells and undergoes dephosphorylation in serum-stimulated cells probably through its interaction with RasGAP. Analysis of G3BP deletion mutants has shown that the binding of the protein to its RNA target and its cleavage activity can be separated. G3BP harbours a carboxyl terminal RNA binding domain (an RRM domain) that dictates the binding of the protein to the consensus sequence ACCCA(U/C)(A/C) (C/G)G(C/A)AG implying that its RNA targets comprise a subset of cellular mRNAs containing this high affinity binding sequence. G3BP has been shown to cleave specifically c-*myc* mRNA at several sites within its 3'UTR in an *in vitro* reaction assay. Furthermore, in mouse embryonic fibroblasts containing a defective G3BP isoform, c-*myc* mRNA decay was found to be delayed implying that this regulated endoribonuclease may execute the task of regulated mRNA decay conveyed through RasGAP signalling in response to external stimuli.

The fact that the c-*myc* mRNA CRD instability element could also serve as a target for endonucleolytic cleavage was first demonstrated by J. Ross and colleagues that used an *in vitro* mRNA decay system recapitulating many aspects of the degradation pathways in intact cell [Bernstein *et al.,* 1992]. They postulated that CRD may function as a target for an RNA-binding protein involved in c-*myc* mRNA decay regulation and supplemented the decay system with excess of an RNA competitor corresponding to the CRD region. In the presence of the competitor the c-*myc* mRNA was 8-fold less stable and this was proven to be due to endonucleolytic cleavage within the CRD region, implying that the protein(s) binding to the region acts as a shield protecting the message from cleavage. The protein recognising specifically the CRD region was named CRD-BP, purified to homogeneity [Prokipcak *et al,* 1994] and subsequently cloned [Doyle *et al.,* 1998].

Apparently, the mode of action of the c-*myc* CRD differs from that of the c-*fos* and Interferon-beta coding region instability elements. Whereas c-*fos* and Interferon-beta coding region elements trigger the rapid deadenylation [Shyu *et al.,* 1991; Paste *et al.,* 2003], like the AREs, the c-*myc* CRD provides an endonucleoase sensitive target and thus may represent the substrate used to switch between the two alternative decay pathways. The activation of alternative decay pathways, initiating *via* endonucleolysis within specific regions, has been shown to be pivotal for the rapid clearance of certain mRNAs in response to alterations of the internal milieu or to external stimuli [Stoeckle 1992; Binder *et al.,* 1994; Ioannidis *et al.,*

1996; Cunningham *et al.*, 2001]. Based on the observation that the c-*myc* CRD region is highly enriched in rare codons the following model has been proposed [Lemm and Ross, 2002]. Transient ribosome pausing within the 5' segment of the CRD region may leave the 3' segment of the CRD exposed to endonucleolysis unless protected by CRD-BP/IMP1 protein binding. A schematic representation summarizing the pathways and factors involved in c-*myc* mRNA decay is shown in Figure 1.

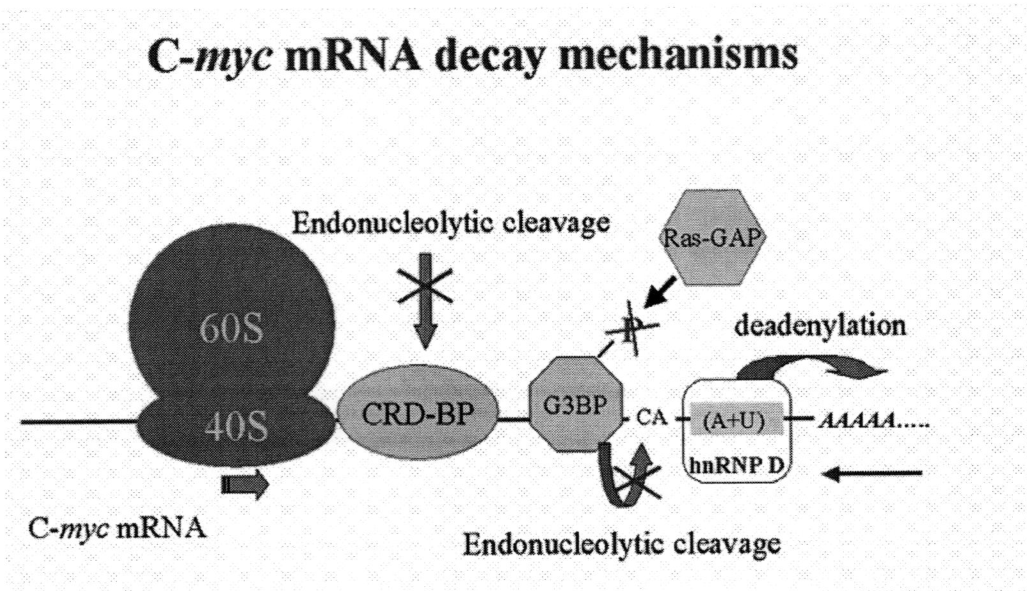

Fig1. C-myc mRNA decay pathways and regulatory proteins. C-*myc* mRNA contains a Class I ARE (A+U) that directs a biphasic pattern of decay. Prior to the degradation of the main body of the message, synchronous deadenylation leads to the generation of poly(A) shortened intermediate whose half life could be twice as long as that of their polyadenylated counterparts. The AU-binding protein hnRNP D protein binds to the c-*myc* ARE and promotes the degradation of c-*myc* mRNA in an *in vitro* decay system. C-*myc* mRNA may follow alternative pathways of decay initiating *via* endonoucleolytic cleavage not necessarily requiring the prior removal of the poly(A) tail. Two regions, the 3' UTR and the CRD have been shown to serve as targets for endoribonucleolysis. G3BP a phosphorylation dependent endoribonuclease regulated by RasGAP, recognizes the consensus sequence ACCCA(U/C)(A/C) (C/G)G(C/A)AG and cleaves specifically c-*myc* mRNA at several sites within its 3'UTR.The c-*myc* mRNA CRD instability element, that is recognized by CRD-BP/IMP1, could also serve as a target for endonucleolytic cleavage. Transient ribosome pausing within the 5' segment of the CRD region, which is highly enriched in rare codons, may leave the 3' segment of the CRD exposed to endonucleolysis unless protected by CRD-BP/IMP1 binding.

IFG-II Leader 3' mRNA and H19 RNA: In addition to c-*myc* mRNA, CRD-BP/IMP1 specifically binds *IFG-II* leader 3' mRNA and *H19* RNA. The expression of these two CRD BP/IMP1 targets is linked. The maternally expressed *H19* gene is located 90 kilobase pairs downstream from the paternally expressed *insulin-like growth factor II* (*IGF-II*) gene on human chromosome 11 and mouse chromosome 7; and due to their reciprocal imprinting and identical spatiotemporal expression, it is assumed that the two genes are functionally coupled. A 2.4 kb imprinting control region regulates somatic monoallelic expression of the *IGF-II* and *H19* genes. This is achieved through DNA methylation – dependent chromatin insulation and promoter silencing activities on the maternal and paternal chromosomes. In human somatic cells the hypomethylated maternally inherited imprinting control region binds the insulator protein CTCF at four sites, blocking the adjacent *IGF-II* gene transcription by insulating the promoter from its distal enhancers [Ishihara *et al.*, 2002]. On the other hand, the *H19* imprinting results from the methylation of the 7-9 kb domain surrounding the corresponding paternal allele [Drewell *et al.*, 2000]. Thus, the *H19* gene is exclusively expressed from the maternal chromosome and this also depends on CTCF binding which has a positive effect upon its transcription [Drewell *et al.*, 2000; Scoenherr *et al.*, 2003]. Down regulation of CTCF expression results in increased methylation of the *H19* differentially methylated domain (DMD) and decreased developmental competence of CTCF-deficient oocytes [Fedoriw *et al.*, 2004]. The *H19* gene product is a 12.3 kb long RNA spliced and polyadenylated. The gene transcribed in both humans and mice consists of five exons separated by very short introns. No translation product has been detected even though at least in humans a putative ORF exists which would encode for a 26 kD protein. *H19* RNA is an oncofetal RNA expressed at high levels in placenta and several fetal tissues. After birth *H19* expression declines and drops to undetectable levels in most tissues except in skeletal muscle, heart and mammary cells [Ohana *et al.*, 1999]. It is also expressed *de novo* in tumors derived from embryonic tissues. No function has been assigned to the *H19* gene product, which clearly seems to play its role as RNA. The loss or inactivation of the maternal copy, which normally is in an active chromatin conformation, has been found to be associated with several tumors, including Wilms' tumor. Its *de novo* activation in certain tumor types and on the other hand, the loss of maternal expression in others, implies a tumor suppressor function [Drewell *et al.*, 2000; Cui *et al.*, 2002; Bock *et al.*, 2003].

Insulin-like growth factor II is a fetal growth factor with auto- and paracrine modes of action [Stewart and Rotwein, 1996]. IGF-II can act as a cell survival and mitogenic factor and it can also modify metabolism. In the mouse, lack of IGF-II results in a small but apparently normal progeny [Baker *et al.*, 1993], whereas increased supply of embryonic IGF-II frequently results in disproportionate overgrowth associated with heart enlargement, oedema and fetal death, and may account for the large offspring syndrome. In humans, increased levels of IGF-II are associated with the Beckwith-Wiedemann syndrome, which is characterized by a disproportionate overgrowth and malformations [Weksberg *et al.*, 1993; Eggenschwiler *et al.*, 1997; Sun *et al.*, 1997]. Overexpression of IGF-II occurs commonly in tumors, functioning possibly as an autocrine factor [Reviewed: LeRoith *et al.*, 1995; Toresky *et al.*, 1996; Zumkeller and Burdach, 1999; Yu and Rohan, 2000].

IGF-II expression is also largely regulated by posttranscriptional mechanisms. The human IGF-II gene is a complex transcriptional unit containing four promoters that are active in a developmental stage-dependent and tissue-specific fashion [Sussenbach, 1989]. Transcription from these different promoters yields a population of transcripts ranging in size from 2.2. kb to 6.0 kb. The various transcripts differ in their 5'-UTR sequences and are translated with different efficiencies. All transcripts share the 552 nt long IGF-II coding region, and apart from the minor 2.2 kb transcript, which is polyadenylated at an alternative upstream polyadenylation site, the 4 kb long 3'-UTR is common to all transcripts. The leader 4' mRNA with a 5' UTR of 100 nucleotides is constitutively translated. On the other hand, the more abundant leader 3' mRNA, which contains a long 1,170-nucleotide cytidine rich 5' UTR, is stored in a 100S particle and not translated except in growing cells [Nielsen *et al.*,

1990]. Moreover, the mouse homologue *IGF-II* leader 3' mRNA ceases to be translated between embryonic day 11.5 (E11.5) and E12.5 [Newell *et al.,* 1994]. CRD-BP/IMP-1 associates specifically with the 5' UTR of the human 6.0-kb *IGF-II* leader 3' mRNA and its overexpression was shown to repress translation of a chimeric leader 3'-luciferase mRNA *in vivo* [Nielsen *et al.,* 1999].

***Tau* mRNA:** *Tau* mRNA encodes for a microtubule-associated protein (MAP) that is localized primarily in the cell body and the axon of neurons. Neurons are highly polarized cells elaborating morphologically distinct axons and dendrites that are characterized, among other things, by the different polarity of their microtubules and the composition of their associated MAPs [Ginsburg, 1991]. Accumulating evidence indicates that mRNA pre-localization is critical for the creations of these asymmetries. Messenger RNAs travel as motile RNPs (RNA-protein particles) across dendrites and axons and are locally translated in response to external stimuli and growth factors [Schuman E.M., 1999; Snee *et al.,* 2002]. *Tau* mRNA traffics in a microtubule dependent manner from the cell body to the distal parts of the axon in developing neurons and its axonal localization is controlled by a *cis* element located in the 3' UTR of *tau* mRNA, known as the axon localization signal (ALS) [Aronov *et al.,* 1999]. HuD, an Elav protein that stabilizes several neuronal mRNAs, targeted to the axon, binds to the ALS element [Aranda-Abreu *et al.,* 1999]. Atlas *et al.* [2004] using as a bait HuD protein isolated *Tau* RNP particles and analysed their protein composition. Besides HuD, two cytoplasmic proteins were also identified, CRD-BP/IMP1 and G3BP. Furthermore, upon using the G3BP protein as bait, HuD and CRD-BP/IMP1 co-precipitated. The association of the three proteins was proven RNase sensitive implying that complexes containing all three or any two of the isolated proteins must form in an RNA-depended manner[2]

CRD-BP/IMP1 Protein Characteristics and Protein Relatives

Cloning of CRD-BP/IMP1 cDNA [Doyle *et al.,* 1998, Nielsen *et al.,* 1999] predicted a protein sequence of 577 amino acids containing three characteristic RNA-binding protein motifs: four copies of the KH (hrRNP K homology) domain at the C-terminus, an RGG motif and two RRM motifs (RNA recognition motifs) in the N-terminus. CRD-BP/IMP1 belongs to a highly conserved family of RNA-binding proteins that in humans is represented by three homologues: CRD-BP/IMP1, IMP2 and IMP3/KOC. Other known members of this family are the *Xenopus* Vg1 RNA-binding protein (xVg1-RBP/Vera) and the chicken zipcode binding protein (cZBP-1). Proteins in this family are closely related exhibiting 70-95% amino acid identity [Doyle *et al.,* 1998; Yaniv and Yisraeli, 2002]. Vg1-RBP/Vera binds with high affinity to the vegetal localization element (VLE) of the *Xenopus Vg1* RNA [Schwartz *et al.,*

---

[2]. In addition CRD-BP/IMP1 was recently shown to interact with the FMRP protein isoform 18 [Rachham & Brown, 2004] and to cooperate with the poly(A) binding protein PABP and the UNR protein [Patel *et al.,* 2005] to form heteromeric RNP complexes. These findings imply that CRD-BP/IMP1 may participate in multiprotein regulatory "factors" controling various aspects of post-transcriptional regulation of more than a few mRNAs.

1992] whereas cZBP1 binds to the localization elements present in the 3'UTR region of beta-actin mRNA that form a sort of zipcode [Ross *et al.,* 1997]. The acronym VICKZ has been proposed for this protein family where V stands for Vg1-RBP/Vera, I for IMP, C for CRD-BP, K for KOC and Z for ZBP-1 [Yaniv and Yisraeli, 2002]. The different homologues from the various species can be classified in three subfamilies each represented by a different human homologue. A putative *Drosophila* homolog has been identified that lacks the two N-terminal RRMs but contains the four KH domains at the C-terminus, exhibiting a 47% sequence similarity within this region with all individual members of the VICKZ family [Nielsen *et al.,* 2000]. Vertebrates, with the apparent exception of *Xenopus*, exhibit three members the VICKZ gene family that might have arisen from two gene duplications, shortly before their segregation during evolution.

The KH domain, first found in the hnRNP K protein, is present in one or more copies in many RNA binding proteins and shows preference for single stranded RNA [Ostarek-Lederer *et al.*, 1998]. The RGG motif, usually found in RNA binding proteins along with other RNA-binding domains, was first identified in the hnRNP U protein and is formed by a variable number of Arg-Gly-Gly (RGG) repeats. The RRM motif consists of about 90 amino acids and contains two short sequences (RNP1 and RNP2) that have been shown to form the RNA binding domain [Burd and Dreyfuss, 1994]. The first CRD-BP/IMP1 RRM motif contains typical RNP1 and RNP2 sequences whereas in the second one the RNPs diverge from the consensus sequences. Electrophoretic mobility-shift analysis of deletion mutants of CRD-BP/IMP1, using as target a high affinity 173 nucleotide segment of the *H19* RNA, revealed that the four KH domains, but not the two RRMs, are sufficient for RNA binding [Nielsen *et al.,* 2002]. Mutants lacking either the KH1-2 or the KH3-4 modules exhibited no binding activity, implying that the entire KH segment is the binding region. UV cross-linking experiments using as target a 600 bp nucleotide fragment from *tau* mRNA 3'UTR also revealed that the 4 KH domain containing portion of the protein, and not the RRM 1-2 N-terminal region, mediates RNA binding to this target as well [Atlas *et al.,* 2004]. However, the lower binding affinity of the four KH domain segment alone, for both targets, compared to the entire CRD-BP/IMP1 protein, implies that the two RRMs may contribute towards a lower K_D.

The specificity of CRD-BP/IMP1 for its known RNA targets has been analysed by gel shift and UV cross linking assays as well as with functional assays using the *in vitro* decay system. The K_D of CRD-BP/IMP-1 for c-*myc* CRD and *IGF-II* leader 3' regions have been estimated to 0.7 nM and 0.1 nM respectively [Prockipcak *et al.,* 1994, Nielsen *et al.,* 1999, Ioannidis *et al.,* 2001]. The difference probably reflects the presence of multiple attachment sites to the *IGF-II* leader 3' RNA compared to the one binding site in the c-*myc* CRD region. Four binding sites with K_D values ranging from 0.4 to 1.3 nM were identified in the 3' terminal segment of *H19* RNA encompassing exons 4 and 5 [Runge *et al.,* 2000]. Whereas IMP1 is able to bind to these RNA regions with K_D at the subnanomolar range, no binding has been detected to a number of other *in vitro* synthesised RNAs (i.e. beta-globin, 5' UTR of the leader 4' *IGF-II* RNA or prokaryotic RNA templates*)* showing that this protein recognizes specific RNA targets and does not exhibit any non-specific RNA binding activity. CRD-BP/IMP1 has been shown to bind also to the region of the N-*myc* mRNA that is homologous to the c-*myc* CRD, albeit with 5 fold lower affinity. This segment of both mRNAs encodes the highly conserved helix-loop-helix motifs of the proteins and exhibits 68% similarity at the nucleotide level. However, upon using a 1000 fold excess of this N-*myc* RNA region as a competitor in the *in vitro* decay system, no effect was observed upon the stability of RNA transcripts containing the c-*myc* CRD element. Apparently, the lower affinity for the N-*myc* RNA region could not titrate away CRD-BP/IMP1 from c-*myc* mRNA [Prokipcak *et al.,* 1994]. Furthermore, the 27 nt zipcode beta-actin mRNA element, that was used to purify the cZBP1 protein, is not able to gel shift the recombinant mouse CRD-BP/IMP1 protein, implying the high specificity of these closely related proteins (mCRD-BP/IMP1 and cZBP1 exhibit 77.9% identity and 83.8% similarity in their protein sequence) for their respective RNA targets [Doyle *et al.,* 1998]. No significant sequence similarity exists among the so far identified CRD-BP/IMP1 RNA binding sites and therefore it is not possible to predict other putative RNA targets on the basis of their sequence. Furthermore, probing with ribonoucleases of the *H19* RNA segment that exhibits the higher affinity for CRD-BP/IMP1 indicated the absence of a higher order structure [Runge *et al*, 2000]. Thus specificity of target recognition may rely upon protein protein – protein interactions and this is supported by the following observations: 1) There are indications that the KH domain may have the ability to oligomerize and to mediate protein –protein interactions [Chen *et al.,* 1997]. 2) Some of the CRD-BP/IMP1 targets exhibit multiple attachment sites. 3) Other members of the IMP family exhibit binding activity for the same RNA segments, and 4) other proteins, besides IMPs, bind to the same or adjacent RNA regions as has been exemplified by the cooperative association of CRD-BP/IMP1, G3BP and HuD with *tau* mRNA [Atlas *et al.,* 2004]. CRD-BP/IMP1 and HuD bind to distinct regions in the 3' UTR region of the message and binding is mediated by the CRD-BP/IMP1, 1-4 KH and the HuD RRM3 domains. RNase treatment abolishes HuD-CRD-BP/IMP1 association indicating that these proteins do not interact in the absence of an RNA molecule, nevertheless, it is possible that binding of one of these proteins to its RNA target sequence may facilitate the binding of the other. It is noteworthy that an RNA-dependent association of HuD and CRD-BP/IMP1 was also observed in the human embryonic kidney HEK-293 cell line that does not express the *tau* mRNA implying the existence of additional shared targets besides this message.

CRD-BP/IMP1 Putative Functions and the Pattern of its Expression in Development

Prockipcak *et al.* [1994] purified CRD-BP/IMP1 from polysomes isolated from the K562 human erythroleukemia cell line and using an *in vitro* RNA decay system proposed a role for this protein in the regulation of c-*myc* mRNA decay. Nielsen *et al.* [1999] purified CRD-BP/IMP1 from the human rabdomyosarcoma cell line RD and showed that the overexpression of this protein in mouse NIH-3T3 cells inhibited the translation of reporter mRNAs containing the leader 3', but not the leader 4' *IGF-II* mRNA segments.

Indications for alternative or additional roles of CRD-BP/IMP1 came from the functions ascribed to the xVg1-RBP/Vera and cZBP-1 proteins. Vg1-RBP protein binds specifically to the VLE sequence of *Vg1*mRNA and mediates its association with microtubules *in vitro* [Yisraeli *et al.,* 1990; Mowry and Melton, 1992; Gautreau *et al.,* 1997]. Thus, it appears to be a trans acting factor for the localization of *Vg1* mRNA [Deshler *et al.,* 1998; Havin *et al.,* 1998]. Furthermore, Vg1-RPB localizes asymmetrically to extended processes in migrating neural crest cells [Yaniv *et al.,* 2003]. The *trans*-acting factor ZBP1 binds directly to the zipcode of beta-actin mRNA and colocalizes with the mRNA to the leading lamellae of Chicken embryo fibroblasts and to the neuronal growth cone [Ross *et al.,* 1997]. The formation of a complex involving both β–actin mRNA and ZBP1 is a prerequisite for the increases in β-actin mRNA levels and the formation of the growth cone in neurotrophin-stimulated neurons [Zhang *et al.*, 2001]. Such complex is also necessary for cellular polarity, lamellar localization of ZPB1 and the persistent movement of migrating embryo fibroblasts [Kislauskis *et al.,* 1997; Farina *et al.,* 2003].

Immunocytochemistry has revealed that CRD-BP/IMP1 is predominantly cytoplasmic exhibiting either a localized concentration in perinuclear regions and lamellipodia or a completely delocalised pattern [Nielsen *et al.,* 1999]. Nielsen *et al.* [2002] have used GFP-full length CRD-BP/IMP1 as well as GFP-truncated CRD-BP/IMP1 fusion proteins to study the intracellular distribution and behaviour of CRD-BP/IMP1. The delocalised distribution initially observed in the majority of resting NIH-3T3 cells turned to a localized pattern upon seeding the cells on laminin coated plates that stimulate their migration. CRD-BP/IMP1 present in large granules exhibited a characteristic movement towards the leading edge into the cortical region of extending lamellipodia at an average speed estimated to 0.12 μm/sec. Subsequently they have shown that the CRD-BP/IMP1 containing granules associate with microtubules in the cytoplasm and with microfilaments in the lamellipodia. The GFP-truncated CRD-BP/IMP1 fusion constructs revealed that the protein region containing the four KH domains was sufficient for cytoplasmic trafficking whereas deletion of either KH1 or KH4 domains abolished both movement and lamellipodia anchoring. These findings imply that the same four KH domain-containing region probably mediates both intracellular trafficking and RNA binding, since it is indispensable for RNA recognition.

CRD-BP/IMP1 contains two putative nuclear export signals (NES) and a putative nuclear localization signal (NLS), suggesting the facilitated transport to the nucleus of this ~67 kDa protein. Whereas immunocytochemistry revealed a predominant cytoplasmic localization, two observations imply that CRD-BP/IMP1 can also enter the nucleus: 1) Nuclear staining has been observed in cells overexpressing CRD-BP/IMP1, upon transfection of the gene under the control of the strong CMV promoter, and 2) human adult testis exhibit nuclear immunoreactivity in spermatogenic cells unlike the other CRD-BP/IMP1 positive tissues and cells studied. Analysis of deletion mutants fused to the GFP fluorescent protein has led to the identification of two NES situated one in the KH2 and the other in the KH4 domain, whereas no simple NLS has been so far identified [Nielsen *et al,* 2003]. However a full-length mutant with two substitutions of lysines with glutamates in the GxxG loop regions of KH1-3 domains undergoes rapid nuclear translocation as revealed by FRAP analysis. Since this mutant exhibits reduced RNA binding capacity, it has been postulated that this protein cannot participate in granule formation in the cytoplasm and rapidly enters to the nucleus and its nuclear retention implies that in the absence of RNA-binding fails to be exported. The above observations indicate that CRD-BP/IMP1 may attach to its targets in the nucleus and co-export with them to determine their cytoplasmic fate.

In differentiated P19 neuronal cells CRD-BP/IMP1, HuD and G3BP-1 colocalize primarily in the cell body and not in neurites [Atlas *et al.,* 2004]. The CRD-BP/IMP1 protein levels are upregulated early following the induction for cell differentiation and decline at day 8 when cells exhibit the characteristics of fully differentiated cells. G3BP-1 is constituvely expressed, while *tau* mRNA and HuD levels rise at days 2-4 and remain high thereafter. Analysis of protein and RNA content from co-immunoprecipitated material indicated that CRD-BP/IMP1 and G3BP-1 are coassociated from day 0 and throughout the differentiation process, despite the fact that *tau* mRNA is not present until day 2, implying the shared recognition of additional targets. *Tau* mRNA is associated from day 2 (the day it first appears) whereas HuD co-associates from day 6, when neurites are visible. Besides P19 cells, the co-association of CRD-BP/IMP1, HuD and G3BP-1 proteins is also demonstrated in mouse brains of day E13 embryos that is the developmental stage in which CRD-BP/IMP1 expression increases in the expanding brain and neural tract.

In all organisms, so far studied, CRD-BP/IMP1 expression was found to be restricted exclusively to embryonal tissues, with the exception of spermatogenic cells in human adult testis [Leeds *et al.,* 1997; Runge *et al.,* 1999; Ioannidis *et al.*, 2001; Nielsen *et al.,* 2003]. Low levels of expression in mouse adult small and large intestines have been reported by Hansen *et al.* [2004] but not in normal colon tissue in humans [Ross *et al.*, 2001]. Nevertheless, the expression detected in adult mouse intestines is substantially lower than that observed in the corresponding embryonal tissue. Analysis of total RNA from mouse embryos indicated a peak of CRD-BP/IMP1 expression at embryonic day 12.5 (E 12.5) followed by a decline towards birth and its disappearance in neonatal mice shortly after birth [Runge *et al.,* 2000]. At day E12.5 the *H19* RNA and *IGF-II* mRNA are in the initial phase of their expression that peaks at E17.5. Besides the temporal similarities between CRD-BP/IMP1, *IGF-II* mRNA and *H19* RNA expression patterns, it is noteworthy that at the same time c-*myc* becomes indispensable for the developing organism, as c-*myc* null mice die around day 10.5 [Davis *et al.,* 1993]. The spatiotemporal pattern of expression of CRD-BP/IMP1 in development was elucidated by monitoring in transgenic mice the expression of a CRD-BP/IMP1-β-galactosidase-neomycin fusion molecule, generated by a gene trap insertion in the second intron of the gene [Hansen *et al.,* 2004]. Galactosidase staining was observed in E10 in the fore- and hindbrain, the snout, branchial arches, limb bunds and the tail. At day E12.5 its expression increased in the expanding fore- and hindbrain and in the neural tract and appeared also in the developing eye, tongue, heart, lung and liver. In the limb buds at day 14.5 its expression became restricted to the future tenonts, whereas in day 17.5 its expression generally decreased with the exception of intestines, developing tubules of the kidney and the liver. Its expression was detected in both mesenchymal and parenchymal cells in the internal organs of the mouse embryo.

CRD-BP/IMP1 null mice exhibited dwarfism and high perinatal lethality [Hansen *et al.,* 2004]. Only 40% of homozygous mice survived into adulthood and their body sizes were 40% smaller compared to that of the wild type or heterozygous mice. Growth kinetics showed that until day E14.5 there were no differences in the body weight between wild type, heterozygous or null mice. From day E17.5 growth retardation increased progressively, reached its maximum a week following birth, remained stable thereafter and thus null mice never reached the body size of their normal litter mates. The reduced Proliferating Cell

Nuclear Antigen (PCNA) staining in developing organs, as well as the reduced proliferation of Mouse Embryo Fibroblast isolated from null mice, indicated that the smaller size resulted mainly from hypoplasia. With the exception of minor differences in the external appearance, null mice were proportionally reduced in size. However, significant histological abnormalities were observed in the intestine, kidneys, and to a small extent in the liver. The small intestine was hypoplastic and with reduced villus size, whereas in the colon the crypts were short and irregular and the kidneys exhibited an immature phenotype. Heterozygous c-*myc* or *IGF-II* mice exhibit dwarfism, whereas lack of *H19* RNA is associated with increased growth of the embryo [Trumpp *et al.*, 2001; DeChiara *et al.,* 1990; Leighton *et al.*, 1995]. Analysis of c-*myc* and *IGF-II* mRNA and *H19* RNA expression in day E12.5 revealed no significant changes in their levels between knock out and heterozygous embryos. However, sucrose gradient fractionation of polysomes and Northern blot analysis of *IGF-II* mRNA in the fractions, revealed a redistribution of both leader 3' and leader 4' *IGF-II* mRNAs from polysomes to RNPs, indicating a reduced translational efficiency of these mRNA species in CRD-BP/IMP1 null mice. Microarray analysis of the global expression pattern revealed downregulation in a number of genes in the gut, mainly encoding proteins associated with the extracellular matrix[3]

CRD-BP/IMP1 and Cancer

CRD-BP/IMP1, initially identified and purified from two cancer cell lines, was subsequently found to be expressed in practically all established cell lines studied, despite the fact that most have originated from adult organisms. The crucial role of c-*myc* in the process of tumorigenesis, the autocrine role of IGF-II in various cancers, as well as the abnormalities in the expression of *H19* in some tumor types, prompted a number of laboratories to monitor for CRD-BP/IMP1 abnormal expression in tumor specimens. Initially Doyle *et al.* [2000] have detected CRD-BP/IMP1 gene locus amplification in 14 out of 40 breast cancer tumors studied. CRD-BP/IMP1 is localized in the short arm of chromosome 17, at 17q21.3 region, adjacent to genes like BRCA1, p53 and Her-2/neu, whose abnormal expression is frequently encountered in cancer. One of the most common abnormalities in breast carcinomas is the amplification of the Her-2/neu gene [Reviewed: Wang and Hung, 2001]. Thus CRD-BP/IMP1 locus may be modified as part of a larger amplicon not necessarily CRD-BP/IMP-1 amplification in itself being causally associated to cancer development or cancer cell phenotype. Nevertheless, Doyle *et al.*, [2000] have detected the amplification of CRD-BP/IMP1 without the concomitant amplification of the Her-2/neu in half of the cases studied, implying an independent role for this gene in cancer. In this first study no attempts have been made to associate CRD-BP/IMP1 amplification with the expression of the gene. However, due to the fact that this gene is normally expressed only in foetal tissues, the detection of its expression in tumor samples has been proven the most pertinent approach for monitoring its

[3] A stem/progenitor cell pattern of expression was recently described for the CRD-BP/IMP1 gene in human umbilical cord blood cells [Ioannidis *et al.*, 2005] and in gonads [Hammer *et al.*, 2005] suggesting a normal role for this protein in the biology of pluripotent cells in development and possibly in tissue renewal, at least for certain adult tissues.

de novo activation in malignancies. In all studies no CRD-BP/IMP expression could be detected in normal adult human tissues, even upon using the RT-PCR methodology, rendering this highly sensitive and simple approach the suitable method to monitor CRD-BP/IMP1 expression in tumors. The *de novo* activation of CRD-BP/IMP1 gene has been detected in a large number of primary tumors from various types of human malignancies, with frequencies ranging from 12.5% in prostate to 81% in colon carcinomas (Table 1). Besides malignant tumors the activation of CRD-BP/IMP1 has also been detected in premalignant or benign lesions, albeit to a lower extend [Ioannidis *et al.*, 2001; Ioannidis *et al.*, 2004 and unpublished observations]. Furthermore, these studies imply the existence of additional to gene amplification, mechanisms of activation of CRD-BP/IMP1 in human tumors. In breast carcinomas approximately 60% of tumors studied were found positive for CRD-BP/IMP1 *de novo* expression, whereas CGH analysis detected genomic alterations in the CRD-BP/IMP1 gene locus in only 20% and in sarcomas that express CRD-BP/IMP1 at a frequency of 70%, none of the karryotyped tumors exhibit alterations at the 17q21-q22 region [Ioannidis *et al.*, 2001; Ioannidis *et al.*, 2003].

Table 1. CRD-BP/IMP1 Expression In Human Neoplasias

Tumor type	CRD-BP/IMP1 Aberration	Reference
	Samples expressing *de novo* CRD-B/IMP1	
BENIGN		
Breast Fibroadenomas	4/14 (28%)	Unpublished
Meningiomas (Grade I)	8/14 (57%)	Ioannidis *et al.*, 2004
Mesenchymal Solid Tumors	4/10 (40%)	Ioannidis *et al.*, 2001
MALIGNANT		
Colon Carcinomas	17/21 (81%)	Ross *et al.*, 2001
Breast Carcinomas	69/118 (58.5%)	Ioannidis *et al.*, 2003
Sarcomas	24/33 (73%)	Ioannidis *et al.*, 2001
Prostate Carcinomas	3/24 (12.5%)	Unpublished
Non Small Cell Lung Cancer	4/16 (25%)	Ioannidis *et al.*, 2004
Meningiomas (Grade II)	2/3	Ioannidis *et al.*, 2004
Chronic Myelogenous Leukemias	0/9	Ioannidis *et al.*, 2001
	Overexpression compared to normal tissue	
Ovarian Carcinomas	CRD-BP/IMP1 levels: Higher in ovarian cancer and adenomas compared to normal ovaries	Gu *et al.*, 2004
	CRD-B/IMP1 gene amplification	
Breast Carcinomas	14/40 (35%)	Doyle *et al.*, 2000
Breast Carcinomas	11/60 (18.3%)	Ioannidis *et al.*, 2003
	Autoantibody response to CRD-BP/IMP1	
Breast Carcinomas	5/64 (7.8%)	Zhang *et al.*, 2003
Lung Cancer	4/56 (7.1%)	
Colorectal Cancer	6/45 (13.3%)	
Gastric Cancer	15/91 (16.5%)	
Hepatocellular Carcinomas	10/65 (15.4%)	
Prostate Cancer	18/206 (8.7%)	

In ovarian carcinomas CRD-BP/IMP1 expression was highly elevated compared to normal tissue specimens and furthermore CRD-BP/IMP1 expression levels were found to associate with advanced clinical stage and poor prognosis [Gu *et al.* 2004]. In human breast carcinomas the CRD-BP/IMP1 *de novo* expression was found to be associated with the absence of estrogen receptors and advanced grade (Grade III) both representing features of less differentiated and aggressive tumor types. Furthermore, the concomitant presence of 8q24 locus amplification, the most common alteration affecting c-*myc* expression in breast tumors, and the *de novo* activation of CRD-BP/IMP1 expression cooperate towards a less differentiated tumor phenotype [Ioannidis *et al.*, 2003]. The above-mentioned observations in breast and ovarian carcinomas, link the activation of this gene with certain cancer phenotypic characteristics, and imply that its ectopic expression in malignancies may promote more aggressive tumors.

The fact that CRD-BP/IMP1 can act as an oncogene has been definitively demonstrated by Tessier *et al.* [2004] who developed a transgenic mouse expressing CRD-BP/IMP1 in adult mammary tissue. A CRD-BP/IMP1 mini gene comprising the whole coding sequences was placed downstream of the mouse whey acidic protein (WAP) promoter that targets transgene expression to the mammary epithelium and is induced during pregnancy and lactation. Two lines of animals expressing high and low levels of CRD-BP/IMP1 during pregnancy were isolated and monitored for tumor induction. In both lines the activation of CRD-BP/IMP1 induced the formation of palpable mammary tumors, albeit with different frequencies. Ninety-five percent of high expressing and sixty percent of low CRD-BP/IMP1 expressing mice developed tumors by an average age of 53 and 60 weeks respectively, demonstrating that the *de novo* activation of CRD-BP/IMP1 in adult mammary tissue is capable of initiating tumorigenesis. No association of CRD-BP/IMP1 expression and tumor histological type was observed as the whole spectrum of pathological entities from ductal carcinoma *in situ* to anaplastic adenocarcinomas were induced, in agreement with the observations made in the human breast tumor study. Although most animals developed only a single palpable tumor in one gland, nearly all glands exhibited hyperplastic areas. As previously mentioned the expression of CRD-BP/IMP1 in human tumors has also been detected in pre-neoplastic or benign entities such as benign mesenchymal tumors, Grade I meningiomas, benign neuroepithelial tumors and breast fibroadenomas. The above imply that *de novo* activation of CRD-BP/IMP1 likely affects normal cell proliferation and thus provides a platform – i.e. a cancerogenic field – upon which additional genetic alterations are required to induce malignant transformation. Such a mode of action is supported by the relatively long latency and the fact that most mice develop only one tumor despite the presence of extensive hyperplasia in all glands.

Analysis of the expression of CRD-BP/IMP1 known RNA targets, at day 12 of the transgenic mice first pregnancy, i.e. well before the onset of significant preneoplastic or neoplastic changes, revealed a 100 and 8 fold increase in the levels of *IGF-II* mRNA and *H19* RNA respectively. No alterations have been observed in c-*myc* mRNA and protein levels as well as in the IGF-II intracellular protein levels. These findings were unexpected since the overexpression of *H19* RNA and IGF-II in human breast tumors has been shown to characterize mainly the stromal and not the malignant epithelial cells [Andriaenssens *et al.*, 1998; Giani *et al.*, 1996] and furthermore, the aberrant c-*myc* expression is a common characteristic of malignant epithelial cells [Deming *et al.*, 2000]. Analysis of *c-myc* mRNA levels in CRD-BP/IMP1 expressing and non expressing human tumors has shown statistically significant differences in sarcomas but not in breast tumors and this discrepancy was attributed to the inherent cellular heterogeneity in the latter tumor type. However, in light of the findings in the WAP-CRD-BP/IMP1 transgenic mice, it is likely that CRD-BP/IMP1 may not affect c-*myc* expression in this tissue and if so, its effects may be obscured by the activation of other transcriptional or postrascriptional compensatory mechanisms acting to maintain c-*myc* mRNA at constant levels.

Besides CRD-BP/IMP1, p62/IMP2 and KOC/IMP3 have also been found to be abnormally expressed in human malignancies [Lu *et al.*, 2001; Mueller *et al.*, 2003; Wang *et al.*, 2003]. Furthermore, the expression of all three IMP family members in tumors could induce autoantibody response in the sera of cancer patients (Zhang et *al.*, 2003). Actually, p62/IMP2 was initially identified as an autoantigen in human hepatocellular carcinoma [Zhang *et al.*, 1999] whereas KOC/IMP3 was initially identified in a large-scale screen for genes differentially expressed in pancreatic cancer [Mueller-Pillasch *et al.*, 1997]. Both genes are normally expressed only in development and are *de novo* activated in a variety of human tumors. Transgenic mice expressing KOC/IMP3 in adult tissues, under the control of the metallothionein promoter that is usually active in multiple organs, exhibit morphological alterations mainly in exocrine pancreas [Wagner *et al.*, 2003]. Pancreata of KOC/IMP3 transgenic mice exhibit fetal morphologic features implying that the re-expression of this gene, which is highly expressed in fetal pancreas, may recapitulate a developmental program active in embryogenesis and impede the maintenance of a fully differentiated adult phenotype. Thus, it is likely that all three IMP members play a crucial role in development and their *de novo* activation in adult tissues may impart adult cells with fetal cell characteristics, like augmented proliferative capacity, and/or impede differentiation. Each IMP family member may exhibit different tissue specificities, however a general mode of their action in tumorigenesis could be the generation of a "platform" consisting of cells with fetal like features upon which the acquisition of additional mutations can open the road to malignant transformation.

Concluding Remarks

CRD-BP/IMP1 protein was identified in 1992 and purified in 1994, but following the cloning of its cDNA in 1997 the gene has attained increasing interest mainly because: 1) CRD-BP/IMP1 a gene normally expressed only during development, is *de novo* activated or overexpressed in a variety of human tumors. 2) CRD-BP/IMP1 belongs to a small family of highly homologous proteins, two members of which are known to be involved in the creation of cellular asymmetries *via* localization of specific mRNAs, a mechanism pivotal for development and differentiation, and 3) the thus far identified CRD-BP/IMP1 RNA targets play critical roles in development and in normal and abnormal cell proliferation. Recently, the usage of transgenic mice conclusively demonstrated the important role of this gene in both development and tumorigenesis. Whereas CRD-BP/IMP1 null mice exhibit growth retardation, gut malformation and high perinatal lethality, the de novo activation of CRD-BP/IMP1 in adult mouse mammary tissue, is in itself sufficient to open the road to tumorigenesis. High percentages of CRD-BP/IMP1 *de novo* activation have been recorded in some of the most common human malignancies, and for some tumor types CRD-BP/IMP1 expression alone or in conjunction with other genetic abnormalities has been shown to predict for a more aggressive disease. Thus the *de novo* activation of CRD-BP/IMP1 could be regarded as an abnormality acquired late in the multistep process of tumorigenesis, concurring with or promoting tumor progression. However the data from the transgenic mice, which upon the activation of CRD-BP/IMP1 develop generalized mammary hyperplasia, as well as the findings that CRD-BP/IMP is activated in human benign and premalignant lesions, imply that the activation CRD-BP/IMP1 is probably an early event in the transformation process that upon tumor progression predisposes for an aggressive phenotype. If so, and as different studies have shown that tumor cells remain dependent on the initiating mutations, CRD-BP/IMP1 may provide a target for therapeutic intervention having the attractive feature as well, to be normally confined in development.

On the other hand, these two animal studies have not shed light but rather perplexed the issue of CRD-BP/IMP1 function. Being initially identified as a gene stabilizing c-*myc* mRNA and thus considered to upregulate its expression, no alterations in c-*myc* mRNA levels have actually been detected in the early stages of mammary transformation in the transgenic mouse with targeted CRD-BP/IMP1 expression neither in the tissues of CRD-BP/IMP1 null mice. Transfection with CRD-BP/IMP1 constructs of the NIH-3T3 cell line, which expresses low CRD-BP/IMP1 levels, leads to the translational impediment of leader 3' *IGF-II* mRNA. Surprisingly, in the null mouse not only leader 3' but leader 4' *IGF-II* mRNA too, are redistributed from polysomes to RNPs, indicating their reduced translational efficiency. In mice expressing *de novo* CRD-BP/IMP1 in mammary tissue, a 100-fold increase in the levels of *IGF-II* mRNA was recorded albeit not accompanied by a concomitant elevation in the *IGF-II* intracellular protein levels. All the above indicate that the action of CRD-BP/IMP1 upon its RNA targets is more complex than initially thought. Accumulating evidence assigns a role to this gene in RNA trafficking, similar to the one exerted by its "cousins" xVg1-RBP/Vera and cZBP-1. Since the localization process entails protection from degradation and translational repression while the message is transported along the cytoskeleton to its translation site, it can be postulated that shielding from decay and repressing mRNA translation may be secondary to the primary role of CRD-BP/IMP1 as a localizing protein. However in certain instances, as for example following the aberrant activation of CRD-BP/IMP1 in cancer, these functions may become prominent and may explain the translational repression of *IGF-II* mRNA in the CRD-BP/IMP1 overexpressing NIH-3T3 cell. It is noteworthy that a vast excess of CRD-BP/IMP1 molecules (50,000 –100,000 molecules/ cell) has been estimated in the K562 cell line [Prockipcak *et al.,* 1994], and this number exceeds in abundance 1000 fold the c-*myc* mRNA, and must also exceed the levels of all its known RNA targets. Thus, in transformed cells, CRD-BP/IMP1 may exert illegitimate roles due to its overproduction.

Finally, a putative differential localization effect of CRD-BP/IMP1 upon the *IGF-II* mRNA in cell protrusions, may serve the polarized production of this secreted growth factor to selectively affect the surrounding cells. C-*myc* mRNA contains in its 3' UTR an 86 nt perinuclear localisation signal, including also the AUUUA element. Fusion mRNAs containing this signal are preferentially localized in the perinuclear cytoplasm and associate with the cytoskeleton. It has been postulated that this localization, probably ensures the efficient nuclear import of the produced c-MYC protein [Veyrune *et al.,* 1996; Dagleish *et al.*, 1999]. There is no obvious purpose for CRD-BP/IMP1 mediated localization of *c-myc* mRNA unless to provide daughter cells with different c-*myc* levels following cell division and thus predisposing them for different fates.

CRD-BP/IMP1 and KOC/IMP3 and p62/IMP2, all three members of the IMP family, represent examples of RNA binding proteins whose aberrant expression is associated with cellurar malfunction and tumorigenesis.

References

Amati, B; Land H. Myc-Max-Mad: a transcription factor network controlling cell cycle progression, differentiation and death. *Curr. Opin. Genet. Dev.* (1994) 4, 102-108.

Andriaenssens, E; Dumont, L; Lottin, S; Bolle, D; Lepretre, A; Deloballe, A; Bouali, F; Dugimont, T; Coll, J; Curgy, JJ. H19 overexpression in breast adenocarcinoma stromal cells is associated with tumor values and steroid receptor status but independent of p53 and Ki-67 expression. *Am. J. Pathol.* (1998) 153, 1597-1607.

Aranda-Abreu, GE; Behar, L; Chung, S; Furmeaux, H; Ginzburg, I. Embryonic lethal abnormal vision-like RNA binding proteins regulate neurite outgrowth and tau expression in PC12 cells. *J. Neurochem.* (1999) 22, 8827-8837.

Aronov, S; Marx, R; Ginzburg, I. Identification of 3' UTR regions implicated in tau mRNA stabilization in neuronal cells. *J. Mol. Neurosci.* (1999) 12, 131-145.

Ashley, CT Jr; Wilkinson, KD; Reines, D; Warren, ST. FMR1 protein: conserved RNP family domains and selective RNA binding. *Science* (1993) 262, 563-566.

Atlas, R; Behar, L; Elliot, E; Ginzburg, I. The insulin-like growth factor mRNA binding-protein IMP-1 and the Ras-regulatory protein G3BP associate with tau mRNA and HuD protein in differentiated P19 neuronal cells. *J. Neurochem.* (2004) 89, 613-626.

Baker, JJ; Liu, P; Robertson, EJ; Efstratiadis, A. Role of insulin-like growth factors in embryonic and postnatal growth. *Cell* (1993) 75, 73-82.

Bardoni, B; Mandel, JL. Advances in understanding of fragile X pathogenesis and FMRP function, and in identification of X linked mental retardation genes. *Curr. Opin. Genet. Dev.* (2002) 12, 284-293.

Barker, DD; Wang, C; Moore, J; Dickinson, LK; Lehmann, R. Pumilio is essential for function but not for distribution of the Drosophila abdominal determinant Nanos. *Genes Dev.* (1992) 6, 2312-2326.

Baumbach, LL; Stein, GS; Stein, JL. Regulation of human histone gene expression: transcriptional and posttranscriptional control in the coupling of histone mRNA stability with DNA replication. *Biochemistry* (1987) 23, 6178-6187.

Bernstein, PL; Herrick, DJ; Prokipcak, RD; Ross J. Control of c-myc mRNA half-life in vitro by a protein capable of binding to a coding region stability determinant. *Genes Dev.* (1992) 6, 642-654.

Binder, R; Horowitz, JA; Basilion, JP; Koeller, DM; Klausner, RD; Hartford, JB. Evidence that the pathway of transferin receptor mRNA degradation involves an endonucleolytic cleavage within the 3'UTR and does not involve poly(A) tail shortening. *EMBO J.* (1994) 13, 1969-1980.

Blanchard, JM; Piechaczyk, M; Dani, C; Chambard, JC; Franchl, A; Pouyssegur, J; Jeanteur, P. C-myc is transcribed at high rate in G0 arrested fibroblasts and is post-transcriptionally regulated in response to growth factors. *Nature* (1985) 317, 443-445.

Bock, O; Schlue, J; Kreipe, H. Reduced expression of H19 in bone marrow cells from chronic myeloproliferative disorders. *Leukemia* (2003) 17, 815-816.

Bonnieu, A; Roux, P; Marty, L; Cuny, M; Jeanteur, P; Piechaczyk, M. AUUUA motifs are dispensable for rapid degradation of the mouse c-myc mRNA. *Oncogene* (1990) 5, 1585-1588.

Brewer G. An A+U-Rich element RNA-binding factor regulates c-myc mRNA stability in vitro. *Mol. Cell. Biol.* (1991) 11, 2460-2466.

Brewer, G; Saccani, S; Sarkar, S; Lewis, A; Petska S. Increased interleukin-10 mRNA stability in melanoma cells is associated with decrease levels of A+U-rich element binding factor AUF1. *J. Interf. Cytok. Res* (2003) 23, 553-564.

Burd, CG; Dreyfuss, G. Conserved structures and diversity of functions of RNA-binding proteins. *Science* (1994) 265, 615-621.

Casey, JL; Koeller, DM; Ramin, VC; Klausner, RD; Harford, JB. Iron regulation of transferrin receptor mRNA levels requires iron responsive elements and a rapid turnover determinant in the 39 untranslated region of the mRNA. *EMBO J.* (1989) 8, 3693–3699.

Castagnetti, S; Hentze, MW; Ephrussi, A; Gebauer, F. Control of oskar mRNA translation by Bruno in a novel cell-free system from Drosophila ovaries. *Development* (2000) 127, 1063-1068.

Chen, CY; Shyu, AB. AU-rich elements: characterization and importance in mRNA degradation. *Treds. Biochem. Sci.* (1995) 20, 465-470.

Chen, T; Damaj, BB; Herrera, C; Lasko, P; Richard, S. Self-association of the single-KH-domain family members Sam68, GRP33, GLD-1, and Qk1: role of the KH domain. *Mol. Cell. Biol. (1997)* 10, 5707-5718.

Chkheidze, AN; Lyakhov, DL; Makeyev, AV; Morales, J; Kong, J; Liebhaber, SA. Assembly of the -globin mRNA stability complex reflects binary interaction between the pyrimidine-rich 3'-untranslated region determinant and poly(C) binding protein CP. *Mol. Cell. Biol.* (1999) 19 4572–4581.

Clegg, JB; Weatherall, DJ; Milner, PF. Haemoglobin Constant Spring: a chain termination mutant? *Nature* (1971) 234, 337–340.

Coller, HA; Grandori, C; Tamayo, P; Colbert, T; Lander, ES; Eisenman, RN; Golub, TR. Expression analysis with oligonucleotide microarrays reveals that MYC regulates genes involved in growth, cell cycle, signaling and adhesion. *Proc. Natl. Acad. Sci. USA* (2000) 97, 3260-3265.

Cui, H; Onyango, P; Brandenburg, S; Wu, Y; Hsieh, CL; Feinberg, AP. Loss of imprinting in colorectal cancer linked to hypomethylation of H19 and IGF2. *Cancer Res* (2002) 62, 6442-6446.

Cunningham, KS; Hanson, MN; Schoenberg, DR. Polysomal ribonuclease I exists in a latent form on polysomes prior to estrogen activation of mRNA decay. *Nucl. Acids Res.* (2001) 29, 1156-1162.

Dalgeish, G; Veyrene, JL; Accornero, N; Blanchard, JM; Hesketh J. Localization of a reporter transcript by the c-myc 3'UTR is linked to translation. *Nucleic Acids Res.* (1999) 27, 4363-4368.

Dang, CV. C-Myc target genes involved in cell growth, apoptosis, and metabolism. *Mol. Cell. Biol.* (1999) 19, 1-11.

Dani, C; Blanchard, JM; Piechaczyk, M; El Sabouty, S; Marty, L; Jeanteur, P. Extreme instability of myc mRNA in normal and transformed human cells. *Proc. Natl. Acad. Sci. USA* (1984) 81, 7046-7050.

Dani, C; Mechti, N; Piechaczyk, M; Lebleu, B; Jeanteur, P; Blanchard, JM. Increased rate of degradation of c-myc in interferon treated Daudi cells. *Proc. Natl. Acad. USA* (1985) 82, 4896-4899.

Darnell, JC; Jensen, KB; Jin, P; Brown, V; Warren, ST; Darnell, RB. Fragile X mental retardation protein targets G quartet mRNAs important for neuronal function. *Cell* (2001) 107, 555-557

Davis, AC; Wims, M; Spotts, GD; Hann, SR; Bradley, A. A null c-myc mutation causes lethality before 10.5 days of gestration in homozygotes and reduced fertility in heterozygotes female mice. *Genes Dev.* (1993) 7, 671-682.

DeChiara, TM; Efstratiadis, A; Robertson, EJ. A growth deficiency phenotype in heterozygous mice carrying an insulin-like growth factor II gene disrupted by targeting. *Nature* (1990) 345, 78-80.

Decker, CJ; Parker, R. Mechanisms of mRNA degradation in eukaryotes. *TIBS* (1994) 19, 336-340.

Deming, SL; Nass, SJ; Dickson, RB; Trock, BJ. C-myc amplification in breast cancer: a meta-analysis of its occurrence and prognostic relevance. *British J. Cancer* (2000) 83, 1688-1695.

Denkert, C; Weichert, W; Pest, S; Koch, I; Kobel, M; Reles, A; Sehouli, J; Dietel, M; Hauptmann, S. Overexpression of the embryonic-lethal abnormal vision-like protein HuR in ovarian carcininoma is a prognostic factor and is associated with increased cyclooxygenase 2 expression. *Cancer Res.* (2004) 64, 189-195.

Deshler, JO; Highett, MI; Schnapp, B J. A highly conserved RNA binding protein for cytoplasmic mRNA localization in vertebrates. *Curr. Biol.* (1998) 8, 489-496.

Dixon DA, Tolley ND, King PH, Nabos, LB; McIntyre TM; Zimmerman, GA; Prescott, SM. Altered expression of the mRNAstability factor HuR promotes cyclooxygenase-2 expression in colon cancer cells. *J. Clin. Invest.* 2001, 108, 1657-1665.

Doyle, GAR; Betz, NA; Leeds, PF; Fleisig, AJ; Prokipcak, RD; Ross, J. The c-*myc* coding region determinant-binding protein: a member of a KH domain RNA-binding proteins. *Nucleic Acids Res.* (1998) 26, 5036-5044.

Doyle, GA; Bourdeau-Heller, JM ; Couthard, S; Meisner, LF; Ross, J. Amplification in human breast cancer of a gene encoding a c-*myc* mRNA binding protein. *Cancer Res.* (2000) 60, 2756-2759.

Drewell, RA; Brenton, JD; Ainscough, JF; Barton, SC; Hilton, KJ; Arney, KL; Dandolo, L; Surani, MA. Deletion of a silencer element disrupts H19 imprinting independently of a DNA methylation epigenetic switch. *Development* (2000) 127, 3419-3428.

Driever, W; Nusslein-Volhard C.The bicoid protein determines position in the Drosophila embryo in a concentration-dependent manner. *Cell* (1988) 54, 95-104.

Driever W, Thoma G, Nusslein-Volhard C. Determination of spatial domains of zygotic gene expression in the Drosophila embryo by the affinity of binding sites for the bicoid morphogen. *Nature* (1989 a) 340, 363-367.

Driever, W; Nusslein-Volhard C. The bicoid protein is a positive regulator of hunchback transcription in the early Drosophila embryo. *Nature* (1989 b) 337, 138-143.

Driever, W; Siegel,V; Nusslein-Volhard C. Autonomous determination of anterior structures in the early Drosophila embryo by the bicoid morphogen. *Development* (1990) 109, 811-820.

Dubnau, J; Struhl, G. RNA recognition and translational regulation by a homeodomain protein. *Nature* (1996) 379, 694-699.

Eggenschwiler, J; Ludwig, T; Fisher, P; Leighton, PA; Tilghman, SM; Efstratiadis, A. Mouse mutant embryos overexpressing IGF-II exhibit phenotypic features of the Beckwith-Wiedemann and Simpson-Golabi-Behmel syndromes. *Genes Dev.* (1997) 11, 3128-3142.

Evan, G; Littlewwod, T. A matter of life and death. *Science* (1998) 281, 1317-1322.

Farina, KL; Huttelmaier, S; Musunuru, K; Darnell, R; Singer, RH.Two ZBP1 KH domains facilitate beta-actin mRNA localization, granule formation, and cytoskeletal attachment. *J.Cell Biol.* (2003) 160, 77-87.

Fedoriw, AM; Stein, P; Svoboda, P; Schultz, RM; Bartolomei, MS. Transgenic RNAi reveals essential function for CTCF in H19 gene imprinting *Science* (2004) 303, 238-240.

Felser, DW; Bishop, JM. Reversible tumorigenesis by MYC in hematopoietic lineages. *Mol Cell* (1999) 4, 199-207.

Fernandez, PC; Frank, SR; Wang, L; Schroeder, M; Liu, S; Greene, J; Cocito, A; Amati, B. Genomic targets of the human c-MYC protein. *Genes Dev.* (2003) 17, 1115-1129.

Ferrandon, D; Koch, I; Westhof, E; Nusslein-Volhard, C. RNA-RNA interaction is required for the formation of specific bicoid mRNA 3' UTR-STAUFEN ribonucleoprotein particles. *EMBO J.* (1997) 16, 1751-1758.

Gallouzi, I-E; Parker, F; Chebli, K; Maurier, F; Labourier, E; Barlat, I; Capony, JP; Toque, B; Tazi, J. A novel phosphorylation dependent RNase activity of GAP-SH3 binding protein: a potential link between signal transduction and RNA stability. *Mol. Cell. Biol.* (1998) 18, 3956-3965.

Gautreau, D; Cote, CA; Mowry, KL. Two copies of a subelement from the Vg1 RNA localization sequence are sufficient to direct vegetal localization in Xenopus oocytes. *Development* (1997) 124, 5013-5020.

Giani, CKJ; Cullen, D; Rasmussen A. IGF-II mRNA and protein are expressed in the stroma of invasive breast cancers; an in situ hybridization and immunochemistry study. *Br. Cancer Res. Treat.* (1996) 41, 43-50.

Gibson, TJ; Thompson, JT; Heringa, J. The KH domain occurs in a diverse set of RNA-binding proteins that include the antiterminator NusA and is probably involved in binding to nucleic acid. *FEBS Lett.* (1993) 324, 361–366.

Ginsburg, I. Neuronal polarity: Targeting of microtubule components into axons and dendrites. *TIBS* (1991) 16, 257-261.

Grandori, C; Cowley, SM; James, LP; Eisenman, RN. The MYC/MAX/MAD network and the transcriptional control of cell behavior. *Ann. Rev. Dev. Biol.* (2000) 16, 653-699.

Gu, L; Shigemasa, K; Ohama, K. Increased expression of IGF-II mRNA-binding protein mRNA is associated with an advanced clinical stage and poor prognosis in patients with ovarian cancer. *Int. J. Oncol.* (2004) 24, 671-678.

Hammer, NA; Hansen, TO; Byskov, AG; Rajpert-de Meyts, E; Grondahl, ML; Bredkjaer, HE; Wewer, UM; Christiansen, J; Nielsen, FN. Expression of IGF-II mRNA –binding proteins (IMPs) in gonads and testicular cancer. *Reproduction* (2005), 130, 203-212.

Hansen, VO, Hammer, NA, Nielsen, J, Madsen, M, Dalbaeck, Ch; Wewer, UM; Christiansen, J; Nielsen, FC. Dwarfism and impaired gut development in Inslulin-like growth factor II mRNA-binding protein 1-deficient mice. *Mol. Cell. Biology* (2004) 24, 4448-4464.

Havin, L; Git, A; Elisha, Z; Oberman, F; Yaniv, K; Schwartz, SP; Standart, N; Yisraeli, JK. RNA binding protein conserved in both microtubule and microfilament based RNA localization. *Genes Dev* (1998) 12, 1593-1598.

Heintz, N; Sive, H; Roeder, RG. Regulation of human histone gene expression: kinetics of accumulation and changes in the rate of synthesis and in the half-lives of individual histone mRNAs during the HeLa cell cycle. *Mol. Cell. Biol.* (1983) 3, 539-550.

Hentze, MW; Kühn, LC. Molecular control of vertebrate iron metabolism: mRNA-based regulatory circuits operated by iron, nitric oxide, and oxidative stress. *Proc. Natl. Acad. Sci. U. S. A.* (1996) 93, 8175-8182.

Holcik, M; Liebhaber, SA. Four highly stable eukaryotic mRNAs assemble 3'-untranslated region RNA-protein complexes sharing cis and trans components. *Proc. Natl. Acad. Sci. USA* (1997) 94, 2410–2414.

Inoue, SB; Siomi, MC; Siomi, H. Molecular mechanisms of fragile X syndrome. *Med. Invest.* (2000) 47, 101-107.

Ioannidis, P; Havredaki, M; Courtis, N; Trangas, T. In vivo generation of 3' and 5' truncated species in the process of c-myc mRNA decay. *Nucleic Acids Res.* (1996) 24, 4969-4977.

Ioannidis, P; Courtis, N; Havredaki, M; Michailakis, E; Tsiapalis, CM; Trangas T. The polyadenylation inhibitor cordycepin (3'dA) causes a decline in c-myc mRNA levels without affecting c-MYC protein levels. *Oncogene* (1999) 8, 117-125.

Ioannidis, P; Trangas, T; Dimitriadis, E; Samiotaki, M; Kyriazoglou, I; Tsiapalis, CM; Kittas, Ch; Agnantis, N; Nielsen, FC; Christiansen, J; Pandis, N., C-*myc* and IGF-II mRNA-binding protein (CRD-BP/IMP-1) in benign and malignant mesenchymal tumours. *Int. J. Cancer* (2001) 94, 480-484.

Ioannidis, P; Mahaira, L; Papadopoulou, A; Teixeira, MR; Heim, S; Andersen, JA; Evangelou, E; Dafni, U; Pandis, N; Trangas, T. 8q24 copy number gains and expression of the c-myc mRNA stabilizing protein CRD-BP in primary breast carcinomas. *Int. J. Cancer* (2003) 104, 54-59.

Ioannidis, P; Kottaridi, Ch; Dimitriadis, E; Courtis, N; Mahaira, L; Talieri, M; Giannopoulos, A; Iliadis, K; Papaioannou, D; Nasioulas, G; Trangas, T. Expression of the RNA-binding protein CRD-BP in brain and non-small cell lung tumors. *Cancer Lett.* (2004) 209, 245-250.

Ioannidis, P; Mahaira, LG; Perez, SA; Gritzapis, AD ; Kavalakis, GJ; Antsaklis, AI; Baxevanis, CN; Papamichail M. CRD-BP/IMP1 expression characterizes cord blood

CD34 (+) stem cells and affects c-*myc* and *IGF-II* expression in MCF-7 cancer cells. *J. Biol. Chem.* (2005), 280, 20086-20093.

Iritani, BM; Eisenman, RN. C-myc enhances protein synthesis and cell size during B lymphocyte development. *Proc. Natl. Acad. Sci. USA* (1999) 96, 13180-13185.

Ishihara, K; Sasaki, H. An evolutionarily conserved putative insulator element near the 3' boundary of the imprinted Igf2/H19 domain. *Hum Mol Genet* (2002) 1, 1627-1636.

Jones, TR; Cole, MD. Rapid cytoplasmic turnover of c-myc mRNA: requirement of the 3' untranslated sequences. *Mol Cell. Biol.* (1987) 7, 4513-4521.

Karn, J; Watson, JV; Lowe, AD; Green, SM; Verdekis, W. Regulation of cell cycle duration by c-myc levels. *Oncogene* (1989) 4, 773-787.

Keene, JD. Ribonucleotide infrastructure regulating the flow of genetic information between the genome and the proteome. *Proc. Natl. Acad. Sci. USA* (2001) 98, 7018-7024.

Kiledjian, M; Wang, X; Liebhaber, SA. Identification of two KH domain proteins in the - globin mRNP stability complex. *EMBO J* (1995) 14, 4357–4364.

Kislauskis, EH; Zhu, X-C, Singer, RH. ß-actin messenger RNA localization and protein synthesis augment cell motility. *J. Cell Biol.* (1997) 136, 1263–1270.

Klausner, RD; Rouault, TA; Harford, JB. Regulating the fate of mRNA: the control of cellular iron metabolism. *Cell* (1993) 72, 19-28.

Knight, Jr. E; Fahey, AD; Friedland, BK; Jonak, GJ. Interferon regulates c-myc gene expression at the post-transcriptional level. *Proc. Natl. Acad. Sci. USA* (1984) 81, 7046-7050.

Koeller, DM; Horowitz, JA; Casey, JL; Klausner, RD; Harford JB. Translation and the stability of mRNAs encoding the transferring receptor and c-*fos*. *Proc. Natl. Acad. Sci. USA* (1991) 88, 7778–7782.

Laggerbauer, B; Ostareck, D; Keidel, EM; Ostareck-Lederer, A; Fischer, U. Evidence that fragile X mental retardation protein is a negative regulator of translation. *Hum. Mol. Genet.* (2001) 4, 329-338.

Laird-Offringa, IA; Elfferich, P; van der Eb, AJ. Rapid c-myc mRNA degradation does not require (A+U)-rich sequences or complete translation of the mRNA. *Nucleic Acids Res.* (1991) 19, 2387-2394.

Lanenu, A; Pistoi, S; Pournin, S; Babinet, C; Morello, D. Both coding exons of the c-myc gene contribute to its posttranscriptional regulation in the quiescent liver and regenerating liver and after protein synthesis inhibition. *Mol. Cell. Biol.* (1995) 15, 4410-4419.

Leeds, P; Kren, BT; Boylan, JM; Betz, NA; Steer, CJ; Gruppuso, PA; Ross, J. Developmental regulation of CRD-BP, an RNA-binding protein that stabilizes c-myc in vitro. *Oncogene* (1997) 14, 1279-1286.

Lehmann, R; Nusslein-Volhard, C. The maternal gene nanos has a central role in posterior pattern formation of the Drosophila embryo. *Development* (1991) 112, 679-691.

Leighton, PA; Ingram, RS; Eggenschwiler, J; Efstratiadis, A; Tilghman, SM. Disruption of imprinting caused by deletion of the H19 gene region in mice. *Nature* (1995) 375, 34-39.

Lemm, I; Ross, J. Regulation of c-myc mRNA decay by translational pausing in a coding region instability determinant. *Mol. Cell. Biol.* (2002) 22, 3959-3969.

LeRoith, D; Baserga, R; Helman, L; Roberts, CT Jr. Insulin-Like growth factors and cancer. *Ann. Intern. Med.* (1995) 122, 54-59.

Levens, DL. Disentangling the MYC web. *Proc Natl. Acad. Sci USA* (2002) 99, 5757-5759.

Levens, DL. Reconstructing MYC. *Genes Dev.* (2003) 17, 1071-1077.

Levine, B J; Chodchoy N; Marzluff, WF; Skoultchi, A. Coupling of replication type histone mRNA levels to DNA synthesis requires the stem-loop sequence at the 3' end of the mRNA. *Proc. Natl. Acad. Sci. USA* (1987) 84, 6189-6193.

Lu, M; Nakamura, RM; Dent, ED; Zhang, JY; Nielsen, FC; Christiansen, J; Chan, EK; Tan, EM. Aberrant expression of fetal RNA-binding protein p62 in liver cancer and liver cirrhosis. *Am. J. Pathol.* (2001) 159, 945-953.

Menssen, A; Hermeking, H. Characterization of the c-myc regulated transcriptome by SAGE: Identification and analysis of c-myc target genes. *Proc. Natl. Acad. Sci. USA* (2002) 99, 6274-6279.

Morales, J; Russell, JE; Liebhaber, SA. Destabilization of human -globin mRNA by translation anti-termination is controlled during erythroid differentiation and is paralleled by phased shortening of the poly(A) tail. *J. Biol. Chem.* (1997) 272, 6607–6613.

Mowry, KL; Melton DA. Vegetal messenger RNA localization directed by a 340-nt RNA sequence element in Xenopus oocytes. *Science* (1992) 255, 991-994.

Mueller, F; Bommer, M; Lacher, U; Ruhland, C; Stagge, V; Adler, G; Gress, TM; Staufferlein, T. KOC is a novel molecular indicator of malignancy. *British J. Cancer* (2003) 88, 699-701.

Mueller-Pillasch, F; Lacher, U; Wallrap, C; Micha, A; Zimmerhackl, F; Hameister, H; Varga, G; Friess, H; Buchler, M; Beger, HG; Vila, MR; Adler, G; Gress, TM. Cloning of a gene highly expressed in cancer coding for a novel KH domain containing protein. *Oncogene* (1997) 14, 2729-2733.

Mullner, EW. Kuhn, LC. A stem-loop in the 39 untranslated region mediates iron-dependent regulation of transferrin receptor mRNA stability in the cytoplasm. *Cell* (1988) 53, 815–825.

Muslimov, IA; Santi, E; Homel, P; Perini, S; Higgins, D; Tiedge H. RNA transport in dendrites: a cis-acting targeting element is contained within neuronal BC1 RNA. *Neurosci.* (1997) 17, 4722-4733.

Nesbit, CE; Tersak, JM; Prochownik, EV. MYC oncogenes and human neoplastic disease. *Oncogene* (1999) 18, 3004-3016.

Newell, S; Ward, A; Graham, C. Discriminating translation of insulin-like growth factor-II (IGF-II) during mouse embryogenesis. *Mol. Reprod. Dev.* (1994), 39, 249-258.

Nielsen, FC; Gammeltoft, S; Christiansen. J. Translational discrimination of mRNAs coding for human insulin-like growth factor II. *J. Biol. Chem.* (1990) 265, 13431-13434.

Nielsen, J; Christiansen, J; Lykke-Andersen, J; Johnsen, A H; Wewer, UM; Nielsen, FC. A family of Insulin-Like Growth Factor II mRNA- binding proteins represses translation in late development. *Mol. Cell. Biol.* (1999) 19,1262-1270.

Nielsen, J; Nielsen, FC; Jacobsen, KR; Christiansen, J. The biphasic expression of IMP/Vg1-RBP is conserved between vertebrates and Drosophila. *Mech. Dev.* (2000) 96, 129-132.

Nielsen, FC; Nielsen, J; Kristensen, MA; Koch, G; Christiansen, J. Cytoplasmin trafficking of IGF-II mRNA-binding protein by conserved KH domains. *J. Cell. Sci.* (2002) 115, 2087-2097.

Nielsen, J; Adolf, SK; Rajpert-De Meyts, E; Lykke-Andersen, J; Koch, G; Christiansen, J; Nielsen, FC. Nuclear transit of human zipcode-binding protein IMP1. *Biochem. J.* (2003) 376, 383-391.

Ohana, P; Kopf, E; Bibi, O; Ayesh, S; Schneider,T; Laster, M; Tykocinski, M; de Groot, N; Hochberg, A. The expression of the *H19* gene and its function in human bladder carcinoma cell lines. *FEBS Lett.* (1999) 454, 81–84.

Ostarek-Lederer, A; Ostarek, DH; Hentze, MW. Cytoplasmic regulatory functions of the KH-domain proteins hnRNPs K and E1/E2. *Trends Biochem. Sci.* (1998) 23, 409-411.

Pandey, NB; Marzluff, WF. The stem-loop structure at the 3' end of histone mRNA is necessary and sufficient for regulation of histone mRNA stability. *Mol. Cell.Biol.* (1987) 7, 4557-4559.

Pantopoulos, K. Iron metabolism and the IRE/IRP regulatory system: An update. *Ann. N.Y. Acad. Sci.* (2004) 1012, 1-13.

Paste, M; Huez, G; Kruys, V. Deadenylation of interferon-beta mRNA is mediated by both tha AU-rich element in the 3'-untranslated region and an instability sequence in the coding region. *Eur. J. Biochem.* (2003) 270, 1590-1597.

Patel, GP; Ma, S; Bag, J. The autoregulatory translational control element of poly(A)-binding protein mRNA forms a heteromeric ribonucleoprotein complex. *Nucleic Acids Res.* (2005) 33, 7074-7089.

Pelengaris, S; Rudolph, B; Littlewood, T. Action of Myc in vivo-proliferation and apoptosis. *Curr. Opin. Genet. Dev.* (2000) 10, 100-105.

Pelengaris, S; Khan, M; Evan, G. C-myc: More than just a matter of life and death. *Nature Rev.* (2002) 2, 764-776.

Peng, SS; Chen, CY; Shyu, AB. Functional characterization of a non-AUUUA AU-rich element fronm the c-jun proto-oncogene mRNA: evidence for a novel class of AU-rich elements. *Mol. Cell. Biol.* (1996) 16, 1490-1499.

Plumb, M; Stein, J; Stein G. Coordinate regulation of multiple histone mRNAs during the cell cycle in HeLa cells. *Nucleic Acids Res.* (1983) 11, 2391-2410.

Prokipcak, RD; Herrick, DJ; Ross, J. Purification and properties of a protein that binds to the C-terminal coding region of human c-*myc* mRNA. *J. Biol. Chem.* (1994) 269, 9261-9269.

Rackham, O; Brown, CM. Visualization of RNA-protein interactions in living cells: FMRP and IMP1 interact on mRNAs. *EMBO J.* (2004) 23, 3346-3355.

Ramsay, G; Evan, GI; Bishop, JM. The protein encoded by the human proto-oncogene c-myc. *Proc. Natl. Acad. Sci.* (1984) 81, 7742-7746.

Rivera-Pomar, R; Niessing, D; Schmidt-Ott, U; Gehring, WJ; Jackle, H. RNA binding and translational suppression by bicoid. *Nature* (1996) 379, 746-749.

Ross, AF; Oleynikov, Y; Kislauskis, E; Taneja, K; Singer, R. Characterization of a ß-actin mRNA zipcode-binding protein. *Mol. Cell. Biol.* (1997) 17, 2158-2165.

Ross, J; Lemm, I; Berberet, B. Overexpression of an mRNA-binding protein in human colorectal cancer. *Oncogene* (2001) 20, 6544-6550.

Runge, S; Nielsen, FC; Nielsen, J; Lykke-Andersen, J; Wewer, UM; Christiansen, J. H19 RNA binds four molecules of Insulin-like Growth Factor II mRNA-binding protein. *J. Biol. Chem.* (2000) 275, 29562-29569.

Sachs, AB. Messenger RNA degradation in eukaryotes. *Cell* (1993) 74, 413-421.

Schaeffer, C; Bardoni, B; Mandel, JL; Ehresmann, B; Ehresmann, C; Moine, H. The fragile X mental retardation protein binds specifically to its mRNA via a purine quartet motif. *EMBO J.* (2001) 20, 4803-4813.

Schoenenberger, CA; Andres, AC; Groner, B; van der Valk, M; LeMeur, M; Gerling, P. Targeted c-myc gene expression in mammary glands of transgenic mice induces mammary tumours with constitutive milk protein gene transcription. *EMBO J.* (1988) 7, 777-786.

Schoenherr, CJ; Levorse, JM; Tilghman, SM. CTCF maintains differential methylation at the Igf2/H19 locus. *Nat Genet.* (2003) 33, 66-69.

Schuldiner, O; Benvenisty, N. A DNA microarray screen for genes involved in c-MYC and N-MYC oncogenesis in human tumors. *Oncogene* (2001) 20, 4984-4994.

Schuman EM. mRNA trafficking and local protein synthesis at the synapse. *Neuron* (1999) 23, 645-648.

Scwartz, SP; Aisenthal, L; Elisha, Z; Oberman, F; Yisraeli, JK. A 62 kDa RNA binding protein from Xenopus oocyte recognizes a common motif in two vegetally localized maternal mRNAs. *Proc. Natl. Acad. Sci. USA* (1992) 89, 11895-11899.

Shaw, G; Kamen, R. A conserved AU sequence from the 3' untranslated region of GM-CSF mRNA mediates selective mRNA degradation. *Cell* (1986) 46, 659-667.

Shyu, AB; Belasco, JG; Greenberg, ME. Two distinct destabilizing elements in the c-fos message trigger deadenylation as the first step in rapid mRNA decay. *Genes Dev.* (1991) 5, 221-231.

Skoda, RC; Tsai, SF; Orkin, SH; Leder, P. Expression of C-MYC under the control of GATA-1 regulatory sequences causes erythroleukemia in transgenic mice. *J. Exp. Med.* (1995) 181, 1603-1613.

Smibert, CA; Wilson, JE; Kerr, K; Macdonald, PM. Smaug protein represses translation of unlocalized nanos mRNA in the Drosophila embryo *Genes Dev.* (1996) 10, 2600-2609.

Snee, M; Kidd, GJ; Munro, TP; Smith, R. RNA trafficking and stabilization elements associate with multiple brain proteins. *J. Cell Sci.* (2002) 115, 4661-4669.

Spencer, CA; Groudine, M. Control of c-myc regulation in normal and neoplastic cells. *Adv. Cancer Res.* (1991) 56, 2-48.

Stauber, C; Luscher, B; Eckner, R; Lotscher, E; Schumperli, D. A signal regulating mouse H4 mRNA levels in a mammalian cell cycle mutant and sequences controlling RNA3' processing are both contained within the same 80 bp fragment. *EMBO J.* (1986) 5, 3297-3303.

Stauber, C; Schumperli D. 3' processing of pre-mRNA plays a major role in proliferation-dependent regulation of histone gene expression. *Nucleic Acids Res.* (1988) 16, 9399-9414.

Stewart, CE; Rotwein, P. Growth, differentiation, and survival: multiple physiological functions for insulin-like growth factors. *Physiol. Rev.* (1996) 76, 1005-1026.

Stoeckle, MY. Removal of a 3' non-coding sequence is an initial step in degradation of gro-alpha mRNA and is regulated by interleukin-1. *Nucleic Acids Res.* (1992) 20, 1123-1127.

Struhl, G., P. Johnston, and P. A. Lawrence. Control of *Drosophila* body pattern by the hunchback morphogen gradient. *Cell* (1992). 69: 237 –249.

Sun, FL; Dean, WL; Kelsey, G; Allen, ND; Reik, W. Transactivation of Igf2 in a mouse model of Beckwith-Wiedemann syndrome. *Nature* (1997) 389, 809-815.

Sussenbach, JS. The gene structure of the insulin-like growth factor family. *Prog. Growth Factor Res.* (1989) 1, 33-48.

Swartwout, SG; Preisler, H; Guan, W; Kinninburg, AJ. Relative stable population of c-myc mRNA that lacks long poly(A). *Mol. Cell. Biol.* (1987) 7, 2052-2058.

Swartwout, SG; Kinniburgh, AJ. c-myc mRNA degradation in growing and differentiating cells: Possible alternative pathways. *Mol. Cell. Biol.* (1989) 9, 288-295.

Tessier, CR; Doyle, GA; Clark, BA; Pitot, HC; Ross J. Mammary tumor induction in transgenic mice expressing an RNA-binding protein. *Cancer Res.* (2004) 64, 209-214.

Toque, B; Delumeau, I; Parker, F; Maurier, F; Multon, MC; Schwighoffer, F. Ras-GTPase activating protein (GAP): A putative effector of Ras. *Cell Signal.* (1997) 9, 153-158.

Totetsky, A; Helman, LJ. Involvement of IGF-II in human cancer. *J. Endocrinol.* (1996) 149, 367-372.

Tourriere, H; Gallouzi, I-E; Chebli, K; Capony, JP; Moukaikel, J; van der Geer, P; Tazi, J. RasGAP-associated endoribonuclease G3BP: Selective RNA degradation and phosphorylation dependent localization. *Mol. Cell. Biol.* (2001) 21, 7747-7760.

Tourriere, H; Chebli, K; Tazi, J. mRNA degradation machines in eukaryotic cells. *Biochimie* (2002) 84, 821-837.

Trumpp, A; Refaeli, Y; Oskarsson, T; Gasser, S; Murphy, M; Martin, GR; Bishop, JM. c-Myc regulates mammalian body size by controlling cell number but not cell size. *Nature* (2001) 414, 768-773.

Veyrene, JL; Campbell, GP; Wiseman, J; Blanchard, JM; Hesketh J. A localization signal in the 3' untranslated region of c-myc mRNA targets c-myc mRNA and beta globin reporter sequences to the perinuclear cytoplasm and cytoskeletal-bound polysomes. *J. Cell Sci* (1996) 109, 1185-1194.

Wagner, M; Kunsch, S; Duerschmied, D; Beil, M; Adler, G; Mueller, F; Gress, TM. Transgenic overexpression of the oncofetal RNA binding protein KOC leads to remodeling of the exocrine pancreas. *Gastroenterology* (2003) 124, 1901-1914.

Wang, X; Kiledjian, M; Weiss, IM; Liebhaber, SA. Detection and characterization of a 3'-untranslated region ribonucleoprotein complex associated with human-globin mRNA stability. *Mol. Cell. Biol.* (1995) 15, 1769–1777.

Wang, Z; Kiledjian, M. The poly(A)-binding protein and an mRNA stability protein jointly regulate an endoribonuclease activity. *Mol. Cell. Biol.* (2000) 20, 6334–6341.

Wang, SC; Hung, MC. HER2 overexpression and cancer targeting. *Semin. Oncol.* (2001) 58, 115-124.

Wang, T; Fan, L; Watanabe, Y; McNeill, PD; Bangur, C; Fanger, GR; Okada, M; Inoue, Y; Persing, DH; Reed, SG. LS23S, an RNA-binding protein as a poteintial therapeutic target for lung cancer. *British J. Cancer* (2003) 88, 887-894.

Watson, JD; Oster, SK; Shago, M; Khosravi, F; Penn, LZ. Identifying genes regulated in a Myc- dependent manner. *J.Biol. Chem.* (2002) 277, 36921-36930.

Weksberg, R.; Shen, DR; Fei, YL; Song, QL; Squire J. Disruption of insulin-like growth factor 2 imprinting in Beckwith-Wiedemann syndrome. *Nat. Genet.* (1993) 5, 143-150.

Whitfield, ML; Zheng, LX; Baldwin, A; Ohta, T; Hurt, MM; Marzluff, WF. Stem-loop binding protein, the protein that binds the 3′ end of histone mRNA, is cell cycle regulated by both translational and posttranslational mechanisms. *Mol. Cell. Biol.* (2000) 20, 4188-4198.

Wisdom, R; Lee, W. The protein coding region of c-myc mRNA contains a sequence that specifies rapid mRNA turnover and induction by protein synthesis inhibitors. *Genes Dev* (1991) 5, 232-243.

Wreden, C; Verrotti, AC; Schisa, JA; Lieberfarb, ME; Strickland, S. Nanos and pumilio establish embryonic polarity in Drosophila by promoting posterior deadenylation of hunchback mRNA. *Development* (1997) 124, 3015-3023.

Wu, LH; Lengyel, JA. Role of caudal in hindgut specification and gastrulation suggests homology between Drosophila amnioproctodeal invagination and vertebrate blastopore. *Development* (1998) 125, 2433-2442.

Xu, N; Chen, CY; Shyu, AB. Modultion of the fate of cytoplasmic mRNA by AU-rich elements: Key sequence features controlling mRNA deadenylation and decay. *Mol Cell. Biol.* (1997) 17, 46-11-4621.

Yaniv, K; Yisraeli, JK. The involvement of a conserved family of RNA binding proteins in embryonic development and carcinogenesis. *Gene* (2002) 287, 49-54.

Yaniv, K; Fainsod, A; Kalcheim, C; Yisraeli, JK. The RNA-binding protein Vg1 RBP is required for cell migration during early neural development. *Developmen.* (2003) 130, 5649-5661.

Yelding, NM, Procopio, WN; Rehman, MT; Lee, WM. C-myc mRNA is down regulated during myogenic differentiation by accelerated decay that depends on translation of regulatory coding elements. *J. Biol. Chem.* 272 (1998) 273, 15749-15757.

Yisraeli, JK; Sokol, S; Melton, DA. A two-step model for the localization of maternal mRNA in Xenopus oocytes Involvement of microtubules and microfilaments in the translocation and anchoring of Vg1 mRNA. *Development* (1990) 108, 289-298.

Yu, H; Rohan, T. Role of the Insulin-Like growth factor family in cancer development and progression. *J. Natl. Can. Inst.* (2000) 92, 1472-1489.

Zalfa, F; Giorgi, M; Primerano, B; Moro, A; Di Penta, A; Reis, S; Oostra, B; Bagni, C. The fragile X syndrome protein FMRP associates with BC1 RNA and regulates the translation of specific mRNAs at synapses. *Cell* (2003) 112, 317-327.

Zhang, JY; Chan, EKL; Peng, XX; Tan, EM. A novel cytoplasmic protein with RNA-binding motifs is an autoantigen in human hepatocellular carcinoma. *J. Exp. Med.* (1999) 189, 1101-1110.

Zhang, HL; Eom, T; Oleynikov, Y; Shenoy, SM; Liebelt, DA; Dictenberg, JB; Singer, RH, Bassell, GJ. Neurotrophin-induced transport of zipcode binding protein with ß-actin mRNA increases ß-actin levels and stimulates growth cone motility. *Neuron* (2001) 31, 261–275.

Zhang, JY; Casiano, CA; Peng, XX; Koziol, JA; Chan, EKL; Tan, EM. Enhancement of antibody detection in cancer using panel of recombinant tumor-associated antigen. *Cancer Epidem. Biomark. Prev.* (2003) 12, 136-143.

Zheng, L; Dominski, Z; Yang, XC; Elms, P; Raska, CS; Borchers, CH; Marzluff, WF. Phosphorylation of stem-loop binding protein (SLBP) on two threonines triggers degradation of SLBP, the sole cell cycle-regulated factor required for regulation of histone mRNA processing, at the end of S phase. *Mol. Cell. Biol.* (2003) 23, 1590-1601.

Zubiaga, AM; Belasco, JG; Greenberg, ME. The nonamer UUAUUUAUU is the key AU-rich sequence motif that mediates mRNA degradation. *Mol. Cell. Biol.* (1995) 15, 2219-2230.

Zumkeller, W; Burdach, S. The Insulin-Like growth factor system in normal and malignant hematopoietic cells. *Blood* (1999) 11, 3653-3657.

In: Trends in RNA Research
Editor: Peter A. McNamara, pp. 73-94

ISBN: 1-59454-506-5

Chapter IV

Ribozyme against HPV16 E6 mRNA Suppresses Cervical Cancer Cell Growth and Sensitizes Cells to Apoptosis and Anti-Tumor Therapies

Yanfang Zheng[1] *and Jiren Zhang*[2]

[1] Laboratory of Molecular Microbiology, National Institute of Allergy and Infectious Diseases, National Institutes of Health, Rockville, Maryland, USA 20852

[2] Department of Oncology, Zhujiang Hospital, First Military Medical University, Guangzhou. P. R. China, 510282

Abstract

Human Papillomavirus (HPV) is related to more than 90% of cervical cancer. The virus is shown to be essential for the induction and maintenance of the malignant phenotype in cervical cancer. In this chapter, we designed a hammerhead ribozyme to specifically target the HPV16 E6 mRNA, and our results demonstrated that the ribozyme can cleave HPV16 E6 mRNA effectively and with high specificity. When transfected into a HPV16 positive cervical cancer cell, CaSKi, the ribozyme reduced the expression of HPV16 E6 mRNA, and inhibited cell growth both *in vitro* and *in vivo*. The percentage of apoptosis cells was also increased. We found that the anti-HPV16E6 ribozyme reduced the expression of the viral E6 gene, and cellular c-myc, bcl-2 genes, but increased the expression of p53. It is likely that anti-HPV16E6 ribozyme inhibited cervical cancer cell growth by reducing the expression of the HPV16 E6 gene, which may alter the expression of p53, c-myc and bcl-2, and led to apoptosis in cancer cells. We also found that CaSKi cells transfected with the anti-HPV16E6 ribozyme showed increased sensitivity to cisplatin and radiation. Furthermore, the anti-HPV16E6 ribozyme increased expression of HLA-2, B7-1 and B7-2 molecules on CaSKi cell, and

[1] Correspondence to: Twinbrook II Facility Rm201, 12441 Parklawn Drive, NIAID, NIH, Rockville, MD, USA 20852. Fax: 301-480-2618; E-mail: yazheng@niaid.nih.gov

enhanced its antigenic property to immune therapy. Our study demonstrated the potential of anti-HPV16E6 ribozyme for treating cervical cancer, and the possibility of combining ribozyme approach with anti-tumor therapies including chemotherapy, radiotherapy or immunotherapy.

Key words: Ribozyme; Human papillomavirus; Cervical cancer; Gene therapy; apoptosis

Introduction

The discovery in the 1980s that some RNA molecules, named Ribozymes, are capable of catalyzing chemical reactions was a breakthrough in biology [1]. It was soon proposed that they could be used as catalytic antisense tools for the disruption of viral RNAs [2]. Ribozymes catalyze highly sequence-specific reactions determined by RNA-RNA interactions between the ribozyme and its substrate molecules. The capacity of ribozyme to specifically cleave RNAs has made it a promising molecular tool and potential gene suppressor with numerous applications [3]. Naturally exiting ribozymes include hammerhead, hairpin, hepatitis delta virus (HDV) and Varkud Satellite (VS) RNA, group I and II introns [4], and the RNA subunit of RNase P [5]. The hammerhead ribozyme is probably the most widely studied and the most commonly used ribozyme for gene inactivation assay due to its small size and catalytic efficiency. The hammerhead ribozyme has conserved structure which contains two functional modules, a catalytic core that cleaves the target RNA and two flanking regions that direct the ribozyme core to a specific target site by virtue of complementarities [6]. Hammerhead ribozymes can cleave any triplet of NUX (N = any nucleotide, X= A, C, or U) if the flanking sequence is changed appropriately. Highly specific activities of hammerhead ribozymes were shown to inhibit the expression of specific genes by targeting their mRNAs [7]. Ribozymes were shown to be effective against various oncogenes and the drug resistance genes [8]. The results reported along the last two decades, clearly demonstrated the potential applicability of ribozymes as specific gene silencers and their utility as therapeutic and biotechnological tools. Furthermore, many efforts have been dedicated during the recent years towards intracellular delivery and expression of ribozymes [9].

Cervical cancer is a leading cause of cancer-related death in women in developing countries and the second most common cancer among women worldwide [10]. Human papillomavirus (HPV) is one of the causative agents of cervical cancer. The virus is a group of nonenveloped double-stranded DNA virus with over 150 genotypes. Those that are found in cervical and oral cancers are often referred as "mucosal high risk" type, among which HPV-16 and HPV-18 are probably the most carcinogenic [11]. The oncogenic property of HPV has been assigned primarily to two early genes, E6 and E7, which are expressed in HPV-containing cells and carcinomas, and act as transforming genes both in vitro and in vivo [12]. There has been abundant experimental evidence indicating that the E6 and E7 oncogenes of HPV play a major role in the development and maintenance of the malignant phenotype of cervical carcinoma [13]. E6 and E7 of HPV16 induce malignance by degradation of the cellular anti-tumor proteins, p53 and retinoblastoma (Rb) respectively in cervical cells. The p53 gene is present in all cells and is known as the house-keeping gene or cell cycle regulator. One of the important functions of p53 is to recognize DNA damage to arrest cell growth in the G1 phase of the cell cycle for DNA repair and, if repair is not possible, to promote apoptosis. Mutation or deletion of p53 is one of the most common genetic abnormalities in all cancers. The E6 protein of the "high risk" HPVs binds to the p53 protein, and abrogates the transcriptional activation function of p53 by stimulating its degradation via ubiquitin-mediated proteolysis, resulting in tumor progression. The E7 protein also binds to a tumor suppressor, the cellular reitnoblastoma gene product (pRb) and inhibits its function [14]. Evidence has suggested that expression of the HPV E6 and E7 genes is necessary for continued growth of most cervical cancer cells. Thus, both genes could serve as ideal targets for gene therapy using approaches such as ribozymes [15].

In this charpter, we used a hammerhead ribozyme designed to specially cleave the HPV16 E6 mRNA from CaSKi cells which carry HPV16. The present study was to investigate effects of the anti-HPV16E6 ribozyme on the growth of CaSKi cells. We found that the anti-HPV16E6 ribozyme, when expressed in the cells from an expression vector, was able to inhibit cell growth and enhance the sensitivity of CaSKi cells to apoptosis and anti-tumor therapies including chemotherapy, radiotherapy and immunotherapy. These phonotypical changes were likely due to reduced expression of HPV16 E6 protein, which changed the expression of p53, c-myc and bcl-2 proteins, and led to apoptosis in cancer cells. The efficacy of the anti-HPV16E6 ribozyme, both in growth inhibition and cell sensitization to chemicals, radiation and immune cells, illustrates that gene therapy combined with chemotherapy, radiotherapy or immunotherapy could be an effective strategy for cervical cancer treatment.

Material and Methods

Cells and Plasmids

The HPV16-containing human cervical cancer cell line, CaSKi, is grown in Dulbecco's modified Eagle's medium (Invitrogen, Carlsbad, CA, USA) supplemented with 10% fetal bovine serum (FBS, Invitrogen, Carlsbad, CA, USA) and 100 international units/ml penicillin, 100 μg/ml streptomycin (Life technologies, Gaithersburg, MD, USA). The oligonucletides corresponding to the ribozyme were synthesized using ABI 3400 DNA synthesizer (Applied Biosystems, Foster City, CA, USA). To facilitate the cloning, *Xba*I and *Kpn*I sites were added to the ends of the synthesized oligonucletides: anti-HPV16E6 ribozyme-1, 5'-CTAGATATCATGTACTGATGAGTCCGTGAGGACGAAAGTTGTTTGG GTAC-3'; anti-HPV16E6 ribozyme-2, 5'-CCAAACAACTTTCGTCCTCACGGACTCAT CAGTACATGATAT-3'. RNA-trimming plasmid pRG523 was from Dr C. Lu (Biochemical Institute of Shanghai, P.R.China), in which three ribozyme genes, 5'-cis-ribozyme, trans-ribozyme, and 3'-cis-ribozyme genes were ligated to pTZ19 vector, in the order from 5' to 3'downstream of the T7 promoter. Cis-ribozymes on both sides can catalyze self-cleavage; thus the trans-ribozyme will be released without non-specific sequences from the plasmid attached [16]. Anti-HPV16E6 ribozyme gene was put between cis-ribozymes instead of the original trans-ribozyme in pRG523 to construct pRGHRz plasmid. To generate the viral target sequence for *in vitro* assay, a 214bp fragment of HPV16 E6 gene was generated by PCR (denaturation at 94°C for 60 seconds, annealing at 60°C for 60 seconds, and extension at 72°C for 60 seconds; 30 cycles) and cloned into pTZ19 vector to form pTZ16E6 plasmid. *Hind*III and *Eco*RI sites were added to the ends of the PCR primers (forward: 5'-AGTCAAGCTTGAGAACTGCAATGTTTCAGG-3', reverse: 5'-CTGAGAATTCTAT CACATACAGCATATGGA-3') to facilitate the cloning. To construct the eukaryotic ribozyme expressing plasmid, pRGHRz was digested by *Hind*III and *Eco*RI to get the fragment including 5'-cis-Rz, anti-HPV16E6 ribozyme and 3'-cis-Rz. This fragment was inserted into pcDNA3 (Invitrogen, Carlsbad, CA, USA) to construct plasmid pcDNAHRz.

In Vitro Transcription and Cleavage Assay

The ribozyme and substrate RNA were prepared from EcoRI-linearized template (pRG16HRz and pTZ16E6 plasmids) with T7 RNA polymerase (Stratagene, La Jolla, CA, USA) with 10 μCi [α-^{32}P] UTP by in vitro transcription. The transcription products were electrophoresed in 10% PAGE gel and detected by autoradiography. The bands containing ribozymes and target RNAs were excised from the gel and RNAs were eluted with 1% sodium dodecyl sulfate (SDS), 0.5 M CH_3COONH_4, and 1 mM ethylenediaminetetraacetic acid (EDTA). To determine relative RNA concentration, radioactivity was measured by scintillation counting. The cleavage reaction was carried out in 20 μl solution containing 50 mM Tris-Hcl, pH 7.5, 20 mM $MgCl_2$, 20 mM NaCl, 0.04 μM ribozyme and 0.01 μM target RNA. The mixture was incubated at temperatures ranging from 30°C-50°C for 3 hours, and then electrophoresed through 10% polyacrylamide gel. Products of the cleavage reaction were detected by overnight exposure to x-ray film. The cis-ribozymes and anti-HPV16E6-ribozyme were used in the cleavage assay to react with the target RNA respectively.

Expression of Ribozyme in Cervical Cancer Cell Line

The anti-HPV16E6-ribozyme plasmid pcDNAHRz and the control empty vector pcDNA3 were transfected into CaSKi cells by Lipofectamine 2000 reagent (Invitrogen, Carlsbad, CA, USA) according to the manufacturer's protocol. After transfection, stably transfected cells were selected in growth medium containing 800 μg/ml of G418 (geneticin) (Sigma, St Louis, MS, USA) for 4 weeks. The G418-resistant cells with ribozyme or pcDNA3 were cultured and named as CaSKi-R, CaSKi-P, respectively. Total cellular RNAs of CaSKi, CaSKi-R and CaSKi-P cells were extracted by the RNA easy kit (Qiagen, Hilden, Germany). The expression of ribozyme in transfected CaSKi-R cells was monitored by dot blotting. Total cellular RNAs of CaSKi-R and CaSKi-P cells were blotted onto nitrocellulose membrane and hybridized with a ^{32}P-labelled oligonucleotide ribozyme probe (5'-CCAAACAACTTTCGTCCTCACGGACTCATCAGTACATGATAT-3'). The ribozyme obtained from pRGHRz transcribed *in vitro* was used in the RNA dot blotting as positive control.

Northern and Western Blots

To measure the level of HPV16 E6 mRNA, total cellular RNAs of CaSKi, CaSKi-R and CaSKi-P cells were extracted. 2.5 μg of total RNA was denatured, electrophoresed on a 1% formaldehyde-agarose gel, transferred onto Nytran Supercharge membrane, UV cross-linked to the membrane with a Stratalinker (Stratagene, La Jolla, California, USA) and hybridized with ^{32}P-labelled HPV16 E6 probe (5'-AGTCAA GCTTGAGAACTGCAATGTTTCAGG-3') and ^{32}P-labelled β-actin oligonucleotide probe (5'-TGACGGGGTCACCCACACTGTGC CCATCTA-3'). The level of gene expression was measured by densitometry. Expression of HPV16 E6 protein in CaSKi, CaSKi-R and CaSKi-P cells was measured by Western blotting assay. Cells were harvested on ice. Cell extracts were prepared in lysis buffer (0.1 M Tris-cl, pH 8.0, 10 mM EDTA, 0.1% SDS and 2 mM DTT), and protein concentration was determined using the BCA assay (Pierce, Rockford, IL, USA). 5 μg of total protein was resolved by electrophoresis in sodium dodecyl sulfate-polyacrylamide gels, transferred onto Immobilon-P membranes and blocked in 5% milk-TBST buffer (5% nonfat milk, 25 mM Tris-HCl [pH 8.0], 125 mM NaCl, and 0.5% Tween 20). The membrane was probed with an antibody specific for HPV16 E6 (Amersham, Chicago, IL, USA). The membrane was washed in TBST, incubated with species-specific horseradish peroxidase-conjugated donkey antibody (Amersham, Chicago, IL, USA), and washed again in TBST. The membrane was then incubated with ECL+ (Amersham, Chicago, IL, USA), and the signals were detected with Hyperfilm (Amersham, Chicago, IL, USA).

Measurement Of Cancer Cell Growth In Vitro And In Vivo

Cell growth rates of CaSKi, CaSKi-R and CaSKi-P were measured by seeding cells at a concentration of 1×10^4 / well in duplicate into 24-well plates. Cells were harvested daily, and counted in an automated cell counter. All experiments were carried out in triplicates and results represent the average. A soft agar colony formation assay was performed according to Kuroki [17]. CaSKi, CaSKi-R and CaSKi-P cells were plated in triplicate at 5×10^3 / 35-mm dish onto 0.3% agarose supplemented with 20% FBS. Colonies of the cells were counted at 2 weeks after seeding. BALB/c nu/nu mice (female, healthy, 4 weeks old) were injected with 1×10^6 CaSKi cells or transfected cells in Hank's-buffered saline (Invitrogen, Carlsbad, CA, USA). Thirty mice were divided into 3 groups; each group was injected with CaSKi or CaSKi-P or CaSKi-R cells respectively. After inoculation, mice were kept in pathogen free environment for 8 weeks. The growth of the tumors was followed over 8 weeks. Diameters of each tumor were measured once a week. The size was calculated by the equation, V = 0.5 x A x B^2, in which A and B are the experimental measurements in mm for length and width, respectively.

Measurement of Apoptosis and Intracellular Protein by Flow Cytometer

Apoptosis was measured by Annexin-V/ PI staining. Cells were harvested by trypsinization and spun down at 1000 x g for 10 min. Subsequently, cells were washed with Sterofundin (Braun, Melsungen, Germany) supplemented with 10 mM HEPES (PH 8.0) and recentrifuged, The cell pellets were resuspended in Annexin-V-APC (Bender MedSystems, Vienna, Austria) staining solution and incubated for 15 min at 4 ^{0}C in the dark. Cells were then washed again in Sterofundin / HEPES and centrifuged. Supernatants were discarded and the pellets were then resuspended in 50 μl of Sterofundin/ HEPES solution containing 4 μg/ml propidium iodide (Sigma, St Louis, MS, USA) with 40 KU/ml of DNase-free RNase (Stratagene, La Jolla, CA) and analyzed by flow cytometer (BD Pharmingen, San Diego, CA, USA). The intracellular expression of proteins, including HPV16 E6, c-myc, bcl-2, p53 and Fas, was detected by permeablizing cells with 1% Triton X-100, then stained with 20 μl of monoclonal antibodies against HPV16 E6, c-myc, bcl-2, p53 and fas (Amersham, Chicago, IL, USA) for 30 min at room temperature. After washing, cells were incubated with 100 μl of an FITC-labeled sheep anti-mouse IgG (Amersham, Chicago, IL, USA) for 20 min at room temperature, washed and re-suspended in PBS for flow cytometry analysis. The cell surface immune antigens, including HLA-1, HLA-2, B7-1 and B7-2, were also detected by FITC-labeled monoclonal antibodies (Amersham, Chicago, IL, USA).

Measuring Growth Inhibition by MTT Assay

Growth inhibition of CaSKi, CaSKi-R and CaSKi-P cells after being treated with different concentrations of cisplatin (in DMEM + 10% FBS) was detected by MTT [3- (4, 5-methylthiazol-2-yl)-2, 5-diphenyl- tetrazolium bromide] assay to analyze cancer cell sensitivity to chemotherapy agents. Briefly, the three types of cells were seeded at a density of 2000-5000 cells/well in 96-well plates, and grew overnight. Different concentrations of cisplatin (1.25, 2.5, 5, 10, 20, 40, 80 μmol / L) were added into experimental wells and DMEM medium was added into control wells. After 24 hours, supernatant was removed and replaced with serum-free DMEM, then MTT was added (50 μg / well) and cells were continued to incubate at 37^{0}C for additional 6 hours. Solubilization of the converted purple formazan dye was accomplished by adding 100 μl of 0.01 N HCl / 10% SDS and incubating overnight at 37^{0}C. The reaction product was quantified by absorbance at 570 nm. Cell inhibition percentage was calculated by the relative absorbance of treated versus untreated cells. Inhibition rate (%) = (1 – OD of experimental group/ OD of control group) x 100%. Experiments were repeated three times with quadruplicate samples for each treatment in each individual experiment.

Radiation Treatment

CaSKi, CaSKi-R and CaSKi-P cells were given different doses (2Gy, 4Gy, 5Gy, 10Gy, 20Gy; 1.0Gy /min) of X-ray radiation at room temperature respectively. Percentage of apoptotic cells was measured by flow cytometry at 48 hours after radiation treatment. MTT assay was carried out at 48 hours after radiation treatment as described above. Cell growth inhibition rate was calculated.

Peripheral Blood Mononuclear Cell (PBMC) Separation and Cell Killing Assay

10 ml human blood from healthy donors was gently mixed with 10 ml Hank's Balanced Salt Solution (HBSS), then laid on top of 14 ml of Lymphocyte Separation Media (LSM) and centrifuged at 2000 rpm for 30 min in room temperature. The white mononuclear cell layer at the interface was collected, washed and resuspended into RPMI 1640 with 10% FBS and cultured at 37^0C with 5% CO_2. In cell killing assay, PBMC were used directly as the effector cells and CaSKi, CaSKi-P and CaSKi-R cells as the target cells. 100 μl of effector cell and 100 μl of target cell were mixed and cultured at 37^0C for 24 hours. The ratio of effector cells to target cells is 10:1. As controls, 100 μl of effector cells or 100 μl of target cells were cultured alone. MTT assay was carried out after culturing for 24 hours and the reaction product was quantified by absorbance at 570 nm. Killing rate = [1- (C-E) / T] x 100%, where "C" is the OD570 of effector cells and target cells in the co-culture; " E" is the OD570 of effector cell cultured with no target cells; "T" is the OD570 of target cell cultured without effector cells. All experiments were carried out in triplicates and results represent the average.

Statistical Analysis

ANOVA and two tailed Student's t-test were used for statistical analysis. Bonferroni correction was used for multiple comparisons among each group. $P<0.05$ was considered statistically significant.

Results

The HPV16 Specific Hammerhead Ribozyme and Its Cleavage Ability Against HPV16 E6 mRNA in Vitro

To obtain an efficient hammerhead ribozyme against HPV16 E6, we designed ribozyme by computer software and tested its ability to cleave HPV16E6 mRNA *in vitro*. Hammerhead ribozyme is able to cleave the 3'-end of the triplet NUX (where N is any base and X is C, A or U) in RNA molecules. The structure of hammerhead ribozyme includes a central core and two flanking sequences that can specifically bind to the target RNA [18]. We checked for all suitable cleavage sites on the HPV16 E6 mRNA [19], and compared all sites to find the one which needs lowest energy to cleave. The secondary structure of the ribozyme was calculated with computer software. The position between 170 to 171nt of the E6 mRNA of HPV16 was selected as the cleavage site. The predicated secondary structure of the hammerhead ribozyme was shown in Figure 1A. In this hammerhead ribozyme, the two flanking sequences provide specific binding to the HPV16 E6 mRNA from position 161 to 169 nt and from position 171 to 179 nt respectively; whereas the central core is responsible for the cleavage at position between 170 and 171nt. To generate this HPV16 E6 specific ribozyme *in vitro*, a RNA-trimming plasmid pRG16HRz was constructed, in which three ribozyme genes, 5'-cis-ribozyme, anti-HPV16E6 ribozyme, and 3'-cis-ribozyme genes were ligated to the pTZ19 vector in the order from 5' to 3'downstream of the T7 promoter. The cis-ribozymes on both sides of anti-HPV16E6 ribozyme can catalyze self-cleavage; thus the anti-HPV16E6 ribozyme will be released with no non-specific flanking sequences attached from the plasmid. In vitro transcription and cleavage reaction was carried out to test the cleavage ability of this anti-HPV16E6 ribozyme. We observed that nearly all the RNA transcribed from pRG16HRz was processed into three fragments: the 66 nt anti-HPV16E6 ribozyme, the 63 nt 5'-cis-ribozyme and the 50 nt 3'-cis- ribozyme (data not shown), demonstrating that cis-ribozymes could cleave on both sides and release the anti-HPV16E6 ribozyme successfully. To generate HPV16 E mRNA target for *in vitro* cleavage, a fragment of the HPV16 E6 gene (+94—+296) was cloned into pTZ19 vector by PCR amplification to form pTZ16E6. A 214 nt of HPV16 E6 mRNA was transcribed from pTZ16E6. The cleavage reaction of the HPV16 E6 RNA fragment by the transcribed ribozymes from linearized pRG16HRz was shown in Figure 1B. It was found that only the anti-HPV16E6 ribozyme (66nt) can cleave the HPV16 E6 RNA (214 nt fragment), whereas the 63nt and 50nt fragments which contain the 5' and 3' cis-ribozymes cannot cut the target RNA. The optimal temperature and time of the reaction at was at 40^0C for 100 minutes (data not shown).

The Ribozyme Stably Expressed in Caski Cells Reduced Expression of HPV16 E6 mRNA and Protein

In order to investigate the effect of anti-HPV16E6 ribozyme on CaSKi cells, pcDNAHRz and the control empty vector, pcDNA3, were transfected into CaSKi cells with lipofectin. After transfection, stably transfected cells were selected in growth medium containing G418 for 4 weeks. The G418-resistant cells with ribozyme or pcDNA3 were cultured and named as CaSKi-R or CaSKi-P respectively. The expression of anti-HPV16E6 ribozyme in transfected CaSKi-R cells was monitored by RNA dot blotting, as shown in Figure 2A. It was seen that anti-HPV16E6 ribozyme expressed stably in CaSKi-R cells. The CaSKi-P cells did not generate non-specific hybridization signal. Total cellular RNAs from CaSKi, CaSKi-R and CaSKi-P cells were extracted and hybridized with ^{32}P-labelled HPV16 E6 and ß-actin oligonucleotide probes by northern blotting analysis. As showed in Figure 2B, expression of HPV16 E6 mRNA was much lower in CaSKi-R than that in CaSKi cells ($p<0.05$), while there was no difference of expression of HPV16 E6 mRNA in CaSKi cells and that in CaSKi-

P cells (p>0.05). Expression of ß-actin mRNA were no difference among CaSKi, CaSKi-R and CaSKi-P cells (p>0.05). Total cellular proteins were further extracted from CaSKi, CaSKi-R and CaSKi-P cells, resolved by gel electrophoresis and probed with antibody specific for HPV16E6 (Figure 2C). We found much lower levels of HPV16 E6 protein expressed in CaSKi-R cells than in CaSKi and CaSKi-P cells, demonstrating that anti-HPV16E6 ribozyme expressed in CaSKi cells can reduce the expression of both HPV16 E6 mRNA and protein.

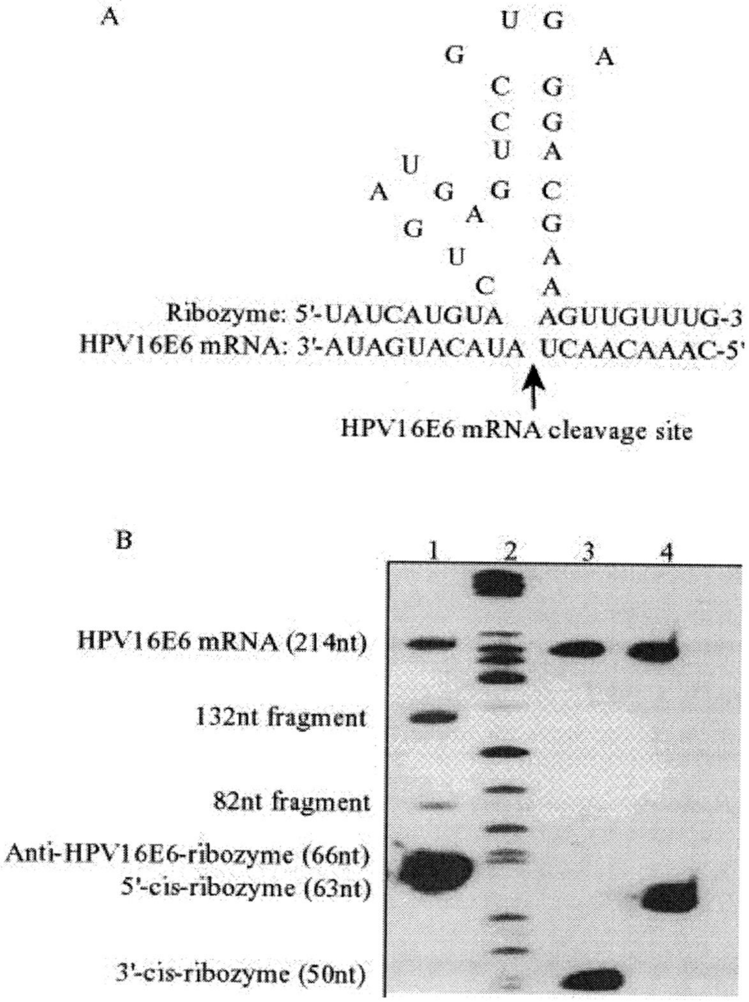

FIGURE 1. Anti-HPV16E6 ribozyme and its cleavage activity in vitro. (A) Predicated secondary structure of the anti-HPV16E6 ribozyme. The central core of the hammerhead is responsible for the cleavage reaction; whereas the flanking sequences mediate specific binding to the HPV16E6 mRNA. (B) Anti-HPV16E6 ribozyme can cleave the target RNA specifically. The cis-ribozymes and anti-HPV16E6 ribozyme were transcribed in vitro, purified from the gel and reacted with HPV16E6 mRNA respectively. The reaction products were electrophoresed in 10% PAGE gel and detected by autoradiography. Anti-HPV16E6 ribozyme (66nt) cleaved the HPV16E6 mRNA (214 nt) into two fragment (132 and 82 nt, lane 1); whereas the 5'-cis-ribozyme (63nt) and 3'-cis-ribozyme (50nt) did not cut the HPV16E6 mRNA (lane3, 4); pBR322/HaeIII marker is in lane2.

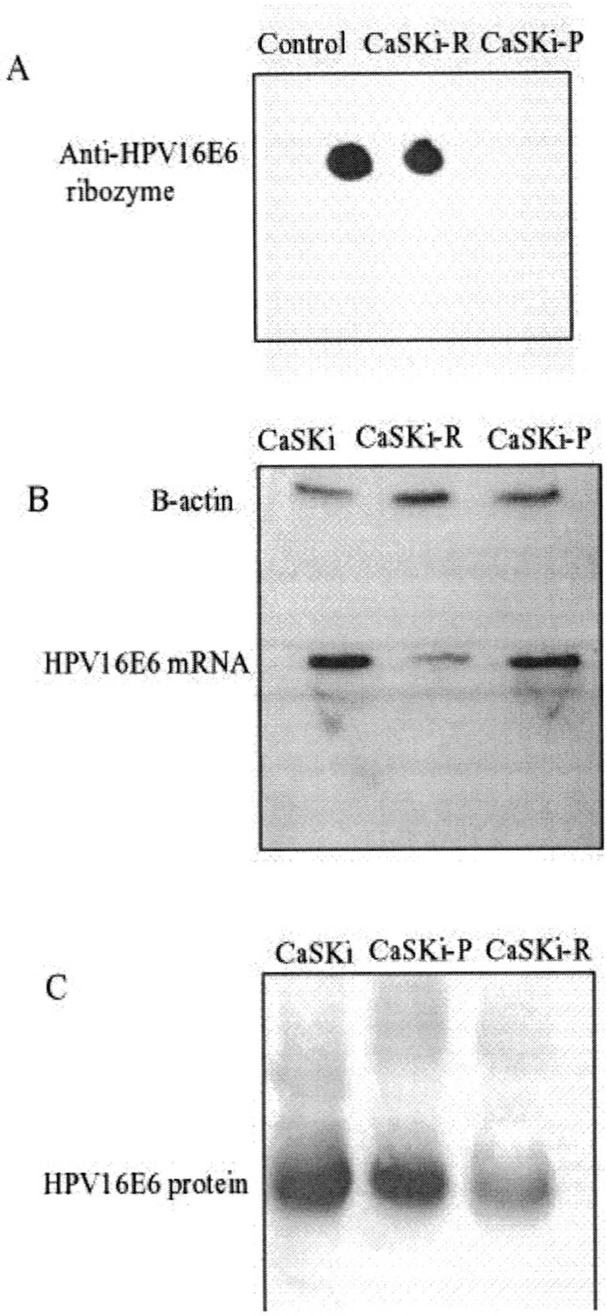

FIGURE 2. Expression of HRz and de-regulation of HPV16 E6 mRNA in CaSKi-R cells. (A) Total cellular RNAs of CaSKi, CaSKi-R and CaSKi-P cells were blotted onto nitrocellulose membrane and hybridized with a ^{32}P-labelled oligonucleotide ribozyme probe. HRz mRNA expressed stably in CaSKi-R cells, but not in CaSKi-P cells. Lane 1 is a positive control. (B) Total cellular RNAs of CaSKi, CaSKi-R and CaSKi-P cells were extracted and hybridized with ^{32}P-labelled HPV16 E6 and ß-actin oligonucleotide probes by northern blotting analysis. E6 mRNA was much lower in CaSKi-R than in CaSKi cells, while there was no difference of E6 mRNA levels between CaSKi and CaSKi-P cells. (C) Total cellular protein was extracted from CaSKi, CaSKi-R and CaSKi-P cells. Five micrograms of each sample were resolved by gel electrophoresis, transferred onto a membrane, and probed with antibody specific for HPV16E6. HPV16E6 protein expressed much less in CaSKi-R cells than in CaSKi and CaSKi-P cells.

Anti-HPV16E6 Ribozyme Inhibits Caski-R Cell Growth in Vitro and in Vivo

In an effort to investigate possible phenotypical changes by transfected anti-HPV16E6 ribozyme, we measured the growth of CaSKi, CaSKi-P and CaSKi-R cells both in cell culture condition and in transferred mouse models. The cell growth curves in cell culture condition were shown in Figure 3. We did not observe difference both in morphology and growth rate between CaSKi and CaSKi-P cells, but we found significant growth decrease in CaSKi-R cells. The colony forming units of CaSKi, CaSKi-P and CaSKi-R cells in 0.3% agarose were 118 (100%), 109 (92%) and 27 (23%) respectively. Statistical analysis confirmed that the soft agar-colony forming rate of CaSKi-P was similar to that of CaSKi cells, while that of CaSKi-R was decreased significantly (a 77% decrease). The tumor forming ability of CaSKi, CaSKi-P and CaSKi-R cells were further compared in a mouse model by transferring these cells into nude mice. Only two tumors (2/10) were formed 25 days after inoculation with CaSKi-R cells; whereas tumors formed in all mice inoculated with CaSKi (10/10) and CaSKi-P cells (10/10). The tumor volume of CaSKi-R cells was significantly decreased (308 mm^3) compared with CaSKi (990 mm^3) and CaSKi-P cells (1015 mm^3) (on day 42 after inoculation, 1 x 10^6 cells per mouse), demonstrating decreased ability of CaSKi-R cells to form tumors in nude mice in comparison with CaSKi and CaSKi-P cells ($p<0.01$). Thus, the anti-HPV16E6 ribozyme can partially reverse malignant phenotype of CaSKi cells by inhibiting cancer cell growth.

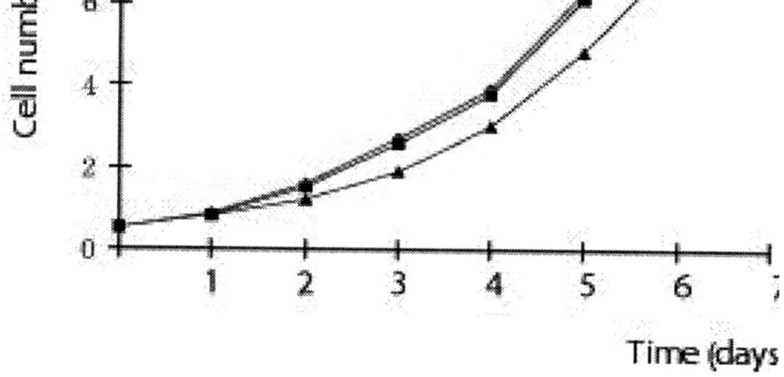

FIGURE 3. Cell growth curve of three cervical cancer cells. Cell growth rates of CaSKi, CaSKi-R and CaSKi-P were measured by seeding cells at a concentration of 1 x 10^4 / well. Cells were harvested and counted daily. All experiments were carried out in triplicates and results represent the average. There was no distinct difference of the growth rate between CaSKi and CaSKi-P, but the growth rate of CaSKi-R decreased significantly.

Anti-HPV16E6 Ribozyme Increases Apoptosis in Caski-R Cells by Changing Expression of Cellular Proteins

To understand mechanisms of anti-HPV16E6 ribozyme on CaSKi cells, apoptosis and expression of apoptosis related proteins were detected by flow cytometry analysis. Apoptosis was measured by Annexin-V/ PI staining. Percentage of apoptotic cells in CaSKi-R cells (22.2%) was significantly higher than that in CaSKi cells (5.2%) and CaSKi-P cells (5.5%, P<0.05). The intracellular expression of proteins, including HPV16 E6, c-myc, bcl-2, p53 and fas, was detected by flow cytometry using FITC-labeled monoclonal antibodies. As shown in Figure 4, analysis shows that expression of HPV16 E6, c-myc and bcl-2 proteins in CaSKi-R cells was much lower than that in CaSKi cells (P<0.05), while expression of these proteins on CaSKi-P cells was no difference with CaSki cells (P>0.05). Expression of p53 protein in CaSKi-R cells increased significantly compared with that in CaSKi and CaSKi-P cells. Expression of fas protein was similar among the three kinds of cells (P<0.05). Anti-HPV16E6 ribozyme, which can cut HPV16 E6 mRNA, also reduced expression of c-myc and bcl-2 proteins on CaSKi-R cells, and increased expression of p53 protein significantly. It was likely that anti-HPV16E6 ribozyme increases apoptosis in CaSKi-R cells by reducing the expression of c-myc and bcl-2 proteins, and increasing the expression of the p53 protein.

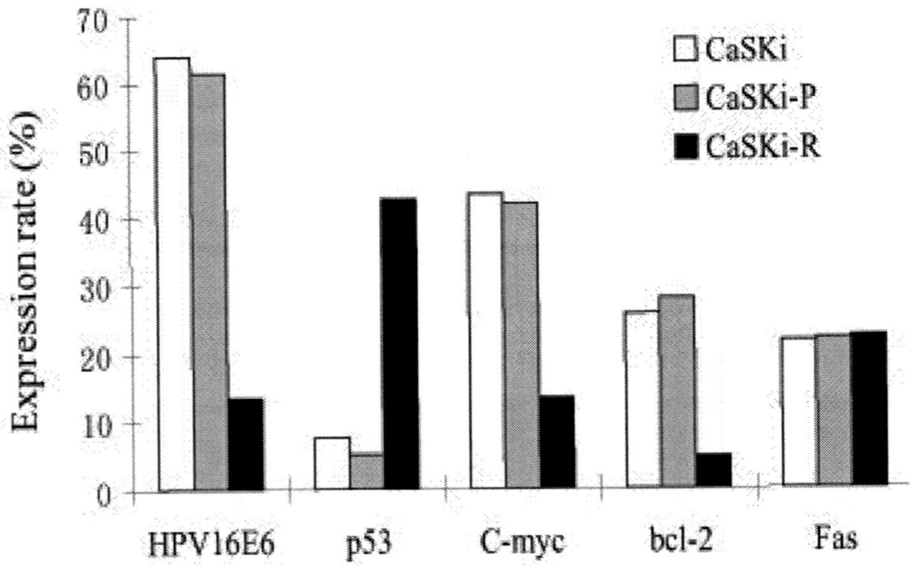

FIGURE 4. Intracellular expression of viral and cellular proteins in CaSKi, CaSKi-P and CaSKi-R cells. The intracellular expression of proteins, including HPV16E6, c-myc, bcl-2, p53 and fas, was detected by flow cytometry using FITC-labeled monoclonal antibodies. Expression of E6, c-myc and bcl-2 proteins were significantly less in CaSKi-R cells than those in CaSKi cells, and p53 expression is higher in CaSKi-R cells than in CaSKi cells. Expression of these proteins in CaSKi-P and CaSKi cells was similar. The expression of fas was similar in all three cells.

Anti-HPV16E6 Ribozyme Sensitizes Caski-R Cells to Cisplatin

In order to compare the sensitivity of CaSKi, CaSKi-R and CaSKi-P cells to different concentrations (1.25, 2.5, 5, 10, 20, 40, 80 μmol/L) of chemotherapy agent, cisplatin, MTT assay was carried out and cell inhibition rate was calculated (1 – OD of experimental group/ OD of control group x 100%). As shown in Figure 5A, cisplatin inhibits the growth of cervical cancer cells in a concentration dependent manner. After treatment by different concentrations of cisplatin for 24 hours, the growth of CaSKi-R cells were inhibited significantly compared with CaSKi cells. However, there was no difference in cell inhibition between CaSKi-P and CaSKi cells ($p>0.05$). When the concentration of cisplatin was 20μmol / L, the cell inhibition rate of CaSKi-R cells reached 52.7% while that of CaSKi and CaSKi-P cells was 38.5% and 40.1% respectively. The IC50 of cisplatin was as follows: CaSKi (31.07 μmol / L), CaSKi-P (30.04 μmol / L), CaSKi-R (13.62 μmol / L). The IC50 of CaSKi-R cells are significantly lower than CaSKi and CaSKi-P cells, indicating CaSKi-R cell is more sensitive to cisplatin. Apoptosis in CaSKi, CaSKi-P and CaSKi-R cells treated with different concentrations of cisplatin for 24 hours were detected by Annexin-V/ PI staining. We observed a direct correlation between the rate of apoptotic cells and the concentration of cisplatin. As well, percentage of apoptosis cells in CaSKi-R was much higher than those in CaSKi and CaSKi-P cells after cisplatin treatment. As shown in Figure 5B, percentage of apoptotic cells in CaSKi-R was 53.6% while that in CaSKi and CaSKi-P cells was 32.1% and 32.9% after being treated with 20 μmol / L cisplatin. The higher sensitivity of CaSKi-R cells to cisplatin was probably due to increase in the percentage of apoptotic cells.

Anti-HPV16E6 Ribozyme Enhances Sensitivity of Caski-R Cells to Radiation

CaSKi, CaSKi-R and CaSKi-P cells were given different doses of X-ray radiation at 2Gy, 4Gy, 5Gy, 10Gy, and 20Gy respectively. We investigated the sensitivity of cervical cancer cells to radiation by MTT and apoptosis assay at 48 hours after radiation treatment. Cell inhibition rate was calculated using result of MTT assay as described above. As shown in Figure 6A, radiation inhibits cervical cancer cells' growth in a dose dependent manner. Cell inhibition rates of CaSKi-R cells increased significantly compared with CaSKi cells after being treated by different doses of radiation for 48 hours. Inhibition rate of CaSKi-R cell went up to 81.7% whereas that of CaSKi-P and CaSKi cells was 53.5% and 52.1% after 20Gy of radiation. However, there was no difference in cell inhibition rates between CaSKi-P and CaSKi cells ($p>0.05$). Percentage of apoptosis cells in these three cervical cancer cells increased significantly 48 hours after radiation treatment as the dose of radiation increased. Percentage of apoptosis cells in CaSKi-R cells was much higher than that in CaSKi or CaSKi-P cells after radiation treatment ($p<0.05$, Figure 6B). When radiation dose is 20Gy, percentage of apoptosis cells in CaSKi-R cells was 80.5%, whereas that of CaSKi-P and CaSKi cells was 55% and 53.2% respectively, demonstrating that CaSKi-R cells were more sensitive to radiation than CaSKi and CaSKi-P cells.

A

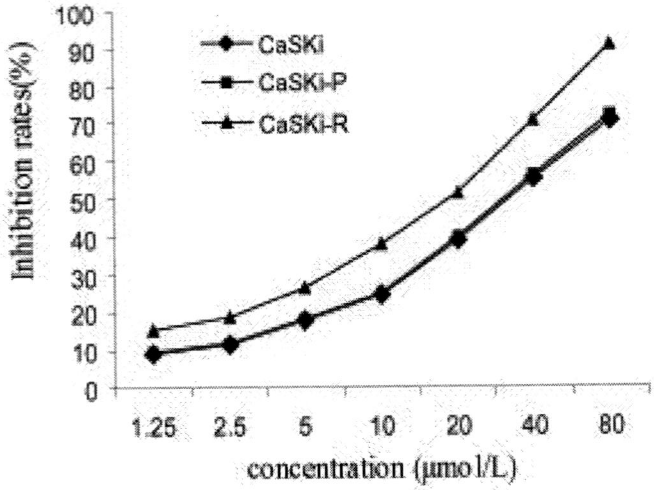

B

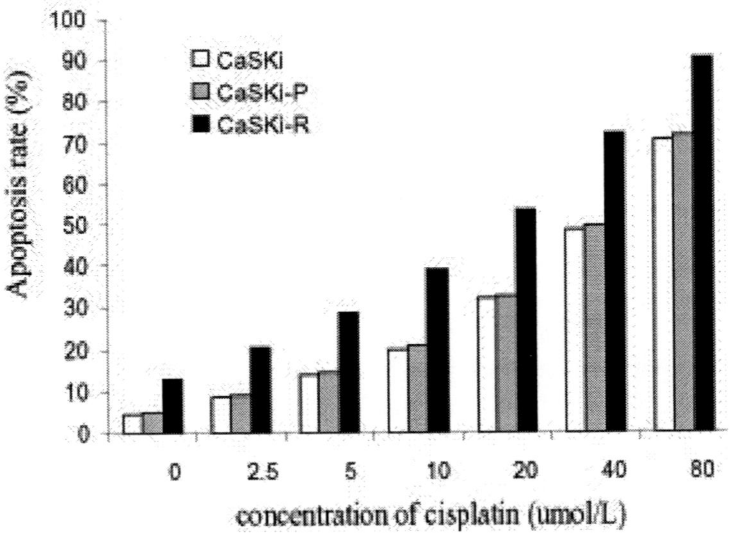

FIGURE 5. CaSKi-R cells were more sensitive to cisplatin than CaSKi and CaSKi-P cells. (A) After treated by different concentration of cisplatin for 24 hours, cell inhibition rates of CaSKi-R cells increased significantly compared with CaSKi cells. There was no difference in cell inhibition rates between CaSKi-P and CaSKi cells. MTT assay was carried out after cisplatin treatment. Cell inhibition rate (%) = (1 – OD of experimental group/ OD of control group) x 100%. (B) Percentage of apoptosis cells in CaSKi-R was much higher than those in CaSKi and CaSKi-P cells after being treated with different concentrations of cisplatin for 24 hours. Apoptosis was measured by Annexin-V/ PI staining and detected by flow cytometry.

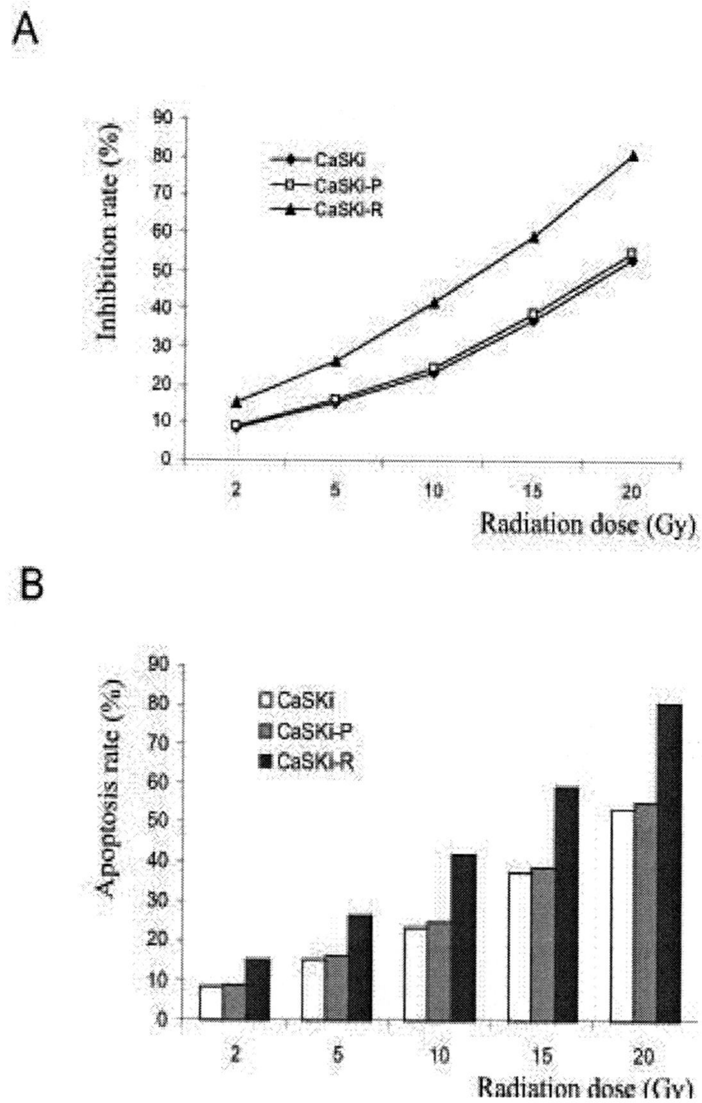

FIGURE 6. CaSKi-R cells were more sensitive to radiation than CaSKi and CaSKi-P cells. (A) Cell inhibition rates of CaSKi-R cells increased significantly compared with CaSKi cells after being treated by different doses of radiation for 48 hours. There was no difference of cell inhibition rates between CaSKi-P and CaSKi cells. Cell inhibition rates were calculated from the result of MTT assay. (B) Apoptosis rates of these three cervical cancer cells increase significantly 48 hours after different doses of radiation treatment. Apoptosis rate in CaSKi-R cells was much higher than in CaSKi and CaSKi-P cells after radiation treatment. Apoptosis was measured by Annexin-V/ PI staining and detected by flow cytometry.

Anti-HPV16E6 Ribozyme Changed Expression of Immune Antigens and Sensitized Cells to PBMC

In an effort to explore the possibility of ribozyme combined with immunotherapy to treat cancer, immune antigens of CaSKi, CaSKi-R and CaSKi-P cells, including HLA-1, HLA-2, B7-1 and B7-2, were detected by flow cytometry using FITC-labeled monoclonal antibodies.

As shown in Figure 7, HLA-1 protein was highly expressed in CaSKi, CaSKi-P and CaSKi-R cells (98.1%, 95.5%, and 98.7% respectively). Expression rates of HLA-2, B7-1 and B7-2 antigens in CaSKi and CaSKi-P cells were very low. There was no difference between CaSKi and CaSKi-P cells ($P>0.05$). Expression rates of HLA-2, B7-1 and B7-2 antigens in CaSKi-R cells were significantly higher than those of CaSKi and CaSKi-P cells ($P<0.05$). To measure the antigenic properties of these cells, human PBMC were prepared from blood of healthy donors and used as effector cells. CaSKi, CaSKi-P and CaSKi-R cells as target cells and co-cultured with PBMC at 37^0C with the ratio of 10:1. MTT assay was carried out after culturing for 24 hours. The average killing rates of PBMC to CaSKi cells and CaSKi-P cells are 21.7% and 22.5% respectively. There is no statistic difference between them ($p>0.05$). The killing rate of PBMC to CaSKi-R cells (38.9%) is significant higher than those of CaSKi and CaSKi-P cells ($p<0.05$). The killing effect of PBMC is mainly due to cytotoxic T lymphocytes (CTLs) and natural killer (NK) cells.

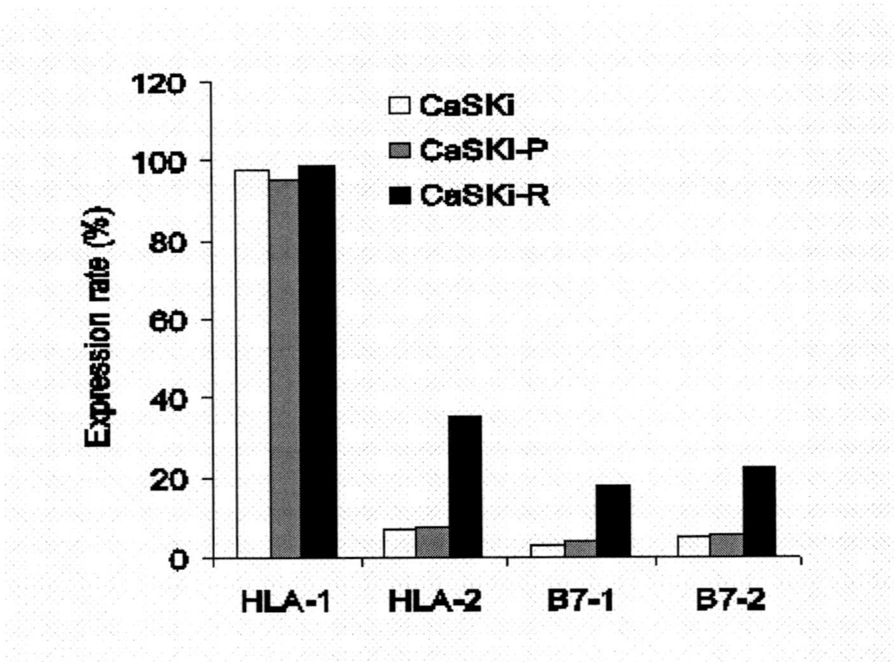

FIGURE 7. Anti-HPV16E6 ribozyme increased the expression of HLA-2, B7-1, B7-2 molecules on CaSKi cells. Expression of HLA-1, HLA-2, B7-1 and B7-2 molecules in cervical cancer cells was detected by flow cytometry using FITC-labeled monoclonal antibodies. HLA-1 protein was highly expressed in CaSKi, CaSKi-P and CaSKi-R cells (98.1%, 95.5% and 98.7%). Expression rates of HLA-2, B7-1 and B7-2 antigens in CaSKi and CaSKi-P cells was low. Expression of HLA-2, B7-1 and B7-2 antigens in CaSKi-R cells were significantly higher than those of CaSKi and CaSKi-P cells ($P<0.05$).

Discussion

Gene therapy of human cancers is most effective when tumor-specific expression of therapeutic genes is achieved. The human papillomavirus is present in many cervical and oral cancers, and is one of the presumed etiological agents; thus, specific targeting of the virus would provide an ideal strategy for the treatment of cervical cancer [20]. In this chapter, we used a HPV16 specific ribozyme to target the viral E6 gene since ribozymes can effectively cleave the viral mRNA and inhibit its expression [21]. Prior to testing the ribozyme in cancer cells, we investigated its ability to cleave the target sequence of HPV16 E6 mRNA in a cell-free system. Ribozymes usually have long non-related plasmid sequences attached when transcribed from plasmid. These long non-related sequences affect the cleavage efficiency of hammerhead ribozymes [16]. To avoid this problem, we cloned the anti-HPV16E6 ribozyme to a vector with cis-ribozymes on both sides which can catalyze their self-cleavage. The ribozyme is released without flanking non-specific sequences. We have shown that the anti-HPV16E6 ribozyme generated by this way is highly effective *in vitro*. It catalyzed site-specific cleavage of the target HPV16E6 mRNA [22]. These *in vitro* data demonstrated the possibility to use ribozymes to inhibit the expression of the HPV16E6 genes.

Further to demonstrate the specificity and efficiency of the ribozyme, we chose the CaSKi cells which contain 500-600 copies of integrated HPV16 genome, compared with carcinoma SiHa cells which contain only 1-2 copies of HPV16. The DNA of both HPV16 and HPV18 can transform cells *in vitro* to an immortal phenotype. The transformation of keratinocytes requires stable levels of expression of the E6 and E7 genes. Continuous expression of E6 and E7 genes may be essential to retain the malignant phenotype of HPV-related cervical cancers [23]. When compared, we found that the growth rate and colony forming ability of the anti-HPV16E6 ribozyme transfected CaSKi-R cells were significantly decreased compared with un-transfected CaSKi and empty vector transfected CaSKi-P cells. The tumor forming ability of the CaSKi-R cells in nude mice was also inhibited by anti-HPV16E6 ribozyme in comparison with that of CaSKi-P and CaSKi cells. These data demonstrated that the anti-HPV16E6 ribozyme could partially reverse the malignant phenotype of CaSKi cells. It is likely that expression of the ribozyme decreased the HPV16 E6 mRNA and affected subsequent expression of downstream cellular genes in cervical cancer cells.

Specific down-regulation of target genes involved in neoplastic transformation and progression has been extensively studied. However, most studies were concentrated on the direct down-regulation of target gene expression and have assumed that down-regulation of the target gene is the direct cause of tumor growth inhibition. Little attention has been paid to the intermediate steps between target gene down-regulation and tumor growth inhibition. E6 and E7 were reported to regulate cell proliferation by reducing the activity of the tumor suppressor protein p53 and pRb respectively. It has been demonstrated that the E6 protein forms complexes with p53 and can target p53 for degradation through the ubiquitin pathway. Apoptosis is a process of single-cell deletion. Recent studies have suggested that alteration in the rate of cell death may be equally important to tumorgenesis [24]. Deregulation of cell death pathways is an important feature of carcinogenesis [25]. The tumor suppressor gene p53 is crucial to the control of apoptosis. In the presence of cellular DNA damage, expression of p53 blocks cell cycle progression, allowing time for the cell to repair its DNA before cycling. Overexpression of p53 would initiate a pathway ending in apoptotic cell death [26]. The HPV E6 protein that binds to p53 can block its function, allowing continued cell division in spite of unrepaired DNA damage, and setting the stage for the development of malignant phenotype [27]. Thus, ribozymes targeting HPV E6 would be expected to affect the cell cycle and apoptosis. In this study, we found that anti-HPV16E6 ribozyme affected the viability and morphology of the cervical carcinoma CaSKi cells and induced apoptosis. Since there was evidence that the bcl-2 protein can inhibit apoptosis [28], and a recent report demonstrated that HPV16 E6 increased the expression of the bcl-2 protein during carcinogenesis [29]. In our study, we also found that when HPV16 E6 expression was decreased, bcl-2 expression was increased, consistent with these previous findings. C-myc is an oncogene essential for cell cycle in normal cells. In many cancers, c-myc is overexpressed and its overexpression induced pathways that contribute to apoptosis [30]. Recent studies have showed that HPV E6 protein can increase c-myc transcription in a p53-dependent fashion [31], and expression of c-myc was markedly inhibited by repression of HPV E6 [32]. In this study, we observed decreased expression of c-myc after transfection of the anti-HPV16E6 ribozyme. It is likely that the ribozyme induced apoptosis by reducing the expression of E6 protein, increasing the expression of p53, and decreasing the expression of c-myc and bcl-2.

There was limited information on the combined effect of ribozyme therapy with chemotherapy, radiotherapy or immunotherapy for cancer treatment [33]. Our study was also intended to explore and expend this area. Combination of different therapies has been widely used and proven effective in cancer treatment for enhancing efficacy, reducing toxicity, and preventing or delaying development of resistance. However, as the difference between normal cells and cancer cells is usually trivial, combination of conventional therapies often cause intolerably high levels of toxicity in most cancer patients. Gene therapy uses genetically encoded function to attack molecular events involved in the development and maintenance of malignancies, providing a molecular approach to eradicate cancer cells and to protect normal cells from the toxicity of conventional therapies [34]. Since synergistic effects can be achieved by combining gene therapy, chemotherapy, radiotherapy and immunotherapy, it may be an effective strategy for human cancer treatment. In this study, we found that the inhibition rate of CaSKi-R cells to cisplate or to radiation was significantly higher than that of CaSKi and CaSKi-P cells, which illustrated that CaSKi-R cells were more sensitive to chemotherapy agents or radiation treatment. Interactions between the immune system and malignant cells play an important role in tumor genesis. Failure of the immune system to detect and reject transformed cells leads to cancer development. Tumors use multiple mechanisms for escaping from immune surveillance, which represents hurdles for effective cancer immunotherapy [35]. Such immune escape mechanisms may include apoptosis resistance, impaired antigen presentation, secretion of immunosuppressive factors, and other strategies. We observed low expression of HLA-2, B7-1 and B7-2 antigens in CaSKi cells, which may be one of mechanisms involved in immune escape of cervical cancer cells [36]. Many immunotherapeutic approaches may fail because tumors use multiple mechanisms to become resistant to apoptosis or to counterattack the immune system [37]. Anti-HPV16E6 ribozyme sensitizes cervical cancer cells to apoptosis and increases expression of HLA-2, B7-1 and B7-2 antigens, thus sensitizes cancer cells to immune cells. All these findings demonstrated that ribozyme gene therapy against HPV16E6 may enhance the therapeutic efficiency of subsequent chemotherapy, radiotherapy or immunotherapy in the treatment of cervical cancer.

The hammerhead ribozyme was first discovered as a catalytic motif in different plant pathogen RNAs. All hammerhead ribozymes share a characteristic secondary structure [35]. They are easy to design, synthesize and have predictable specific affinities for the targeted disease genes. Our data demonstrated the therapeutic potential of anti-HPV16E6 hammerhead ribozyme. If targeted against a tumor-specific gene and placed under the control of a tissue-specific promoter, ribozymes would offer minimal toxicity and maximal efficiency in gene therapy [36]. The successful use of ribozymes as therapeutic agents depends upon many factors, e.g., the relative amounts of active ribozymes in cells, their co-localization with target RNAs, the structural features of transcripts that influence accessibility to specific target sites, catalytic efficiency, the interaction of target RNAs with proteins, and the intracellular stability of targeted RNAs and ribozymes [37]. The most effective approach to maximize ribozyme function in a complex intracellular environment is to understand the intracellular fate of RNA that is targeted [38]. Much progress has been achieved recently. The relative amounts of ribozymes inside cells and their localization can be controlled by using different promoters to drive ribozyme synthesis [39]. The catalytic performance of ribozymes can be improved by sequence modifications [40]. Greater RNA stability can be achieved by chemical modification [41]. Clinical trials to analyze the efficacy and validity of ribozymes as therapeutic agents still have been limited and unsatisfactory so far, and the production of drugs based on catalytic nucleic acids remains a remote goal. The lack of suitable carriers represents a formidable obstacle to the clinical use not only of ribozymes, but also of gene therapy in general [42]. Although limitations still exist to the widespread application of ribozyme based gene therapy, the strategy has already showed therapeutic promise and remains an interesting field to be explored.

Acknowledgement

We thank Lianghu Qu for his help in experiment design and Yuntao Wu for critical readings of the manuscript.

Reference

[1] Cech TR. Ribozymes and their medical implications. *JAMA* 1988; 260: 3030-34.

[2] Sarver N, Cantin EM, Chang PS, et al. Ribozymes as potential anti-HIV-1 therapeutic agents. *Science* 1990; 247: 1222-25.

[3] Puerta-Fernandez E, Romero-Lopez C, Barroso-delJesus A, Berzal-Herranz A. Ribozymes: recent advances in the development of RNA tools. *FEMS Microbiol Rev* 2003; 27(1): 75-97.

[4] Young KJ, Gill F, Gtrasby JA. Metal ions play a passive role in the hairpin ribozyme catalysed reaction. *Nucleic Acid Res* 1997; 25: 3760-66.

[5] Cobeleda C, Sanchez-Garcia I. In vivo inhibition by a site-specific catalytic RNA subunit of RNase P designed against the BCR-ABL oncogenic products: A novel approach for cancer treatment. *Blood* 2000; 95: 731-7.

[6] Peracchi A. Prospects for antiviral ribozymes and deoxyribozymes. *Rev Med Virol* 2004; 14: 47-64.

[7] Tsuchida T, Kijima H, Hori S, Oshika Y, Tokunaga T, Kawai K, Yamazaki H, Ueyama Y, Scanlon KJ, Tamaoki N, and Nakamura M. Adenovirus-mediated anti-K-ras ribozyme induces apoptosis and growth suppression of human pancreatic carcinoma. *Cancer Gene Ther* 2000; 7 (3): 373-83.

[8] Tokunaga T, Tsuchida T, Kijima H, Okamoto K, Oshika Y, Sawa N, Ohnishi Y, Yamazaki H, Miura S, Ueyama Y and Nakamura M. Ribozyme-mediated inactivation of mutant K-ras oncogene in a colon cancer cell line. *Br J Cancer* 2000; 83 (6): 833-9.

[9] Lustig B, Jeang KT. Biological applications of hammerhead ribozymes as anti-viral molecules. *Curr Med Chem* 2001; 8: 1181-1187.

[10] Zur Hausen H. Virus in human cancers. *Science* 1991; 254: 1167-73.

[11] Wolf JK, Franco EL, Arbeit JM, Shroyer KR, Wu TC, Runowicz CD, Tortolero-luna G, Herrero R, Crum CP. Innovations in understanding the biology of cervical cancer. *Cancer* 2003; 98 (9 Suppl): 2064-9.

[12] Duensing S, Munger K. Mechanisms of genomic instability in human cancer: insights from studies with human papillomavirus oncoproteins. *Int J Cancer* 2004; 109 (2): 157-62.

[13] Steller MA. Cervical cancer vaccines: progress and prospects. *J Soc Gynecol Investig* 2002; 9 (5): 254-64.

[14] Alani RM, Munger K. Human papillomaviruses and associated malignancies. *J Clin Oncol* 1998; 16(1): 330-7.

[15] Cho CW, Poo H, Cho YS, Cho MC, Lee KA, Lee SJ, Park SN, Kim IK, Jung YK, Choe YK, Yeom YI, Choe IS, and Yoon DY. HPV E6 antisense induces apoptosis in CaSKi cells via suppression of E6 splicing. *Exp Mol Med* 2002; 34 (2): 159-66.

[16] He YK, Lu CD, Qi GR. In vitro cleavage of HPV16 E6 and E7 RNA fragments by synthetic ribozymes and transcribed ribozymes from RNA-trimming plasmids. *FEBS Letter* 1993; 322 (1): 21-4.

[17] Kuroki T. Agar plate culture and Lederberg-style replicaplating of mammalian cells. *Methods Cell Biol* 1978; 9: 157-78.

[18] Pyle AM. Ribozymes: a distinct class of metalloenzymes. *Science* 1993; 261: 709-14.

[19] Seedorf K, Krammer G, Durst M, Suhai S, Rowekamp WG. Human papillomavirus type16 sequence. *Virology* 1985; 145: 181-5.

[20] Chen Z, Kamath P, Zhang S, St John L, Adler-Storthz K, Shillitoe EJ. Effects on tumor cells of ribozymes that cleave the RNA transcripts of human papillomavirus type 18. *Cancer Gene Ther* 1996; 3(1): 18-23.

[21] Ke A, Zhou K, Ding F, Cate JH, Doudna JA.A conformational switch controls hepatitis delta virus ribozyme catalysis. *Nature* 2004; 429(6988): 201-5.

[22] Zheng Y, Zhang J, Qu L. Effects of anti-HPV16E6-ribozyme on phenotype and gene expression of a cervical cancer cell line. *Chinese Med J* 2002; 115(10): 1501-6.

[23] Zur Hausen H. Papillomaviruses infection—a major cause of human cancers. *Biochim Biophys Acta* 1996; 1288: 55-78.

[24] Evan G, Littlewood T. A matter of life and cell death. *Science* 1998; 281: 1317-22.

[25] Alfandari J, Magal SS, Jackman A, Schlegel R, Gonen P and Sherman L. HPV16E6 oncoprotein inhibits apoptosis induced during Serum-calcium differentiation of foreskin human keratinocytes. *Virology* 1999; 257: 383-96.

[26] Levine AJ. P53, the cellular gatekeeper for growth and division. *Cell* 1997; 88: 323-31.

[27] Ahn WS, Bae SM, Lee KH, Lee JM, Namkoong SE, Chun HJ, Kim CK, and Kim YW. Recombinant adenovirus-p53 gene transfer and cell-specific growth suppression of human cervical cancer cells in vitro and in vivo. *Gynecologic Oncology* 2004; 92: 611-21.

[28] Kelekar A, Thompson CB. Bcl-2 family proteins: The roles of the BH3 domain in apoptosis. *Trends Cell Biol* 1998; 8:324-30.

[29] Du J, Chen GG, Vlantis AC, Chan PK, Tsang RK, van Hasselt CA. Resistance to apoptosis of HPV 16-infected laryngeal cancer cells is associated with decreased Bak and increased Bcl-2 expression. *Cancer Lett* 2004; 205(1): 81-8.

[30] Nilsson JA, Cleveland JL. Myc pathways provoking cell suicide and cancer. *Oncogene* 2003; 22(56): 9007-21.

[31] Kinoshita T, Shirasawa H, Shino Y, Moriya H, Desbarats L, Eilers M, Simizu B. Transactivation of prothymosin alpha and c-myc promoters by human papillomavirus type 16 E6 protein. *Virology* 1997; 232(1): 53-61.

[32] DeFilippis RA, Goodwin EC, Wu L, DiMaio D. Endogenous human papillomavirus E6 and E7 proteins differentially regulate proliferation, senescence, and apoptosis in HeLa cervical carcinoma cells. *J Virol* 2003; 77(2): 1551-63.

[33] Nishizaki M, Meyn RE, Levy LB, Atkinson EN, White RA, Roth JA, and Ji L. Synergistic inhibition of human lung cancer cell growth by adenovirus-mediated wild-type p53 gene transfer in combination with docetaxel and radiation therapeutics in vitro and in vivo. *Clin Cancer Res* 2001; 7: 2887-97.

[34] Fang B, Roth JA. The role of gene therapy in combined modality treatment strategies for cancer. *Curr Opin Mol Ther* 2003; 5 (5): 475-82.

[35] Ko EC, Wang X, Ferrone S. Immunotherapy of malignant diseases. Challenges and strategies. *Int Arch Allergy Immunol* 2003; 132(4): 294-309.

[36] Pawelec G. Immunotherapy and immunoselection – tumor escape as the final hurdle. *FEBS Letter* 2004; 567: 63-6.

[37] Igney FH, Krammer PH. Immune escape of tumors: apoptosis resistance and tumor counterattack. *J Leukoc Biol* 2002; 71(6): 907-20.

[38] Forster AC, Symons RH. Self-cleavage of plus and minus RNAs of a virusoid and a structural model for the active sites. *Cell* 1987; 49: 211-20.

[39] Steele D, Kertsburg A, Soukup GA. Engineered catalytic RNA and DNA: new biochemical tools for drug discovery and design. *Am J Pharmacogenomics* 2003; 3(2): 131-44.

[40] Cech TR. Ribozymes, the first 20 years. *Biochem Soc Trans* 2002; 30(Pt 6):1162-6.

[41] Khan AU, Lal SK. Ribozymes: a modern tool in medicine. *J Biomed Sci* 2003; 10(5): 457-67.

[42] Kato Y, Kuwabara T, Warashina M, Toda H, Taira K. Relationships between the activities in vitro and in vivo of various kinds of ribozyme and their intracellular localization in mammalian cells. *J Biol Chem* 2001; 276: 15378-85.

[43] Esteban JA, Banerjee AR, Burke JM. Kinetic mechanism of the hairpin ribozyme. Identification and characterization of two nonexchangeable conformations. *J Biol Chem* 1997; 272: 13629-39.

[44] Pavco PA, Bouhana KS, Gallegos AM, Agrawal A, Blanchard KS, Grimm SL, Jensen KL, Andrews LE, Wincott FE, Pitot PA, Tressler RJ, Cushman C, Reynolds MA, Parry TJ. Antitumor and antimetastatic activity of ribozymes targeting the messenger RNA of vascular endothelial growth factor receptors. *Clin Cancer Res* 2000; 6(5): 2094-103.

[45] Moon C, Oh Y, Roth JA. Current status of gene therapy for lung cancer and head and neck cancer. *Clin Cancer Res* 2003; 9: 5055-67.

In: Trends in RNA Research
Editor: Peter A. McNamara, pp. 95-119

ISBN: 1-59454-506-5

Chapter V

Transcriptome Analysis of Dendritic Cells

Bradley Messmer and Davorka Messmer
Laboratory of Experimental Immunology, North Shore – Long Island Jewish Research Institute, 350 Community Dr., Manhasset, NY, 11030, USA

Abstract

Dendritic cells (DCs) are potent antigen presenting cells. They initiate and tailor the immune response by sensing immunostimulatory signals and foreign molecular structures. DCs play the role of central processing unit of the immune system, whereby different stimuli lead to a qualitatively different maturation outcome. There is growing evidence that DC subsets can differ in their response based on microenvironmental changes or pathogens and their products. Many of the functional characteristics of these tailored responses have been defined, but the underlying molecular pathways involved in DC maturation are not fully characterized. In the past few years, highly parallel expression analyses by microarrays have begun to elucidate the genetic pathways underlying DC maturation as well as DC subset differentiation. These experiments have identified unexpected cytokine pathways, which play an important role in DC-mediated T cell proliferation, as well as transcription regulators which proved to be critical for the development of certain DCs lineages *in vivo*. Furthermore, global gene expression profiling has revealed that the DC maturation program contains a set of genes that is commonly induced in response to various stimuli as well as stimulus-specific genes, indicating a certain plasticity.

Abbreviations

DC	Dendritic cells
APC	Antigen presenting cells

LPS	lipopolysaccharide
CTL	cytotoxic T lymphocyte
TLR	Toll-like-receptor
IL	Interleukin
TCR	T cell receptor

Introduction

The immune system is the body's primary defense system against pathogens and foreign material. In higher vertebrates, an adaptive immune system has evolved to customize the reaction against the particular antigens encountered. At its core, this system involves the coordinated interaction of B cells, T cells, and Dendritic cells (DCs).

B cells produce antibodies that can bind specifically and with high affinity to extracellular antigens, such as a bacteria or free virus particle. T cells are generally divided into two types, helper and cytotoxic. Helper T cells provide activation signals to B cells or other T cells. Cytotoxic T cells can identify and trigger the destruction of other infected cells. Dendritic cells are the cellular initiators and directors of the adaptive immune response. They are responsible for activating the appropriate T cells, thus polarizing the immune response towards the production of antibodies or cytotoxic T cells.

The power of the adaptive immune system derives from its ability to produce antibodies specific to a particular antigen or cytotoxic T cells that will kill only cells infected with a particular substance. This process begins when a DC internalizes, processes (cleaves into small peptides in the case of protein antigens), and presents antigen to T cells. Antigen presentation is performed on the surface of the dendritic cell by molecules of the major histocompatiblity class (MHC) family. Each T cell has on its surface a unique T cell receptor (TCR) molecule. If the particular TCR binds to the processed antigen displayed on the MHC, a series of interactions between the DC and T cell take place, the consequence of which is an activated helper or cytotoxic T cell. If the antigen was extracellular, then the helper T cell will look for an appropriate B cell. B cells also have a unique receptor of which the antibody molecule is the essential recognition component. The B cell that produces an antibody that binds the particular antigen will also internalize, process, and display that antigen in a manner similar to the DC. The helper T cell will recognize the same processed molecule in context with the MHC, and specifically activate those B cells whose antibodies bind the same antigen recognized by the activating DC. If the antigen was intracellular or in some cases even extracellular, the dendritic cell will activate cytotoxic T cells (CTLs) whose TCR recognize the particular displayed antigen. The CTLs will then kill specific target cells that show the same antigen on their surface.

Dendritic Cell Biology

Dendritic cells (DCs) play a critical role in the initiation and regulation of the adaptive immune responses against pathogens and in tolerance against "self tissue" [1, 2]. They are professional antigen[1] presenting cells (APCs)[2] but have several other functions. Collectively, these can be grouped into environmental sensing, antigen uptake and processing, antigen presentation, and T cell activation. These functions are discrete in both their timing and their physiological settings. DCs are present in most peripheral tissues in a functionally immature state, meaning that they cannot activate T cells. In this state they primarily sense the immune environment and, in particular, detect molecular structures indicative of microbe invasion. Once such structures are encountered, they take up the antigens and migrate to draining lymph nodes. Once there, they will have processed and presented the antigen on their surface, as well as various other molecules that enhance their ability to interact with T cells. At this point the DC is considered mature, and when a T cell with the appropriate TCR sees the antigen presented by the DC, the T cell will receive additional signals from the DC to become activated (Fig.1).

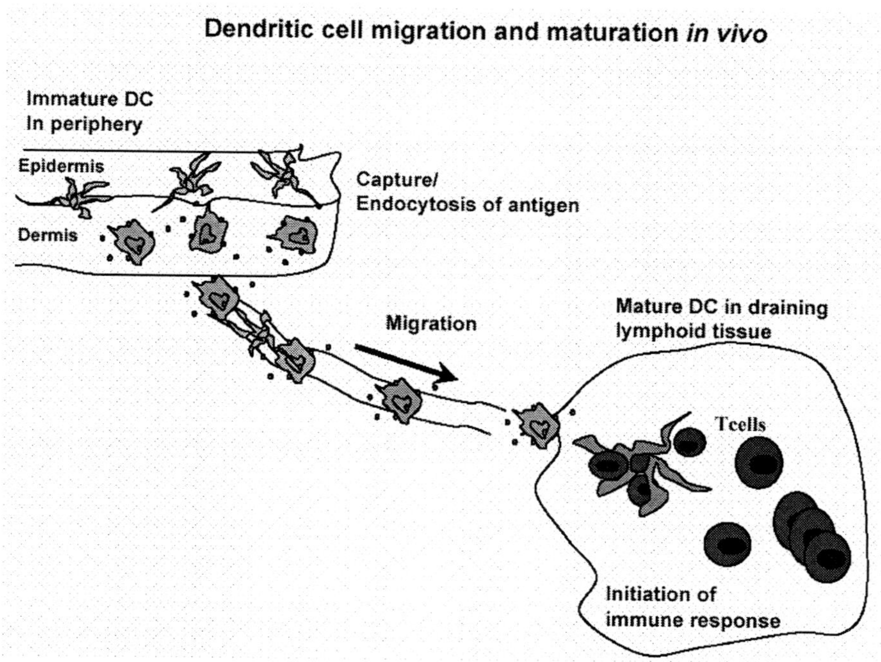

Figure 1

Immature DCs express high levels of receptors involved in antigen uptake (CD206, CD205, FcR), but low levels of MHC class II molecules, adhesion molecules (CD54 and CD58) and co-stimulatory molecules (CD80 and CD86). Immature DCs also secrete low levels of inflammatory cytokines and chemokines. Therefore, immature DCs are very efficient in antigen uptake, but are very weak in their capacity to activate T cells (Fig.2). In contrast mature DCs have increased levels of MHCII, co-stimulatory and adhesion molecules, show enhanced production of inflammatory cytokines and chemokines and are potent in inducing T cell responses against various antigens. The maturation process is irreversible and terminates in apoptotic death of the DC [3].

The DC maturation program can be triggered by the recognition of evolutionarily conserved molecular structures unique to foreign pathogens, such as LPS [4], DNA molecules containing unmethylated CpG motifs [5], peptidoglycans [6] and bacterial lipoproteins [7, 8] by a set of receptors on the cell surface of immature DCs. These Toll-like receptors (TLRs) are critical sensors of infection. Upon ligand binding they induce activation of the immune and inflammatory responses. Eleven TLRs have been identified so far in mammals [9]. Only a few TLRs will be discussed in the context of this chapter. TLR2 mediates recognition of peptidoglycans and bacterial lipoproteins [10-12]. TLR4 is a receptor involved in recognition of gram-negativer lipopolysaccharide (LPS) [13, 14], and TLR9 is described as a receptor of bacterial DNA [15]. Not only can DCs detect microbial components, but they also respond to endogenous signals of cellular distress or damage [16-19]. Endogenous molecules that can induce DC maturation include: CD40L [20, 21], TNF-α [22] heat shock proteins (Hsps) [23-26], and High mobility group box protein 1 (HMGB1) [27].

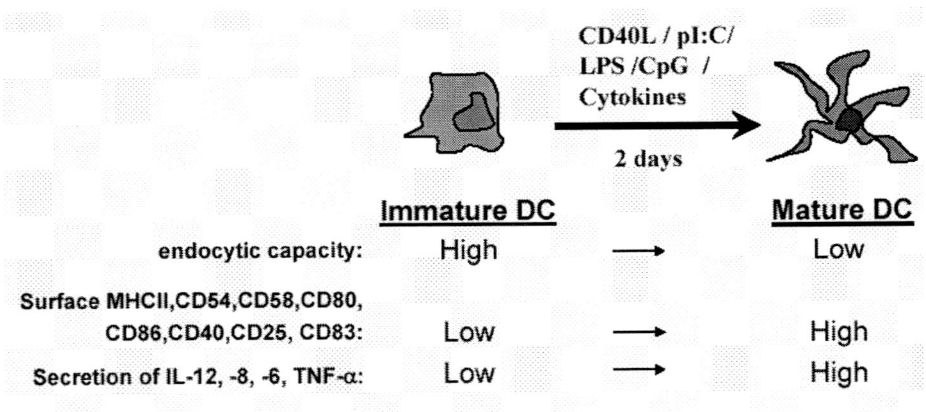

Figure 2

DC Subsets and their Plasticity

While the antigen-presenting function is central to DCs, the concordant immunological effector functions are dependent on additional variables. Some studies suggest that DCs can induce different types of T cell-mediated immune responses, depending on their lineage, activation signal and maturation stage. Distinct DC subsets have been documented and can regulate the immune response differently [28]. There is also growing evidence that DC subsets can differ in their response based on microenvironmental changes or pathogens and their products [29-31]. Furthermore, it has been suggested that DCs play the role of central processing unit of the immune system, whereby different stimuli lead to a qualitatively different maturation outcome [32]. This functional polarization is supported by studies showing that myeloid DCs that encounter different stimuli secrete distinct sets of cytokines and differ in their Th1/Th2-inducing capacity [29, 33].

DCs have been identified in most tissues including peripheral blood on the basis of their morphology, surface phenotype and function. However, only a small number of DCs (<1%), which is insufficient for many experiments, can be isolated directly from tissue or blood. The ability to generate DCs in vitro from progenitors has opened many doors for molecular and cell biological studies. Bone marrow (BM), peripheral blood monocytes ($CD14^+$), and $CD34^+$ stem cells in cord blood preparations (Fig.3) are three sources of such progenitors. The progenitors are cultured with cytokines for several days during which time they differentiate into immature DCs. If they are subsequently exposed to various stimuli (Fig.2) they develop into mature DCs. Their differentiation and maturation process has been studies in vitro [34-37].

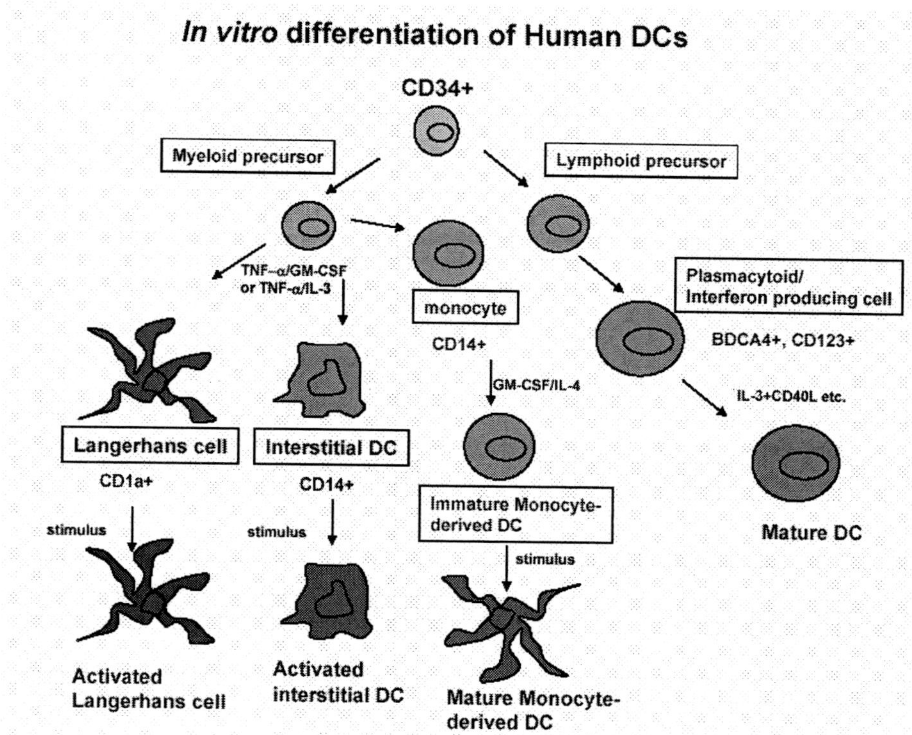

Figure 3

Understanding the molecular pathways of DC maturation could prove valuable in designing better vaccines as well as interfering with undesired autoimmune responses. There are several important questions in DC research, some of which have been addressed using microarray approaches in the past few years. 1) What is the molecular basis of DC amturation? 2) What are the functional relationships and properties of different DC subsets? 3) What is the molecular basis of the functional plasticity of a particular DC subset? 4) What is the molecular basis of DC-mediated tolerance? Insight into changes in gene expression has been made possible through the use of sequential analysis of gene expression (SAGE) [38] and hybridization to DNA immobilized on solid-state microarrays [39-41]. In the past few years highly parallel expression analyses by microarrays have begun to elucidate the genetic pathways underlying DC function. This chapter will discuss some of these findings. The experimental details of most current publications are summarized in Table.1.

Transcriptional Profile in Dcs in Response Cytokine Stimulation and CD40L

An early study addressed the expression of 4110 known genes in immature DCs and mature DCs. Maturation was induced by culturing DCs in a cocktail of inflammatory cytokines: IL-1β, TNF-α, IL-6 and prostaglandin (PG) E-2 for 3 days [42]. The authors showed that immature and mature DCs express a similar number of transcripts and upon maturation the levels of 291 transcripts increased, while the levels of 78 transcripts decreased. They found that transcripts for 10 CD antigens (CD40, CD49d, CD54, CD58, CD61, CD83, CD116, CD124, CD127, and CD162) were higher in mature DCs and levels of 9 CD antigens (CD1a, CD1b, CD11b, CD44, CD49e, CD75, CD115, CDw121b, and CD143) were lower compared to immature DCs. Levels of chemokines and chemokine receptors also changes during maturation. TARC, MCP-1, and MCP-2 were expressed exclusively in mature DCs, and transcripts for STCP1 (MDC-1) were elevated in mature DCs. They also uncovered numerous genes of unknown function as well as genes of known function but previously not investigated in DCs.

Ju et al.[43] have analyzed the molecular components controlling DC differentiation from CD34+ hematopoetic progenitor cells. They investigated transcriptional changes using the human HG-U95A arrays (Affymetrix), which represent 13,000 genes. The investigators compared gene expression in progenitor cells, cells differentiated into DCs, and DCs stimulated with TNF-α for 24 h. A set of CD molecules was highly expressed in TNF-α-treated DCs only, comprising: CD208, CDw197, CD205, CD83, CD200, CD213, CD85k, CD127, CD116, CD87, CD155, and CD166. Another set of genes that are absent or expressed at low levels in progenitor cells and upregulated upon TNF-α stimulation include: CD124, CD141, CD112, CD98, CD123, CD217, CD85c, CD95c, CD54, CD53, CD150, CD131, CD148, CD1b, CD44, CD105, CD121b, CD58, CD1c, CD40, CD13, and CD11c. Furthermore, changes in cytokines and chemokines were observed. Several chemokines were upregulated in response to TNF-α, these include: MCP-1, MIP-1α, MIP-1β, RANTES, MCP-4, TARC, MDC, and IL-8.

Le Naour et al. [44] analyzed gene expression changes in monocyte-derived human DCs after stimulation with TNF-α for 7 days using HuGeneFL Arrays, complementary to 6,300 human genes. Transcripts from ~40% of the 6,300 unique genes assessed were detected in all the cell populations tested. They also identified a subset of 255 genes that differed in their expression levels during DC differentiation and maturation. The number of genes that showed increased expression levels upon differentiation/maturation was similar to the number of genes that showed decreased expression levels. While they found surprisingly few differences in gene expression between immature and mature DCs, this could be explained by the late (day14) timepoint at which they analyzed the DCs. TNF-α is not a sufficient stimulus to induce terminal DC maturation [45] and the cells may have reverted to an immature phenotype by the end of the long culture.

Table.1. Overview of DC microarray experiments

Reference	Species	Cells	Time of Activation	Stimulus	Time-Points (h)	Chip
Dietz et al. [42]	human	Mo-DC	Day 7	(IL-6 + IL-1β + TNF-α + PGE-2)	72	Human UniGem V microarrays
Le Naour et al. [44]	human	Mo-DC	Day 7	TNF-α	7 days	HuGeneFL
Huang [48]	human	Mo-DC	Day 7	E.coli Influenza C. albicans LPS Mannan pI:C	0, 1, 2, 4, 8, 12, 18, 24, 36	HuGeneFL
Messmer [49]	human	Mo-DC	Day 7	LPS (IL-6 + IL-1β + TNF-α + PGE-2), CD40L	24, 48	HG-U95A-E
Tureci et al. [47]	human	Mo-DC	Day 5	CD40L-transfected cells	2, 40	HuGEneFL
Chaussabel et al. [68]	human	Mo-DC	Day 8	*B. malayi* *T. gondii* *L major* *L donovani*	16	HG-U95A
Moschella et al. [48]	human	Mo-DC	Day 6	TNF-α, rec sol CD40L, IFN-γ1b	48	Atlas Human Hematology/ immunoloy cDNA expression arrays
Izmailova et al. [109]	human	Mo-DC	Day 7	HIV-1, Adenovirus expressing *tat* or *LacZ*	*HIV*: 0, 1, 6, 12, 24, 72, 7d, 10d, 14d *Adeno*: 0, 4, 10, 16, 30	Custom build Oracle database ChipDB, HU6800
Ju et al. [43]	human	CD34$^+$-DC	Day 6	TNF-α	24	HG-U95A

Table.1. Overview of DC microarray experiments (Continued)

Reference	Species	Cells	Time of Activation	Stimulus	Time-Points (h)	Chip
Suciu-Foca Cortesini et al. [77]	human	Myelomono-cytic cell line KG-1		Allospecific $CD4^+$ T cells, $CD8^+CD28^-$T cells	12	HG-U95A
Hoon Ahn et al. [99]	human	1) peripheral blood DCs $CD11c^-$, 2) $CD1a^+$ and 3) $CD14^+$ generated from $CD34^+$ 4) Mo-DCs		1) IL-3 + rec. CD40L 2) TNF-α + GM-CSF 3) TNF-α + IL-3 4) MCM + TNF-α	1) 5d 2) 8d, 18d 3) 8d, 18d 4) 2 d	HI380
Hacker et al. [100]	human	$CD1a^+$ DCs generated from $CD34^+$	Day 6	*1) TNF-α* 2) TGF-β	1) 24 2) 4, 16, 36	HG-U95Av2
Granucci et al. [56]	mouse	1) D1 cell line, 2) BMDC	2) Day 14	E. coli	0, 4, 6, 12, 18, 24, 48	Mu11kA+B
Granucci et al. [45]	mouse	D1 cell line		LPS, TNF-α	6, 18	Mu6500
Trottein et al. [67]	mouse	1) D1 cell line, 2) BMDC		*S. mansoni* eggs Schistosomul a	0, 4, 8, 12, 24	MG-U74Av2
Edwards et al. [90]	mouse	Splenic $CD11c^{high}$: $CD4^+CD8^-$, $CD8^+CD4^-$, $CD8^-CD4^-$		No stimulus added	Not cultured	U74Av2
Weighardt et al. [61]	Mouse: IFN-$\alpha/\beta R^{-/-}$ and wt	BMDC	Day 10	Synthetic lipid A	3	430A

HS= human serum, FCS= Fetal calf serum, HP=human plasma, MCM=monocyte-condition medium (18), BMDC= bone marrow-derived DC, $CD34^+$-DC= DCs differentiated from $CD34^+$ cells, Mo-DCs=DCs differentiated from monocytes by culture in GM-CSF and IL-4, rec.= recombinant, sol= soluble (not membrane bound)

Moschella et al. analyzed the gene expression profile in human monocyte-derived DCs stimulated with CD40L, TNF-α or IFNγ-1b [46]. Overall the gene expression patterns of DCs treated with either TNF-α or CD40L were closely related. DCs cultured with either TNF-α or CD40L showed increased expression of MIP1β, MIG, IL-14, TPO, MDC, and TARC. Stimulation with IFN-γ induced a unique set of genes: homeobox pbx3, IRF2, host cell factor c1, spi-1/pu-1, dead box protein 6, ikaros and IRF5, and C/EBP γ. Furthermore, IFN-γ treated DCs maintained a high levels of BCL-6-specific mRNA, whereas CD40L or TNF-α treated DCs showed lower levels. Turegi et al.[47] showed that CD40L-induced stimulation of DCs caused an early wave of inflammatory chemokines (MIP-1α, MIP-1β, MIP-1α, MIP-2β, IL-8, IP10, MIP-1d, and MGSA), followed by a sustained increase in constitutive chemokines (TARC, RANTES, MDC, and ENA78) as well as induction of survival proteins. The highly regulated chemokines expression could play a role in the different migratory capacities of DCs at different stages of differentiation.

Transcriptional Profile in DCs Activated with Pathogens or Pathogen Components

The strongest support for the notion that DCs are plastic in their maturation response came from Huang et al. [48], who performed kinetic analyses of human monocyte-derived DCs stimulated with pathogens and their components: LPS, yeast cell wall-derived mannan, and double stranded (ds) RNA, using an early generation Affymetrix Genechip oligonucleotide microarray which contained probes for ~6800 genes. These investigators demonstrated the existence of common and differential pathogen recognition pathways. They performed a detailed kinetic analysis of the gene expression profiles of DCs in response to *E.coli*, *Candida albicans* (*C.albicans*), and influenza virus as well as to their molecular components and demonstrated distinct gene expression patterns in response to different pathogen-derived stimuli. By comparing the transcriptional response to *E.coli*, *C. albicans* and Influenza, they identified a common set of 166 genes (core response) altered in response to all stimuli. These were divided into immediate early, middle, and late response genes. The early response includes genes associated with pathogenesis and pathogen recognition. Transient increases in genes for cytokines and chemokines were observed at the same time as well as a set of cytoskeletal genes. During the late phase chemokine receptors that mediate responses to lymph node chemokines were upregulated.

A smaller set of genes specifically changed after exposure to *E. coli* or Influenza. E- coli specific changes include innate immune genes, like inflammatory cytokines, neutrophil-and monocyte attracting cytokines, and prostaglandin pathway components. The Influenza-specific response was devoid of genes capable of neutrophil stimulation, but a number of antiviral genes as well as interferon-inducible chemokine genes were strongly induced. No genes were found to be specifically altered in response to *C. albicans*. LPS was able to mimic and account for almost the entire E. coli response, whereas the response to mannan was more similar to the *E. coli* response than to the *C. albicans* response. The response to ds RNA was similar to that induced by *E. coli* and overlapped with responses seen to Influenza. The common response reflects a convergence of pathways from receptors that recognize some of these components and pathogens.

In order to get a deeper understanding into the DC maturation process on the molecular levels as a consequence of exposure to pathogen components as well as inflammatory stimuli, we performed a genome-wide analysis using the HG-U95 series Genechip (Affymetrix) oligonucleotide microarray [49]. These arrays contain >60,000 unique probesets to detect known genes and ESTs, providing nearly complete coverage of the expressed genes in the human genome [50]. Immature human monocyte-derived DCs were induced to mature with a cocktail of inflammatory cytokines (CyC: TNF-α, IL-1β, IL-6, and PGE-2), CD40L, and LPS. Gene expression was analyzed at 24h and 48h after stimulation, detecting mainly late genes. We identified elements of sustained and transient gene expression changes. The sustained transcriptional response to LPS, and CyC was largely overlapping indicating a core maturation response. Interestingly several molecules that could contribute to a controlled and regulated immune response have been identified in the core response. One of these molecules is CD200. CD200 is expressed on DCs in the follicles of spleen and lymph nodes [51]. It provides a negative signal, decreasing macrophage and myeloid cell activity [52], and triggers an immunomodulatory function, which leads to increased allograft survival [53]. The upregualtion of CD200 upon DC maturation in our system could indicate a regulatory mechanism, which puts a brake on the immunostimulatory capacity of a mature DC in order to limit the intensity or duration of a resulting immune response. Thus, the induction of terminal DC maturation might be followed or paralleled by an induction of inhibitory pathways. Further support for this idea is the finding that B7-H1, a co-stimulatory molecule, which may be involved in the negative regulation of cell-mediated immune responses [54, 55] was upregulated in mature DCs. Interaction of DCs with T cells is crucial for the induction of an immune response. Upon DC maturation, several adhesion molecules that contribute to T cell activation are upregulated on the cell surface [1]. Within the core response upregulation of CD54, CD58, integrinβ_8, endothelial differentiation, sphingolipid G-protein-couples receptor 1 and CD42b was observed. However, 15 genes associated with cell adhesion were down-regulated. This might represent an additional inhibitory control for T cell activation or DCs might change their affinity to different environments through their adhesion molecules. A smaller set of genes changed their expression levels in a stimulus-specific manner, suggesting distinct genetic subroutines. These stimuli-specific expression patterns are evidence that DCs process environmental signals and activate different transcriptional programs in response to distinct stimuli. These findings are in concordance with those by Huang et al. [48]. Interestingly, within the transient as well as sustained LPS-specific genes was a high number of chemokines, suggesting that LPS-stimulated DCs have a higher potential for immune cell recruitment and amplification of the immune response.

In the murine system Granucci et al. performed a detailed kinetic analysis of a murine DC cell line (D1) to *Escherichia coli* (*E. coli*) [56] as well as to lipopolysaccharide (LPS) and TNF-α [45]. Studies with this well characterized DC cell line show similar maturation to that observed with fresh splenic or bone marrow-derived DCs [57, 58]. The cell line has the advantage of being homogeneous, since bone marrow-derived DCs are extremely heterogeneous. When comparing the response to LPS and TNF-α, they found that LPS was more potent than TNF-α in activating the expression of genes involved in antigen processing and T cell stimulation, as well as genes that can control inflammatory responses. Terminally differentiated DCs undergo growth arrest and a similar process can be initiated in immature D1 cells when maturation is induced *in vitro*. The authors show that stimulation with TNF-α, unlike LPS, does not induce upregulation of D- type cyclins and anti-proliferative genes such as BTG1, GAS, and RCK, which were exclusively upregulated in response to LPS activation. The gene expression results suggest that only LPS can induce definite growth arrest and the authors supported this by showing that TNF-α-stimulated DCs slow down proliferation, but don't show a complete growth arrest, suggesting that TNF-α induces partial maturation. In addition, LPS but not TNF-a upregulated the PA28 proteasome activator and the TAP-1 mRNA. PA28 increase the spectrum of peptides produced and the efficiency of the 20S proteasome and the TAP-1 molecule is required to transfer proteasome-produced peptides from the cytosol to the ER for loading on MHC molecules [59]. Furthermore IL-6 expression was induced only by LPS and not TNF-a. IL-6 increases the spectrum of peptides presented by DCs [60]. Taken together these data show that only LPS is sufficient to induce a potent T cell stimulatory DC. Another important insight that was derived from their data is that together with the inflammatory response induced by LPS genes that control inflammatory responses are modulated as well. LPS-matured DCs upregulate IL-1β together with the IL-1 receptor antagonist (IL-1RA), which is likely to control inflammation.

Using genome wide analysis, Weighardt et al.[61] identified TRIF as a major regulator of the TLR4-triggered maturation program of DCs. They show that TRIF is necessary for TLR4-induced expression of Th1 type cytokines like IL-12, IL-18, and IL23. They further reveal an interferon-dependent and independent mechanism that regulates TLR4-mediated gene expression changes. The TRIF-dependent genes include IRF-7, Mx1, Mx2, ISG20, IFI204, and TLR3, which are thought to play a role in anti-viral immune responses [62-65]. Their findings showed that TRIF is a major regulator of the TLR-mediated antiviral immune response and TRIF mutant mice are highly susceptible to viral infection

An unsuspected finding arose from gene expression analysis of D1 cells revealing that they produce IL-2 in a tightly regulated time frame after exposure to *E. coli* [56]. IL-2 secretion occurs in two waves; the first being between 4-8 h after bacterial uptake and the second wave is between 14-18 h after activation. They showed that this is a unique feature of DCs, which is not found in macrophages. Using IL-2$^{-/-}$ DCs they showed that IL-2 produced by early bacteria-primed DCs was required for T cell proliferation.

Pathogen-specific functional plasticity of DCs has also been shown in the murine system. The fungus C. albicans stimulates IL-12 production and a Th1 reponse in DCs at the yeast stage and IL-4 production and a Th2 response at the hyphae stage [66]. To understand the different effects different stages of the helminth parasite *Schistosoma* life cycle have on the immune system, their effect on DCs has been investigated by a genome-wide expression study. Trottein et al. [67] have analyzed the transcriptional time-course of mouse D1 cells in response to eggs or schistosomula (SLA), providing molecular insights into the immune evasion mechanism of schistosomula and a role for type I IFN in the innate response to helminth eggs. 283 genes were found differentially expressed after encounter of egg or SLA. Only 30 genes were changed in a similar manner between the two stimuli, whereas the majority of the remaining 253 transcripts were modulated by just one of the two developmental stages of the helminth. Inflammatory cytokines, chemokines, and IFN-inducible proteins were oppositely regulated by the two stimuli. Eggs, but not SLA, induced the expression of IFN-β. IL-1β then through autocrine/paracrine activation of the type 1 interferon receptor caused STAT-1 phosphorylation and the induction of interferon responsive genes: glucocorticoid-attenuated response gene (GARG)-16, GARG-39, GARG-49, IFN-inducible gene (IFI)-1, IFI47, IFN-stimulated gene factor-3γ (ISGF3γ), and IFN-inducible protein (IP)-10.

Further evidence for the notion that DCs are plastic in their response to different pathogens came from Chaussabel et al.[68]. They studied the global transcriptional response in human DCs and macrophages to 5 different pathogens that cause chronic infections in humans. Monocyte-derived DCs were exposed to *Mycobacterium tuberculosis*, and to phylogenetically distinct protozoan: *Leishamia major*, *Leishmania donovani*, *Toxoplasma gondii*, and helminth parasite *Brugaria malayi.* Gene expression profiles were examined 16 h after exposure using the HU95A microarray (Affymetrix). These pathogens induce different types of immune responses in the host. *T gondii* and *Mycobacterium tuberculosis* induce a strong protective Th1 response. In contrast to this *Leishmania* infection may or may not induce a Th1 response depending on the species. Infection with extracellular parasites such as *B. malayi* typically fosters a protective Th2 response. Interestingly immature DCs and macrophages showed very similar gene expression patterns at the baseline before pathogen exposure. 3642 genes were expressed similarly in both cell types, whereas only 130 genes were macrophage-specific and 286 genes were DC-specific. However, those two cell types differed dramatically in their transcriptional response to the same pathogen. The number and complexity of transcriptionally altered genes in DCs was much higher compared to macrophages. As found by other groups with different pathogens Chaussabel et al. [68] demonstrated that DCs show a gene expression profile common to all pathogens tested as well as a pathogen-specific transcriptional profile. Genes that were induced by all pathogens include a large family of genes belonging to the NF-κB signaling pathway and apoptosis regulators. The transcriptional response to *B. malayi* was almost silent. A second group of genes that are induced primarily by *Leishmania* and *M tuberculosis* contains inflammatory mediators (RANTES, MIP-1β, IL-6...) and adhesion molecules. Unique pathogen-specific signatures can be derived by using a set of interferon-induced and "inflammatory" genes.

Transcriptional Profile of "Tolerogenic" DCs

Central tolerance of T cells in the thymus is mediated by thymic DCs [69]. Although evidence suggest that the stage of T cell differentiation rather than the type of DC influences death of those self-reactive T cells, thymic DCs are thought to play a role in tolerizing T cells. DCs also play a role in maintaining peripheral tolerance to self-antigen. The mechanism underlying peripheral tolerance is under debate. Evidence exist to support the hypothesis that immature DCs induce tolerance whereas mature DCs induce immunity [70, 71]. On the other hand it has been shown that mature but quiescent DC induce tolerance, whereas mature and fully activated DCs induce immunity [72, 73]. More recently it has been shown that DCs stimulated with TLR4 and TLR9 ligands reversed regulatory T cell (T_R)-mediated suppression and restored T cell proliferation. This effect was partially mediated by IL-6 [74].

It has been shown that cell-cell interactions between APC and antigen-specific $CD8^+CD28^-$ T suppressor cells (Tc) leads to the development of tolerogenic APC [75, 76]. The molecular changes underlying these events have been analyzed by comparing the gene expression profiles of DCs exposed to either Ts or Th cells [77]. An immature human DCs cell line KG-1 was incubated with Ts or Th for 12 h. The transcriptional changes induced by Ts or Th cells shared certain features. However, specific gene expression changes were observed. Ts cells promoted rather than blocked the maturation of DCs, concordant with the notion that mature, not immature DCs induce tolerance. The specific changes are of interest, because they might reveal the "tolerogenic" features of those DCs. In particular it was observed that the IL-1 receptor agonist (IL-1Ra), which is known to neutralize inflammatory responses [78], was upregulated in DCs cultured with Ts cells. COX-2, which is essential for promoting tolerance [79] was also found upregulated by Ts cells and IL-6 was downregulated.

Taken together these data suggest that tolerogenic DCs have a particular gene expression pattern that is different from both immature and mature DCs. However, one can not exclude that they are a special subset.

Gene Expression Profiling of DC Subsets

To explain the various different functions that are observed in DCs, two models have been proposed. Specialized DCs subsets might represent different activation states of a single lineage and the functional differences would be induced by different environmental signals. We have seen several lines of evidence described above obtained from gene expression profiling that supports this plasticity model. Alternatively, specialized DC subtypes could derive from different lineages [80]. Different DC subsets have been identified in mouse and man.

Mouse DCs

Distinct subtypes of mature mouse DCs have been identified and CD4, CD8 and CD11c expression is used for segregation of the subsets [80, 81]. At least six different DC subtypes have been found in lymphoid tissue [82] four of which are found in the spleen and include $CD11c^{low}$, $CD11c^{high}$ $CD4^-CD8^+$, $CD11c^{high}$ $CD4^+CD8^-$, and $CD11c^{high}$ $CD4^-CD8^-$ DCs [81, 83, 84]. Two additional subtypes are found in the lymph node. Although all the different subtypes of mouse DCs present antigen to T cells and cause cell cycle progression, they differ in the type of T cell response they generate. Their antigen presenting functions differs, in that only $CD8^+$ DCs are effective in uptake and cross-presentation of exogenous antigens on MHCI and stimulate $CD8^+$ T cells [85, 86]. Furthermore, $CD4^+$ DCs do not make bioactive IL-12 p70 [87, 88] and CD8+ DCs express a different set of Toll-like receptors from the other spleen subsets [89]. All other aspects of the mouse DC subsets biology remain unsettled, including their function, antigen presenting capacity, cytokine production and ontogeny. To gain a better understanding of the different subtypes, Edwards et al. [90] performed a global gene expression analysis on $CD11c^{high}$ DC subsets from the spleen. The spleen contains 4 different DCs subsets, three of which express CD11c^{high}. These can be further divided into CD11b^{low} CD8α^{+}, CD11b^{high} CD8α- CD4^{+} and CD11b^{high} CD8α- CD4^{-} [81, 91]. The transcriptional profiles revealed that the CD4^{+} and CD8α^{+} DC subsets showed distinct basal expression profiles differing by >200 genes, whereas the profile of the double negative CD8α- CD4^{-} DCs was similar to the CD4^{+} DCs, suggesting that the CD4^{+} and double negative DCs are phylogenetically closely related. The tanscriptional differences between the CD8α^{+} and CD8α^{-} DCs support the notion that there are derived from more distant lineages. Furthermore, the global gene expression analysis showed that CD1d, CD5, CD22, and CD72 are differentially expressed among CD11c^{high} DCs. CD5, CD22, and CD72 show a restricted distribution and can serve as novel markers fro the CD8α^{-}subset.

Human DCs

Human blood DCs are heterogeneous, which is thought o reflect differences in activation states rather than different lineages [92]. Splenic and tonsillar DCs isolated fresh without additional culture show heterogeneity in expression of CD4, CD11b, and CD11c. However, most studies on the analysis of human DC subtypes have been performed using in vitro generated DCs. Three different precursor-cells have been used to generated human DCs in culture. CD34^{+} cells isolated from bone marrow or umbilical-cord blood. These cells can yield two types of intermediate precursors of DC development depending on the cell culture conditions, one resembling CD1a^{+}CD14^{-} Langerhans cells and the other resembling CD1a^{-} CD14^{+} interstitial DCs [93]. Blood monocytes are also used a precursors to generated human DCs in culture [94]. Plasmacytoid cells represent another precursor of human DC development [28, 95]. They are found in blood and many lymphoid tissues and due to their expression of lymphoid markers it has been suggested that they are of the lymphoid origin [96, 97]. As seen in mouse, the human DCs subtypes show a certain degree of plasticity, which makes lineages categorization difficult. However, monocytes and plasmacytoid cells show different expression patters of toll-like receptors [98], which might suggest different evolutionary/lineage origins of the different DCs subtypes.

The transcriptional profiles of human DC subsets have been investigated using microarray analysis [99]. Four DCs subsets have been examined: one lymphoid DC; CD11c- DCs from peripheral blood, and three myeloid DCs; monocyte-derived DCs from peripheral blood, and CD1a^{+} and CD14^{+} DCs generated from CD34^{+} hematopoetic progenitor cells. Differentially expressed genes were identified in the different DC subsets. As expected from their lineage difference, a strong difference was observed between CD11c^{-} lymphoid and the ex vivo generated myeloid DCs. The interferon regulatory factor-4 (IRF4), as well as dihydropyrimidinase-related protein-2 (DRP-2), and 5-lipoxygenase activating protein (FLAP) were highly expressed in CD11c^{-} DCs. The difference between the myeloid DC subtypes was smaller. Monocyte-derived DCs and CD1a^{+} DCs shown high expression levels of TARC, IRF4, CCR7, and DC-LAMP. No genes were found to be expressed in monocyte-derived DCs only. However, CD1a^{+} DCs and CD14^{+} DCs showed subtype-specific gene expression. CD1a^{+} DCs specific genes include: MMP12, hsp90, FcERI, and GTP-BP. CD14+

DC specific genes include: MCP1, Factor XIII, CCR1, and ARHGDIB. Many of the identified genes have no clear function and are awaiting further analysis.

Using an in vitro culture sysytem to generate cells that resemble Langerhans cells from human CD34+ cell isolated from cord blood, Hacker et al. [100] found that the expression of the transcriptional antagonistic protein Id2 increases during DC differentiation in vitro. Furthermore, they show that Id2 is critical for the development of Langerhans cells and splenic CD8α+ DCs in vivo. TGF-β, which can further increase the frequency of Langerhans cells during differentiation, can induce Id2 expression. In addition some evidence supports the idea that Id2 supressed B cell genes in DCs [100].

Alteration of Gene Expression Patterns in DCs by Viral Proteins

In the context of immunodeficiency viruses, DCs effectively promote virus amplification and spread especially upon encountering $CD4^+$ T cells (reviewed in Ref. [101]). Furthermore, immature DCs may be one of the first leukocytes entrapping and possibly replicating immunodeficiency virus crossing a mucosal surface [102, 103] and then transmitting infection to nearby $CD4^+$ T cells for amplification [104, Stahl-Hennig, 1999 #1361, 105]. Immature DCs account for 65-90% of the infected cells after vaginal exposure to SIV [103]. Thus, a paradox exists between the ability of DCs to present virus for activation of anti-viral immunity versus their ability to exacerbate virus growth. Understanding what drives virus replication in the DC-T cell milieu is critical to identify ways to limit this and favor immune activation.

The viral encoded *tat* is among the first genes expressed during HIV-1 infection [106] and is necessary for viral gene expression and virus production. Tat interacts with host transcriptional regulators, the protein kinase PKR [107] and the transcriptional coactivators p300 and CREB-binding protein (CBP) [108]. Izmailova et al [109] investigated the effect of HIV-1 *tat*-expression on the gene expression program in immature DCs. Several genes were similarly affected by Tat and by HIV-1 infection. These genes encode transcription factors, chemokines, apoptosis-related factors, antiviral proteins and stress-induced proteins. Also Tat induced the expression of interferon-inducible genes, which was not due to interferon being secreted by the DCs. Two of those genes, interferon inducible protein-10 (IP-10) and human monokine induced by interferon-γ (HuMIG), are usually IFN-γ dependent. This might suggest that an interferon independent pathway can induce interferon-inducible genes. However, the benefit of inducing the expression of interferon-inducible genes by an HIV protein for virus replication is not understood. Monocyte chemoattractant protein-2 (MCP-2) also belongs to this group of genes and was found induced by Tat as well as HIV-1 infection. These chemokines induce T cell- and monocyte chemotaxis and it is plausible that the recruitment of these cells to the location of HIV-1 infected DCs might amplify virus replication and transmission.

Concluding Remarks

Collectively, the analyses of gene expression in DCs using microarrays has significantly expanded and deepened the field of DC biology. A product of these experiments has been the identification of a plethora of genes with previously unknown function in DCs, some of which (IL-2) have already been shown to play important roles in DC-mediated T cells activation and the development of special DC subsets (Id2). There are undoubtedly many more testable hypotheses ready to be culled from the many datasets thus far generated.

Future transcriptome analyses of DCs will most likely seek to transcend the limitations of *in vitro* systems to directly assess DC gene expression in various *in vivo* settings. While the paucity of DCs *in vivo* and the difficulty of obtaining some human tissues has been an impediment, improved cell isolation and robust RNA amplification technologies should allow progress. Comparing the *in vitro* results with *in vivo* DC expression profiles in various physiological and immune contexts will be essential. The *in vivo* transcriptional profiles of DCs from patients with various illnesses, including infectious diseases, autoimmune diseases and cancer, will be particularly interesting. Understanding DC malfunctions in disease may allow the development of better vaccines or targeted therapies. In addition, there has been great interest in using DCs directly as immunotherapeutic agents and the monitoring of gene expression in such DCs may prove to be an essential element in their development as therapeutic tools.

Definitions

1. An *antigen* is one of many kinds of substances with which an antibody molecule or T cell receptor may bind. These include sugars, lipids, intermediary metabolites, autocoids, hormones, complex carbohydrates, phospholipids, nucleic acids, and proteins.

2. *Antigen presenting cell*: A cell that can process a protein antigen, break it into peptides, and present it in conjunction with class II MHC molecules on the cell surface where it may interact with appropriate T cell receptors.

References

[1] Banchereau, J., and R.M. Steinman. 1998. Dendritic cells and the control of immunity. *Nature* 392:245-252.

[2] Rescigno, M., C. Winzler, D. Delia, C. Mutini, M. Lutz, and P. Ricciardi_Castagnoli. 1997. Dendritic cell maturation is required for initiation of the immune response. *Journal of Leukocyte Biology* 61:415-421.

[3] Winzler, C., P. Rovere, M. Rescigno, F. Granucci, G. Penna, L. Adorini, V.S. Zimmermann, J. Davoust, and P. Ricciardi-Castagnoli. 1997. Maturation stages of mouse dendritic cells in growth factor-dependent long-term cultures. *J Exp Med* 185:317-328.

[4] De Smedt, T., B. Pajak, E. Muraille, L. Lespagnard, E. Heinen, P. De Baetselier, J. Urbain, O. Leo, and M. Moser. 1996. Regulation of dendritic cell numbers and maturation by lipopolysaccharide in vivo. *J Exp Med* 184:1413-1424.

[5] Hartmann, G., G.J. Weiner, and A.M. Krieg. 1999. CpG DNA: a potent signal for growth, activation, and maturation of human dendritic cells. *Proceedings of the National Academy of Sciences of the United States of America* 96:9305-9310.

[6] Iwaki, D., H. Mitsuzawa, S. Murakami, H. Sano, M. Konishi, T. Akino, and Y. Kuroki. 2002. The extracellular toll-like receptor 2 domain directly binds peptidoglycan derived from Staphylococcus aureus. *J Biol Chem* 277:24315-24320.

[7] Takeda, K., O. Takeuchi, and S. Akira. 2002. Recognition of lipopeptides by Toll-like receptors. *J Endotoxin Res* 8:459-463.

[8] Lee, H.K., J. Lee, and P.S. Tobias. 2002. Two lipoproteins extracted from Escherichia coli K-12 LCD25 lipopolysaccharide are the major components responsible for Toll-like receptor 2-mediated signaling. *J Immunol* 168:4012-4017.

[9] Zhang, D., G. Zhang, M.S. Hayden, M.B. Greenblatt, C. Bussey, R.A. Flavell, and S. Ghosh. 2004. A toll-like receptor that prevents infection by uropathogenic bacteria. *Science* 303:1522-1526.

[10] Schwandner, R., R. Dziarski, H. Wesche, M. Rothe, and C.J. Kirschning. 1999. Peptidoglycan- and lipoteichoic acid-induced cell activation is mediated by toll-like receptor 2. *J Biol Chem* 274:17406-17409.

[11] Takeuchi, O., K. Hoshino, T. Kawai, H. Sanjo, H. Takada, T. Ogawa, K. Takeda, and S. Akira. 1999. Differential roles of TLR2 and TLR4 in recognition of gram-negative and gram-positive bacterial cell wall components. *Immunity* 11:443-451.

[12] Underhill, D.M., A. Ozinsky, K.D. Smith, and A. Aderem. 1999. Toll-like receptor-2 mediates mycobacteria-induced proinflammatory signaling in macrophages. *Proc Natl Acad Sci U S A* 96:14459-14463.

[13] Chow, J.C., D.W. Young, D.T. Golenbock, W.J. Christ, and F. Gusovsky. 1999. Toll-like receptor-4 mediates lipopolysaccharide-induced signal transduction. *J Biol Chem* 274:10689-10692.

[14] Hoshino, K., O. Takeuchi, T. Kawai, H. Sanjo, T. Ogawa, Y. Takeda, K. Takeda, and S. Akira. 1999. Cutting edge: Toll-like receptor 4 (TLR4)-deficient mice are hyporesponsive to lipopolysaccharide: evidence for TLR4 as the Lps gene product. *J Immunol* 162:3749-3752.

[15] Hemmi, H., O. Takeuchi, T. Kawai, T. Kaisho, S. Sato, H. Sanjo, M. Matsumoto, K. Hoshino, H. Wagner, K. Takeda, and S. Akira. 2000. A Toll-like receptor recognizes bacterial DNA. *Nature* 408:740-745.

[16] Gallucci, S., and P. Matzinger. 2001. Danger signals: SOS to the immune system. *Current Opinion in Immunology* 13:114-119.

[17] Gallucci, S., M. Lolkema, and P. Matzinger. 1999. Natural adjuvants: endogenous activators of dendritic cells. *Nature Medicine* 5:1249-1255.

[18] Sauter, B., M.L. Albert, L. Francisco, M. Larsson, S. Somersan, and N. Bhardwaj. 2000. Consequences of cell death: exposure to necrotic tumor cells, but not primary tissue cells or apoptotic cells, induces the maturation of immunostimulatory dendritic cells. *The Journal of Experimental Medicine* 191:423-434.

[19] Basu, S., R.J. Binder, R. Suto, K.M. Anderson, and P.K. Srivastava. 2000. Necrotic but not apoptotic cell death releases heat shock proteins, which deliver a partial maturation signal to dendritic cells and activate the NF-kappa B pathway. *Int Immunol* 12:1539-1546.

[20] Cella, M., D. Scheidegger, K. Palmer-Lehmann, P. Lane, A. Lanzavecchia, and G. Alber. 1996. Ligation of CD40 on dendritic cells triggers production of high levels of interleukin-12 and enhances T cell stimulatory capacity: T-T help via APC activation. *J Exp Med* 184:747-752.

[21] Caux, C., C. Massacrier, B. Vanbervliet, B. Dubois, C. Van Kooten, I. Durand, and J. Banchereau. 1994. Activation of human dendritic cells through CD40 cross-linking. *J Exp Med* 180:1263-1272.

[22] Jonuleit, H., J. Knop, and A.H. Enk. 1996. Cytokines and their effects on maturation, differentiation and migration of dendritic cells. *Arch Dermatol Res* 289:1-8.

[23] Basu, S., and P.K. Srivastava. 2000. Heat shock proteins: the fountainhead of innate and adaptive immune responses. *Cell Stress Chaperones* 5:443-451.

[24] Singh-Jasuja, H., H.U. Scherer, N. Hilf, D. Arnold-Schild, H.G. Rammensee, R.E. Toes, and H. Schild. 2000. The heat shock protein gp96 induces maturation of dendritic cells and down-regulation of its receptor. *European Journal of Immunology* 30:2211-2215.

[25] Kuppner, M.C., R. Gastpar, S. Gelwer, E. Nossner, O. Ochmann, A. Scharner, and R.D. Issels. 2001. The role of heat shock protein (hsp70) in dendritic cell maturation: hsp70 induces the maturation of immature dendritic cells but reduces DC differentiation from monocyte precursors. *Eur J Immunol* 31:1602-1609.

[26] Flohe, S.B., J. Bruggemann, S. Lendemans, M. Nikulina, G. Meierhoff, S. Flohe, and H. Kolb. 2003. Human heat shock protein 60 induces maturation of dendritic cells versus a Th1-promoting phenotype. *Journal of Immunology (Baltimore, Md. : 1950)* 170:2340-2348.

[27] Messmer, D., H. Yang, G. Telusma, F. Knoll, J. Li, B. Messmer, K.J. Tracey, and N. Chiorazzi. 2004. High mobility group box protein 1: an endogenous signal for dendritic cell maturation and Th1 polarization. *J Immunol* 173:307-313.

[28] Rissoan, M.C., V. Soumelis, N. Kadowaki, G. Grouard, F. Briere, R. de Waal Malefyt, and Y.J. Liu. 1999. Reciprocal control of T helper cell and dendritic cell differentiation [see comments]. *Science* 283:1183-1186.

[29] Liu, Y.J., H. Kanzler, V. Soumelis, and M. Gilliet. 2001. Dendritic cell lineage, plasticity and cross-regulation. *Nat Immunol* 2:585-589.

[30] Pulendran, B., J. Banchereau, E. Maraskovsky, and C. Maliszewski. 2001. Modulating the immune response with dendritic cells and their growth factors. *Trends Immunol* 22:41-47.

[31] Luft, T., M. Jefford, P. Luetjens, T. Toy, H. Hochrein, K.A. Masterman, C. Maliszewski, K. Shortman, J. Cebon, and E. Maraskovsky. 2002. Functionally distinct dendritic cell (DC) populations induced by physiologic stimuli: prostaglandin E(2) regulates the migratory capacity of specific DC subsets. *Blood* 100:1362-1372.

[32] Mellman, I., and R.M. Steinman. 2001. Dendritic cells: specialized and regulated antigen processing machines. *Cell* 106:255-258.

[33] Liu, Y.J. 2001. Dendritic cell subsets and lineages, and their functions in innate and adaptive immunity. *Cell* 106:259-262.

[34] Zhou, L.J., and T.F. Tedder. 1996. CD14+ blood monocytes can differentiate into functionally mature CD83+ dendritic cells. *Proc Natl Acad Sci U S A* 93:2588-2592.

[35] Gluckman, J.C., B. Canque, F. Chapuis, and M. Rosenzwajg. 1997. In vitro generation of human dendritic cells and cell therapy. *Cytokines Cell Mol Ther* 3:187-196.

[36] Herbst, B., G. Kohler, A. Mackensen, H. Veelken, R. Mertelsmann, and A. Lindemann. 1997. CD34+ peripheral blood progenitor cell and monocyte derived dendritic cells: a comparative analysis. *Br J Haematol* 99:490-499.

[37] .Lardon, F., H.W. Snoeck, Z.N. Berneman, V.F. Van Tendeloo, G. Nijs, M. Lenjou, E. Henckaerts, C.J. Boeckxtaens, P. Vandenabeele, L.L. Kestens, D.R. Van Bockstaele, and G.L. Vanham. 1997. Generation of dendritic cells from bone marrow progenitors using GM-CSF, TNF-alpha, and additional cytokines: antagonistic effects of IL-4 and IFN-gamma and selective involvement of TNF-alpha receptor-1. *Immunology* 91:553-559.

[38] Velculescu, V.E., L. Zhang, B. Vogelstein, and K.W. Kinzler. 1995. Serial analysis of gene expression. *Science* 270:484-487.

[39] Schena, M., D. Shalon, R.W. Davis, and P.O. Brown. 1995. Quantitative monitoring of gene expression patterns with a complementary DNA microarray. *Science* 270:467-470.

[40] Schena, M., D. Shalon, R. Heller, A. Chai, P.O. Brown, and R.W. Davis. 1996. Parallel human genome analysis: microarray-based expression monitoring of 1000 genes. *Proc Natl Acad Sci U S A* 93:10614-10619.

[41] Shalon, D., S.J. Smith, and P.O. Brown. 1996. A DNA microarray system for analyzing complex DNA samples using two-color fluorescent probe hybridization. *Genome Research* 6:639-645.

[42] Dietz, A.B., P.A. Bulur, G.J. Knutson, R. Matasic, and S. Vuk-Pavlovic. 2000. Maturation of human monocyte-derived dendritic cells studied by microarray hybridization. *Biochemical and Biophysical Research Communications* 275:731-738.

[43] Ju, X.S., C. Hacker, J. Madruga, S.M. Kurz, S. Knespel, G. Blendinger, S. Rose_John, and Z. Martin. 2003. Towards determining the differentiation program of antigen-presenting dendritic cells by transcriptional profiling. *European Journal of Cell Biology* 82:75-86.

[44] Le Naour, F., L. Hohenkirk, A. Grolleau, D.E. Misek, P. Lescure, J.D. Geiger, S. Hanash, and L. Beretta. 2001. Profiling changes in gene expression during differentiation and maturation of monocyte-derived dendritic cells using both oligonucleotide microarrays and proteomics. *J Biol Chem* 276:17920-17931.

[45] Granucci, F., C. Vizzardelli, E. Virzi, M. Rescigno, and P. Ricciardi-Castagnoli. 2001. Transcriptional reprogramming of dendritic cells by differentiation stimuli. *Eur J Immunol* 31:2539-2546.

[46] Moschella, F., A. Maffei, R.P. Catanzaro, K.P. Papadopoulos, D. Skerrett, C.S. Hesdorffer, and P.E. Harris. 2001. Transcript profiling of human dendritic cells maturation-induced under defined culture conditions: comparison of the effects of tumour necrosis factor alpha, soluble CD40 ligand trimer and interferon gamma. *Br J Haematol* 114:444-457.

[47] Tureci, O., H. Bian, F.O. Nestle, L. Raddrizzani, J.A. Rosinski, A. Tassis, H. Hilton, M. Walstead, U. Sahin, and J. Hammer. 2003. Cascades of transcriptional induction during dendritic cell maturation revealed by genome-wide expression analysis. *The Faseb Journal : Official Publication of the Federation of American Societies For Experimental Biology* 17:836-847.

[48] Huang, Q., D. Liu, P. Majewski, L.C. Schulte, J.M. Korn, R.A. Young, E.S. Lander, and N. Hacohen. 2001. The plasticity of dendritic cell responses to pathogens and their components. *Science* 294:870-875.

[49] Messmer, D., B. Messmer, and N. Chiorazzi. 2003. The global transcriptional maturation program and stimuli-specific gene expression profiles of human myeloid dendritic cells. *International Immunology* 15:491-503.

[50] Lander, E.S., L.M. Linton, B. Birren, C. Nusbaum, M.C. Zody, J. Baldwin, K. Devon, K. Dewar, M. Doyle, W. FitzHugh, R. Funke, D. Gage, K. Harris, A. Heaford, J. Howland, L. Kann, J. Lehoczky, R. LeVine, P. McEwan, K. McKernan, J. Meldrim, J.P. Mesirov, C. Miranda, W. Morris, J. Naylor, C. Raymond, M. Rosetti, R. Santos, A. Sheridan, C. Sougnez, N. Stange-Thomann, N. Stojanovic, A. Subramanian, D. Wyman, J. Rogers, J. Sulston, R. Ainscough, S. Beck, D. Bentley, J. Burton, C. Clee, N. Carter, A. Coulson, R. Deadman, P. Deloukas, A. Dunham, I. Dunham, R. Durbin, L. French, D. Grafham, S. Gregory, T. Hubbard, S. Humphray, A. Hunt, M. Jones, C. Lloyd, A. McMurray, L. Matthews, S. Mercer, S. Milne, J.C. Mullikin, A. Mungall, R. Plumb, M. Ross, R. Shownkeen, S. Sims, R.H. Waterston, R.K. Wilson, L.W. Hillier, J.D. McPherson, M.A. Marra, E.R. Mardis, L.A. Fulton, A.T. Chinwalla, K.H. Pepin, W.R. Gish, S.L. Chissoe, M.C. Wendl, K.D. Delehaunty, T.L. Miner, A. Delehaunty, J.B. Kramer, L.L. Cook, R.S. Fulton, D.L. Johnson, P.J. Minx, S.W. Clifton, T. Hawkins, E. Branscomb, P. Predki, P. Richardson, S. Wenning, T. Slezak, N. Doggett, J.F. Cheng, A. Olsen, S. Lucas, C. Elkin, E. Uberbacher, M. Frazier, R.A. Gibbs, D.M. Muzny, S.E. Scherer, J.B. Bouck, E.J. Sodergren, K.C. Worley, C.M. Rives, J.H. Gorrell, M.L. Metzker, S.L. Naylor, R.S. Kucherlapati, D.L. Nelson, G.M. Weinstock, Y. Sakaki, A. Fujiyama, M. Hattori, T. Yada, A. Toyoda, T. Itoh, C. Kawagoe, H. Watanabe, Y. Totoki, T. Taylor, J. Weissenbach, R. Heilig, W. Saurin, F. Artiguenave, P. Brottier, T. Bruls, E. Pelletier, C. Robert, P. Wincker, D.R. Smith, L. Doucette-Stamm, M. Rubenfield, K. Weinstock, H.M. Lee, J. Dubois, A. Rosenthal, M. Platzer, G. Nyakatura, S. Taudien, A. Rump, H. Yang, J. Yu, J. Wang, G. Huang, J. Gu, L. Hood, L. Rowen, A. Madan, S. Qin, R.W. Davis, N.A. Federspiel, A.P. Abola, M.J. Proctor, R.M. Myers, J. Schmutz, M. Dickson, J. Grimwood, D.R. Cox, M.V. Olson, R. Kaul, N. Shimizu, K. Kawasaki, S. Minoshima, G.A. Evans, M. Athanasiou, R. Schultz, B.A. Roe, F. Chen, H. Pan, J. Ramser, H. Lehrach, R. Reinhardt, W.R. McCombie, M. de la Bastide, N. Dedhia, H. Blocker, K. Hornischer, G. Nordsiek, R. Agarwala, L. Aravind, J.A. Bailey, A. Bateman, S. Batzoglou, E. Birney, P. Bork, D.G. Brown, C.B. Burge, L. Cerutti, H.C. Chen, D. Church, M. Clamp, R.R. Copley, T. Doerks, S.R. Eddy, E.E. Eichler, T.S. Furey, J. Galagan, J.G. Gilbert, C. Harmon, Y. Hayashizaki, D. Haussler, H. Hermjakob, K. Hokamp, W. Jang, L.S. Johnson, T.A. Jones, S. Kasif, A. Kaspryzk, S. Kennedy, W.J. Kent, P. Kitts, E.V. Koonin, I. Korf, D. Kulp, D. Lancet, T.M. Lowe, A. McLysaght, T. Mikkelsen, J.V. Moran, N. Mulder,

V.J. Pollara, C.P. Ponting, G. Schuler, J. Schultz, G. Slater, A.F. Smit, E. Stupka, J. Szustakowski, D. Thierry-Mieg, J. Thierry-Mieg, L. Wagner, J. Wallis, R. Wheeler, A. Williams, Y.I. Wolf, K.H. Wolfe, S.P. Yang, R.F. Yeh, F. Collins, M.S. Guyer, J. Peterson, A. Felsenfeld, K.A. Wetterstrand, A. Patrinos, M.J. Morgan, J. Szustakowki, P. de Jong, J.J. Catanese, K. Osoegawa, H. Shizuya, S. Choi, and Y.J. Chen. 2001. Initial sequencing and analysis of the human genome. *Nature* 409:860-921.

[51] Barclay, A.N. 1981. Different reticular elements in rat lymphoid tissue identified by localization of Ia, Thy-1 and MRC OX 2 antigens. *Immunology* 44:727-736.

[52] Barclay, A.N., G.J. Wright, G. Brooke, and M.H. Brown. 2002. CD200 and membrane protein interactions in the control of myeloid cells. *Trends Immunol* 23:285-290.

[53] Gorczynski, R.M., K. Yu, and D. Clark. 2000. Receptor engagement on cells expressing a ligand for the tolerance-inducing molecule OX2 induces an immunoregulatory population that inhibits alloreactivity in vitro and in vivo. *Journal of Immunology (Baltimore, Md. : 1950)* 165:4854-4860.

[54] Dong, H., G. Zhu, K. Tamada, and L. Chen. 1999. B7-H1, a third member of the B7 family, co-stimulates T-cell proliferation and interleukin-10 secretion. *Nat Med* 5:1365-1369.

[55] Freeman, G.J., A.J. Long, Y. Iwai, K. Bourque, T. Chernova, H. Nishimura, L.J. Fitz, N. Malenkovich, T. Okazaki, M.C. Byrne, H.F. Horton, L. Fouser, L. Carter, V. Ling, M.R. Bowman, B.M. Carreno, M. Collins, C.R. Wood, and T. Honjo. 2000. Engagement of the PD-1 immunoinhibitory receptor by a novel B7 family member leads to negative regulation of lymphocyte activation. *J Exp Med* 192:1027-1034.

[56] Granucci, F., C. Vizzardelli, N. Pavelka, S. Feau, M. Persico, E. Virzi, M. Rescigno, G. Moro, and P. Ricciardi-Castagnoli. 2001. Inducible IL-2 production by dendritic cells revealed by global gene expression analysis. *Nat Immunol* 2:882-888.

[57] Rodriguez, A., A. Regnault, M. Kleijmeer, P. Ricciardi-Castagnoli, and S. Amigorena. 1999. Selective transport of internalized antigens to the cytosol for MHC class I presentation in dendritic cells. *Nat Cell Biol* 1:362-368.

[58] Rescigno, M., V. Piguet, B. Valzasina, S. Lens, R. Zubler, L. French, V. Kindler, J. Tschopp, and P. Ricciardi-Castagnoli. 2000. Fas engagement induces the maturation of dendritic cells (DCs), the release of interleukin (IL)-1beta, and the production of interferon gamma in the absence of IL-12 during DC-T cell cognate interaction: a new role for Fas ligand in inflammatory responses. *J Exp Med* 192:1661-1668.

[59] Pamer, E., and P. Cresswell. 1998. Mechanisms of MHC class I--restricted antigen processing. *Annu Rev Immunol* 16:323-358.

[60] Staudt, L.M., and P.O. Brown. 2000. Genomic views of the immune system*. *Annu Rev Immunol* 18:829-859.

[61] Weighardt, H., G. Jusek, J. Mages, R. Lang, K. Hoebe, B. Beutler, and B. Holzmann. 2004. Identification of a TLR4- and TRIF-dependent activation program of dendritic cells. *Eur J Immunol* 34:558-564.

[62] Sato, M., H. Suemori, N. Hata, M. Asagiri, K. Ogasawara, K. Nakao, T. Nakaya, M. Katsuki, S. Noguchi, N. Tanaka, and T. Taniguchi. 2000. Distinct and essential roles of transcription factors IRF-3 and IRF-7 in response to viruses for IFN-alpha/beta gene induction. *Immunity* 13:539-548.

[63] Rolle, S., M. De Andrea, D. Gioia, D. Lembo, L. Hertel, S. Landolfo, and M. Gariglio. 2001. The interferon-inducible 204 gene is transcriptionally activated by mouse cytomegalovirus and is required for its replication. *Virology* 286:249-255.

[64] Horisberger, M.A. 1995. Interferons, Mx genes, and resistance to influenza virus. *Am J Respir Crit Care Med* 152:S67-71.

[65] Espert, L., G. Degols, C. Gongora, D. Blondel, B.R. Williams, R.H. Silverman, and N. Mechti. 2003. ISG20, a new interferon-induced RNase specific for single-stranded RNA, defines an alternative antiviral pathway against RNA genomic viruses. *J Biol Chem* 278:16151-16158.

[66] d'Ostiani, C.F., G. Del Sero, A. Bacci, C. Montagnoli, A. Spreca, A. Mencacci, P. Ricciardi-Castagnoli, and L. Romani. 2000. Dendritic cells discriminate between yeasts and hyphae of the fungus Candida albicans. Implications for initiation of T helper cell immunity in vitro and in vivo. *J Exp Med* 191:1661-1674.

[67] .Trottein, F., N. Pavelka, C. Vizzardelli, V. Angeli, C.S. Zouain, M. Pelizzola, M. Capozzoli, M. Urbano, M. Capron, F. Belardelli, F. Granucci, and P. Ricciardi-Castagnoli. 2004. A type I IFN-dependent pathway induced by Schistosoma mansoni eggs in mouse myeloid dendritic cells generates an inflammatory signature. *J Immunol* 172:3011-3017.

[68] Chaussabel, D., R.T. Semnani, M.A. McDowell, D. Sacks, A. Sher, and T.B. Nutman. 2003. Unique gene expression profiles of human macrophages and dendritic cells to phylogenetically distinct parasites. *Blood* 102:672-681.

[69] Brocker, T., M. Riedinger, and K. Karjalainen. 1997. Targeted expression of major histocompatibility complex (MHC) class II molecules demonstrates that dendritic cells can induce negative but not positive selection of thymocytes in vivo. *The Journal of Experimental Medicine* 185:541-550.

[70] Dhodapkar, M.V., R.M. Steinman, J. Krasovsky, C. Munz, and N. Bhardwaj. 2001. Antigen-specific inhibition of effector T cell function in humans after injection of immature dendritic cells. *The Journal of Experimental Medicine* 193:233-238.

[71] Steinman, R.M., S. Turley, I. Mellman, and K. Inaba. 2000. The induction of tolerance by dendritic cells that have captured apoptotic cells. *The Journal of Experimental Medicine* 191:411-416.

[72] Albert, M.L., M. Jegathesan, and R.B. Darnell. 2001. Dendritic cell maturation is required for the cross-tolerization of CD8+ T cells. 2:1010-1017.

[73] Shortman, K., and W.R. Heath. 2001. Immunity or tolerance? That is the question for dendritic cells. 2:988-989.

[74] Pasare, C., and R. Medzhitov. 2003. Toll pathway-dependent blockade of CD4+CD25+ T cell-mediated suppression by dendritic cells. *Science* 299:1033-1036.

[75] Liu, Z., S. Tugulea, R. Cortesini, S. Lederman, and N. Suciu-Foca. 1999. Inhibition of CD40 signaling pathway in antigen presenting cells by T suppressor cells. *Hum Immunol* 60:568-574.

[76] Li, J., Z. Liu, S. Jiang, R. Cortesini, S. Lederman, and N. Suciu-Foca. 1999. T suppressor lymphocytes inhibit NF-kappa B-mediated transcription of CD86 gene in APC. *J Immunol* 163:6386-6392.

[77] Suciu_Foca_Cortesini, N., F. Piazza, E. Ho, R. Ciubotariu, J. LeMaoult, R. Dalla_Favera, and R. Cortesini. 2001. Distinct mRNA microarray profiles of tolerogenic dendritic cells. *Human Immunology* 62:1065-1072.

[78] Arend, W.P., M. Malyak, C.J. Guthridge, and C. Gabay. 1998. Interleukin-1 receptor antagonist: role in biology. *Annu Rev Immunol* 16:27-55.

[79] Morteau, O. 1999. COX-2: promoting tolerance. *Nat Med* 5:867-868.

[80] Shortman, K., and Y.J. Liu. 2002. Mouse and human dendritic cell subtypes. 2:151-161.

[81] Vremec, D., J. Pooley, H. Hochrein, L. Wu, and K. Shortman. 2000. CD4 and CD8 expression by dendritic cell subtypes in mouse thymus and spleen. *Journal of Immunology (Baltimore, Md. : 1950)* 164:2978-2986.

[82] Henri, S., D. Vremec, A. Kamath, J. Waithman, S. Williams, C. Benoist, K. Burnham, S. Saeland, E. Handman, and K. Shortman. 2001. The dendritic cell populations of mouse lymph nodes. *J Immunol* 167:741-748.

[83] Nakano, H., M. Yanagita, and M.D. Gunn. 2001. CD11c(+)B220(+)Gr-1(+) cells in mouse lymph nodes and spleen display characteristics of plasmacytoid dendritic cells. *J Exp Med* 194:1171-1178.

[84] Asselin-Paturel, C., A. Boonstra, M. Dalod, I. Durand, N. Yessaad, C. Dezutter-Dambuyant, A. Vicari, A. O'Garra, C. Biron, F. Briere, and G. Trinchieri. 2001. Mouse type I IFN-producing cells are immature APCs with plasmacytoid morphology. *Nat Immunol* 2:1144-1150.

[85] Iyoda, T., S. Shimoyama, K. Liu, Y. Omatsu, Y. Akiyama, Y. Maeda, K. Takahara, R.M. Steinman, and K. Inaba. 2002. The CD8+ dendritic cell subset selectively endocytoses dying cells in culture and in vivo. *The Journal of Experimental Medicine* 195:1289-1302.

[86] Schulz, O., and C. Reis e Sousa. 2002. Cross-presentation of cell-associated antigens by CD8alpha+ dendritic cells is attributable to their ability to internalize dead cells. *Immunology* 107:183-189.

[87] Edwards, A.D., S.P. Manickasingham, R. Sporri, S.S. Diebold, O. Schulz, A. Sher, T. Kaisho, S. Akira, and C. Reis e Sousa. 2002. Microbial recognition via Toll-like receptor-dependent and -independent pathways determines the cytokine response of murine dendritic cell subsets to CD40 triggering. *J Immunol* 169:3652-3660.

[88] Hochrein, H., K. Shortman, D. Vremec, B. Scott, P. Hertzog, and M. O_Keeffe. 2001. Differential production of IL-12, IFN-alpha, and IFN-gamma by mouse dendritic cell subsets. *Journal of Immunology (Baltimore, Md. : 1950)* 166:5448-5455.

[89] Edwards, A.D., S.S. Diebold, E.M. Slack, H. Tomizawa, H. Hemmi, T. Kaisho, S. Akira, and C. Reis e Sousa. 2003. Toll-like receptor expression in murine DC subsets: lack of TLR7 expression by CD8 alpha+ DC correlates with unresponsiveness to imidazoquinolines. *Eur J Immunol* 33:827-833.

[90] Edwards, A.D., D. Chaussabel, S. Tomlinson, O. Schulz, A. Sher, and C. Reis_E_Sousa. 2003. Relationships among murine CD11c(high) dendritic cell subsets as revealed by baseline gene expression patterns. *Journal of Immunology (Baltimore, Md. : 1950)* 171:47-60.

[91] Vremec, D., M. Zorbas, R. Scollay, D.J. Saunders, C.F. Ardavin, L. Wu, and K. Shortman. 1992. The surface phenotype of dendritic cells purified from mouse thymus and spleen: investigation of the CD8 expression by a subpopulation of dendritic cells. *The Journal of Experimental Medicine* 176:47-58.

[92] Hart, D.N. 1997. Dendritic cells: unique leukocyte populations which control the primary immune response. *Blood* 90:3245-3287.

[93] Caux, C., B. Vanbervliet, C. Massacrier, C. Dezutter-Dambuyant, B. de Saint-Vis, C. Jacquet, K. Yoneda, S. Imamura, D. Schmitt, and J. Banchereau. 1996. CD34+ hematopoietic progenitors from human cord blood differentiate along two independent dendritic cell pathways in response to GM-CSF+TNF alpha. *J Exp Med* 184:695-706.

[94] Romani, N., D. Reider, M. Heuer, S. Ebner, E. Kampgen, B. Eibl, D. Niederwieser, and G. Schuler. 1996. Generation of mature dendritic cells from human blood. An improved method with special regard to clinical applicability. *Journal of Immunological Methods* 196:137-151.

[95] Grouard, G., M.C. Rissoan, L. Filgueira, I. Durand, J. Banchereau, and Y.J. Liu. 1997. The enigmatic plasmacytoid T cells develop into dendritic cells with interleukin (IL)-3 and CD40-ligand. *The Journal of Experimental Medicine* 185:1101-1111.

[96] Res, P.C., F. Couwenberg, F.A. Vyth_Dreese, and H. Spits. 1999. Expression of pTalpha mRNA in a committed dendritic cell precursor in the human thymus. *Blood* 94:2647-2657.

[97] Spits, H., F. Couwenberg, A.Q. Bakker, K. Weijer, and C.H. Uittenbogaart. 2000. Id2 and Id3 inhibit development of CD34(+) stem cells into predendritic cell (pre-DC)2 but not into pre-DC1. Evidence for a lymphoid origin of pre-DC2. *J Exp Med* 192:1775-1784.

[98] Kadowaki, N., S. Ho, S. Antonenko, R.W. Malefyt, R.A. Kastelein, F. Bazan, and Y.J. Liu. 2001. Subsets of human dendritic cell precursors express different toll-like receptors and respond to different microbial antigens. *The Journal of Experimental Medicine* 194:863-869.

[99] Ahn, J.H., Y. Lee, C. Jeon, S.J. Lee, B.H. Lee, K.D. Choi, and Y.S. Bae. 2002. Identification of the genes differentially expressed in human dendritic cell subsets by cDNA subtraction and microarray analysis. *Blood* 100:1742-1754.

[100] Hacker, C., R.D. Kirsch, X.S. Ju, T. Hieronymus, T.C. Gust, C. Kuhl, T. Jorgas, S.M. Kurz, S. Rose-John, Y. Yokota, and M. Zenke. 2003. Transcriptional profiling identifies Id2 function in dendritic cell development. *Nat Immunol* 4:380-386.

[101] Steinman, R.M., and M. Pope. 2002. Exploiting dendritic cells to improve vaccine efficacy. *The Journal of Clinical Investigation* 109:1519-1526.

[102] Spira, A.I., P.A. Marx, B.K. Patterson, J. Mahoney, R.A. Koup, S.M. Wolinsky, and D.D. Ho. 1996. Cellular targets of infection and route of viral dissemination after an intravaginal inoculation of simian immunodeficiency virus into rhesus macaques. *J Exp Med* 183:215-225.

[103] Hu, J., M.B. Gardner, and C.J. Miller. 2000. Simian immunodeficiency virus rapidly penetrates the cervicovaginal mucosa after intravaginal inoculation and infects intraepithelial dendritic cells. *J Virol* 74:6087-6095.

[104] Zhang, Z., T. Schuler, M. Zupancic, S. Wietgrefe, K.A. Staskus, K.A. Reimann, T.A. Reinhart, M. Rogan, W. Cavert, C.J. Miller, R.S. Veazey, D. Notermans, S. Little, S.A. Danner, D.D. Richman, D. Havlir, J. Wong, H.L. Jordan, T.W. Schacker, P. Racz, K. Tenner-Racz, N.L. Letvin, S. Wolinsky, and A.T. Haase. 1999. Sexual Transmission and Propagation of SIV and HIV in Resting and Activated CD4(+) T Cells. *Science* 286:1353-1357.

[105] Stahl-Hennig, C., R.M. Steinman, P. Ten Haaft, K. Uberla, N. Stolte, S. Saeland, K. Tenner-Racz, and P. Racz. 2002. The simian immunodeficiency virus deltaNef vaccine, after application to the tonsils of Rhesus macaques, replicates primarily within CD4(+) T cells and elicits a local perforin-positive CD8(+) T-cell response. *J Virol* 76:688-696.

[106] Wu, Y., and J.W. Marsh. 2001. Selective transcription and modulation of resting T cell activity by preintegrated HIV DNA. *Science* 293:1503-1506.

[107] Demarchi, F., M.I. Gutierrez, and M. Giacca. 1999. Human immunodeficiency virus type 1 tat protein activates transcription factor NF-kappaB through the cellular interferon-inducible, double-stranded RNA-dependent protein kinase, PKR. *Journal of Virology* 73:7080-7086.

[108] Marzio, G., and M. Giacca. 1999. Chromatin control of HIV-1 gene expression. *Genetica* 106:125-130.

[109] Izmailova, E., F.M. Bertley, Q. Huang, N. Makori, C.J. Miller, R.A. Young, and A. Aldovini. 2003. HIV-1 Tat reprograms immature dendritic cells to express chemoattractants for activated T cells and macrophages. *Nature Medicine* 9:191-197.

In: Trends in RNA Research
Editor: Peter A. McNamara, pp. 121-144

ISBN: 1-59454-506-5

Chapter VI

Post-Transplant Hypertension and Oxidative Stress. Effect of Different Antihypertensive Treatments

Elisa Pagnin Pharm and Lorenzo A Calò[1]

Department of Clinical and Experimental Medicine, Clinica Medica 4, University of Padova, Italy

Abstract

Cyclosporine (CsA) introduction is a milestone in immunosuppresive therapy improving morbidity and mortality of transplanted patients. However, the use of CsA is complicated by side-effects such as hypertension and nephrotoxicity through effects on factors controlling vasomotor tone (endothelin, thromboxane and NO). Nitric oxide (NO) is crucial for maintaining a state of basal vasodilation and inhibition of platelet activation and its impairment by CsA has been advanced as a mechanism for CsA-induced hypertension. However, direct effect of CsA on NO synthesis is not supported as studies in normal human volunteers showed that acute CsA administration produced vasoconstriction and simultaneous increase in endothelial NO release and that CsA produced vasoconstriction in the presence of a simultaneous inhibition of nitric oxide synthase (NOS). Furthermore, incubation of endothelial cells with CsA showed an increased gene expression of endothelial NOS (ecNOS). It is therefore unlikely that CsA is a direct inhibitor of the NO system and decreased NO activity is unlikely to mediate the vasoconstrictive actions of CsA. On the other hand, CsA has been shown to induce superoxide anions production which increase NO metabolism and could represent an link between NO increased metabolism and reduced activity and CsA induced vasoconstriction and hypertension.

---

[1] address for correspondence: Lorenzo A Calò MD, PhD,Department of Clinical and Experimental Medicine,Clinica Medica 4, University of Padova,Via Giustiniani 2, 35128, Padova, Italy,Tel +39/049/8212279-8212179,Fax +39/049/8212151,E-mail: renzcalo@unipd.it

We have evaluated in mononuclear cells of kidney transplanted patients chronically treated with CsA or Tracolimus (FK-506) and post-transplant hypertension, NO system status (ecNOS gene expression and NO metabolites in plasma) and the presence of oxidative stress using a molecular biology approach (gene expression of oxidative stress related proteins such as $p22^{phox}$, heme oxygenase (HO-1), and TGF-β and markers of oxidative stress such has hydroperoxides, peroxynitrites and plasma antioxidant power).
In kidney transplanted patients chronically treated with calcineurin inhibitors and post-transplant hypertension, NO system is upregulated (ecNOS mRNA and NO metabolites are increased in patients). CsA or FK-506 induced oxidative stress as shown by the induction of $p22^{phox}$ and TGF-β expression and of hydroperoxides and peroxynitrites. CsA and FK-506 induced comparable hypertension and degree of oxidative stress with no difference between these two treatments in terms of induction of oxidative stress related proteins, therefore supporting the role of calcineurin inhibitors-induced oxidative stress for the endothelial dysfunction and posttransplant hypertension.
We have also evaluated in kidney transplanted patients chronically treated with CsA or FK-506 and post-transplant hypertension the effect of antihypertensive drugs on oxidative stress related biochemical and molecular markers. We have considered the ACE inhibitor Ramipril, the calcium antagonist Nifedipine and the α1-β blocker Carvedilol. These drugs are equally effective in the reduction of the blood pressure and leave unchanged the upregulation of NO system observed in kidney transplanted patients under chronic CsA or FK-506 treatment. Carvedilol and Nifedipine leave unchanged NO upregulation but only Carvedilol reduced TGF-β and increased HO-1 mRNA production. Ramipril reduced $p22^{phox}$ and TGF-β gene expression while was ineffective on HO-1 mRNA.
The demonstration that in kidney transplant patients chronically treated with CsA or FK-506 and posttransplant hypertension, different antihypertensive drugs induce a comparable blood pressure reduction but different effects on proteins induced by and/or protective from oxidative stress, closely correlated with endothelial dysfunction and fibrogenesis, could give further suggestions on the antihypertensive therapeutic approach to the transplant patient chronically treated with calcineurin inhibitors and with calcineurin inhibitor-induced hypertension besides the need of blood pressure reduction.

Introduction

The introduction of Cyclosporine (CsA) is considered a milestone in immunosuppresive therapy, greatly improving morbidity and mortality of transplanted patients. However, the use of CsA for immunosuppression is complicated by serious side-effects such as hypertension, thromboembolic complications and nephrotoxicity.

Hypertension following renal transplantation is a common occurrence. A variety of causes have been implicated, including rejection episodes, recurrent renal disease and renal artery stenosis [1,2]. The use of calcineurin inhibitors, such as Cyclosporin (CsA) and FK-506, has further increased the prevalence of hypertension associated with transplantation, with approximately 75-80% of these patients having hypertension, which is severe in one fifth of them [2,3]. Hypertension favours post-transplant nephropathy or chronic rejection [4-8] that lead to long term loss of renal function.

Different mechanisms have been proposed to explain the pathophysiology of hypertension mediated by CsA, which, however, has not been completely clarified. The mechanism more frequently proposed is the vasoconstrictor effect at the level of renal afferent arteriole with the consequent alteration of glomerular filtration and sodium excretion.

The pathogenesis of vasocontriction induced by CsA is multifactorial; CsA, in fact, may cause vasoconstriction and hypertension through different mechanisms such as increased sympathetic nervous activity, alteration of intrarenal renin-angiotensin system, increase of the intracellular calcium concentration, endothelin synthesis or release, alteration of prostaglandin metabolism, increase of oxidative stress with superoxide production and consequent increase of nitric oxide (NO) metabolism.

The demonstration that angiotensin II increases production of reactive oxygen species (ROS) by vascular smooth muscle cells [9] and the fact that ROS induce the development of endothelial dysfunction and hypertension [10], have established a link between oxidative stress and hypertension. ROS combined with NO form peroxynitrite, increase NO metabolism and reduce its efficacy as a vasorelaxing factor and as an inhibitor of vascular proliferation, hypertrophy and remodeling [10].

NO is crucial in the maintenance of a state of basal vasodilation and inhibition of platelet activation and its impairment by CsA has thus been advanced as a mechanism for the hypertension associated with CsA. However, in normal human volunteers, acute infusion of CsA produced vasoconstriction and a simultaneous increase in endothelial NO release and in cultured endothelial cells the incubation with CsA increases ecNOS gene expression [11]. The apparent contradiction between NO system upregulation and CsA-induced vasoconstriction is explained by CsA-mediated induction of superoxide anions and reactive oxygen species production, which reacting with NO produce peroxynitrites, reducing the NO bioavailability and, therefore, its vasodilatory action. Calcineurin inhibitors induced oxidative stress could, therefore, represent an attractive link between NO increased metabolism and reduced activity and CsA-induced vasoconstriction and hypertension in transplant patients chronically treated with CsA.

There is always more recognition of an involvment of oxidative stress in cardiovascular pathology and hypertension, therefore the attention has been recently focused on antioxidant strategies to protect the cardiovascular system from the damage induced by reactive oxygen species. The evaluation of antihypertensive drugs has been extended to include their effects on oxidative stress and it has been demonstrated that ACE inhibitors, dihydropiridine calcium channel blocker and Carvedilol, a $\alpha 1$-β blocker, have a potent antioxidant activity.

It has been recently demonstrated that one of the mechanism through which ACE inhibitors increase endothelial function is the reduction of the superoxide radicals formation induced by angiotensin II [12,13]. To confirm this hypothesis, Hornig and colleagues have demonstrated that the beneficial effect of vitamin C, substance with well-known antioxidant properties on endothelium dependent vasodilation in patients with coronary disease was not evident in patients under long-term treatment with ACE inhibitors, therefore suggesting a reduction of the oxidative stress by ACE inhibitors [14]. This demonstration strenghtens the concept that ACE inhibitors have antioxidant effects inducing a reduction of superoxide anions formation and consequently an increase of NO bioavailability. ACE inhibitors could exert their antioxidant effect through different mechanisms such as an increase of extracellular superoxide dismutase activity (EC-SOD). EC-SOD is reduced in patients with coronary disease and it is strongly correlated with endothelium mediated vasodilation, but a four weeks treatment with Ramipril increased strongly the production of EC-SOD which resulted strongly correlated with endothelium dependent vasodilation [14]. The EC-SOD regulation could therefore represent one of the mechanisms of ACE inhibitors mediated antioxidant activity. In addition, angiotensin II (Ang II) induces NADH/NADPH oxidase activity and consequently superoxide anions production; ACE inhibitors reducing the production of Ang II could also reduce the production of superoxide anions.

It has also been demonstrated that several dihydropiridine calcium channel blocker have antioxidant effects. These antioxidant effects were explained with the presence of aromatic rings in the chemical structure of these lipofile compounds which allow them to act as chain-breaker of the radicals reaction [16-21]. In fact, treatment with Nifedipine reduces plasmatic levels of lipoperoxides and isoprostanes and increases plasma antioxidant power and treatment with calcium channel blockers avoids vitamin C effect on vasodilation induced by acetilcolin [20]. These results support the possibility that the increase of endothelial dysfunction due to a restoration of NO bioavaibility induced by Nifedipine, Lacidipine and Amplodipine, is correlated with the intrinsic antioxidant activity of these compounds [22,23].

Carvedilol is a $\alpha 1$-β blocker with several actions. It has been shown, in fact, that Carvedilol is a competitive antagonist of the adrenoreceptors β and a vasodilatant at the same time. This vasodilatant action has been attributed to the antagonistic effect of the Carvedilol at $\alpha 1$-adrenergic receptor levels. Both these activities contribute to the antihypertensive action of this compound [24-26]. In addition, Carvedilol in vitro and in vivo is a scavenger of ROS and it has an antiproliferative effect in vascular smooth muscle cells [27-29]. The antioxidant properties of Carvedilol are greater than vitamin E [27] and animals treated with Carvedilol show delayed atherogenesis, reduced infarct size, and improved survival [29,30].

In our studies we have evaluated "ex vivo" in blood mononuclear cells of kidney transplant patients chronically treated with CsA or Tracolimus (FK-506) and post-transplant hypertension, NO system status and the presence of oxidative stress using a molecular biology approach. In particular, NO systems status has been evaluated considering ecNOS gene expression and NO metabolites in plasma. Oxidative stress has been also evaluated considering oxidative stress-related proteins such as $p22^{phox}$ (NADH/NADPH catalitic subunit critic for superoxides productions), Heme-oxigenase-1 (HO-1) (enzyme induced by and protective from oxidative stress) and TGF-β (fibrogenic cytokine, effector of oxidative stress). We have also evaluated markers of oxidative stress as plasma level of hydroperoxides and peroxynitrites and plasma antioxidant power.

Since it is still unclear whether CsA and FK-506 have the same effects on hypertension and biochemical effects on the oxidative stress markers, we examined in kidney translanted patients the effects of CsA and FK-506 treatment on the above mentioned oxidative stress related markers.

Finally, due to the fact that there is always more recognition of an oxidative stress involvement in cardiovascular pathology and hypertension, in our studies we have evaluated the effect of antihypertensive drugs on the above mentioned biochemical and molecular systems related with oxidative stress in kidney transplant patients chronically treated with CsA or FK-506. In particular, we considered the ACE inhibitor Ramipril, the calcium antagonist Nifedipine and the $\alpha 1$-β blocker Carvedilol.

Patients

We enrolled 41 patients with cadaveric kidney transplant (25 males and 16 females, age range 25-60), at least one year after transplantation and chronically treated either with CsA (Novartis Farma, Italy) or with FK-506 (Fujisawa, Ireland). They had been selected on the basis of the occurrence of post-transplant hypertension, no acute rejection episode for at least six months, and repeatedly absent proteinuria and stable kidney function.

CsA and FK-506 had been mantained in all the patients after transplantation and throughout the study and their doses were adjusted to achieve trough plasma levels between 150 and 200 ng/ml for CsA and between 8 and 12 ng/ml for FK-506. The immunosuppressive protocol included CsA or FK-506 and steroids and CsA or FK-506, steroids and azathioprine.

The diagnosis of post-transplant hypertension was based on the development of hypertension after transplantation for those patients who were normotensive before the transplant, or on the worsening of hypertension associated with CsA or FK-506 treatment despite the tapering of steroids for those patients who were already hypertensive before transplant.

10 healthy subjects (6 males and 4 females, age range 30-46) choose among the staff of Clinica Medica 4 at University of Padova, with normal blood pressure (120/80 mmHg±14/9), plasma electrolytes, renin and aldosterone plasma levels, were used as control group.

Informed consent was obtained from all the study participants, and the study protocol had been approved by our institutional authorities.

After enrollment, the patients were asked to stop any antihypertensive therapy for ten days, were advised to follow their usual lifestyle and to abstain from intense physical activity during the day before the collection of blood samples. All were under a diet containing approximately 150 mmol of sodium per day. Furthermore, a diet with standardized nitrate intake was given for the 48 hours preceding each blood sampling. The patients were also asked to abstain from food, alcohol and caffeine containing drinks for at least 12 hours before before blood sampling. No difference in the usual diet between patients and healthy subjects was revealed by a dietary questionnaire, particularly regarding fruit and vegetable consumption.

The blood pressure was measured in the sitting position with a mercury sphygmomanometer, to the nearest 2 mmHg, by a trained observer, unaware of the study protocol. The mean of three consecutive measurements, two minutes apart, was used for the analysis of the results.

The effect of Ramipril (5mg/die) was evaluated after 2 months of treatment on 16 patients (10 males and 6 females, age range 25-60); the effect of Nifedipine (60mg/die) after 4 months of treatment was evaluated on 10 patients (6 males and 4 females, age range 28-58); the effect of Carvedilol (50mg/die) after 4 months of treatment was evaluated on 15 patients (9 males and 6 females, age range 30-54).

Methods

Peripheral Leukocytes Separation

Peripheral blood monocytes (PBM) were isolated from 35 ml EDTA anticoagulated blood by Ficoll Paque Plus gradient (Amersham Pharmacia Biotech, Sweden) and judged to be 85% pure and functionally alive by cytofluremetric analysis. In brief, whole EDTA anticoagulated blood was layered on Ficoll-Paque Plus solution, polymeric erythrocytes aggregant agent, and centrifugated at 400g for 30 minutes at 20°C. The lymphocytes (PBL) layer were collected after removal of the upper layer of plasma and washed two times using a balanced salt solution (Phosphate Buffer Solution, PBS) and centrifuging at 100g for 10 minutes at 20°C.

RNA Extraction

PBL, collected as described previously, were homogenized adding 1 ml of RNA Ble (Eurobio, France) for ~5 x 10^6 cells. Cells were counted using the vital dye eosin. RNA extraction was obtained adding chloroform (1/10 of the RNABle volume), vortexing for 20 seconds, leaving the tube at 4°C for 5 minutes and centrifuging at 20,000g for 15 minutes at 4°C. The upper aqueous phase was transferred to a new tube and RNA was precipitated using a volume of 2-propanol equal to the aqueous phase. After 15 minutes at 20,000g at 4°C, the RNA pellet was washed twice with ethanol 70% and then recovered in RNAse free water. The extracted RNA had an 280/260 OD ratio between 1.8 and 2.0. RNA integrity was confirmed by 1.4% agarose gel electrophoresis and ethidium bromide as intercalating dye.

Reverse Transcription-Polymerase Chain Reaction (Rt-Pcr)

Gene expression evaluation of ecNOS, TGF-β, HO-1 and $p22^{phox}$ was performed using a semiquantitative analysis of RT-PCR products. Reverse transcription of RNA was performed with Gene Amp RNA PCR Kit (Gene Amp RNA PCR Kit, Applied Biosystems, USA). RNA (~1 μg) was reverse transcribed using Random Hexamers as primers and MuLV reverse transcriptase in a Perkin Elmer 2400 thermalcycler essentially as described by the manufacturer. RT protocol included 10 minutes at 25°C for primer extension, 15 minutes at 42°C for reverse transcription and 5 minutes at 99°C for reverse transcriptase denaturation.

The cDNA obtained in the retrotranscription was used for the amplification of the genes of interest using specific primers and protocols for ecNOS, TGF-β, HO-1 e $p22^{phox}$ as described in Table 1. ecNOS, TGF-β, HO-1 and $p22^{phox}$ primers set were designed by the aid of Primers3 software [31].

Table 1. PCR protocols

Gene of interest	Primers	Amplicone	Primers conc.	$MgCl_2$ Conc.	PCR protocol
ecNOS	F: 5'-GCATCACCAGGAAGAAGACC-3' R: 5'-GAGCCCGAACACACAGAACC-3'	471bp	0.4 μM	1.75mM	Denaturation step: 94°C, 1min Annealing step: 59°C, 1min Extension step: 72°C, 1min n° cycles: touchdown + 20
TGF-β	F: 5'-GCCCTGGACACCAACTATTGCT -3' R: 5'- AGGCTCCAAATGTAGGGGCAGG -3'	161bp	0.4 μM	2.25mM	Denaturation step: 94°C, 45 sec Annealing step: 60°C, 45sec Extension step: 72°C, 2min n° cycles: 26
HO-1	F: 5'-CAGGCAGAGAATGCTGAGTTC -3' R: 5'-GCTTCACATAGCGCTGCA -3'	270bp	0.5 μM	1.5mM	Denaturation step: 94°C, 30sec Annealing step: 58°C, 1min Extension step 72°C, 1min n° cycles: 26
$p22^{phox}$	F: 5'-TGGGCGGCTGCTTGATGGT-3' R: 5'-GTTTGTGTGCCTGCTGGAGT-3'	316bp	0.4 μM	1.5mM	Denaturation step: 94°C, 1min Annealing step: 60°C, 1min Extension step: 72°C, 1min n° cycles: 26

In each experiment AmpliTaqGold (Applied Biosystems, USA) was used as DNA polymerase and β–actin (Clontech, USA) as housekeeping gene.

The numbers of cycles used for the amplification of ecNOS, TGFβ, HO-1, $p22^{phox}$ and β-actin was obtained from the analysis of a kinetic curve set for each gene using an increasing number of cycles from 10 to 40, in the order of 2, to determine the number of cycles corresponding to the exponential phase of the amplification.

PCR products were separated by electrophoresis on polyacrylamide gel and stained by Silver staining.

The identity of PCR products was evaluated at Primm s.r.l., San Raffaele Biomedical Science Park, Milano, Italy, using PRISM Taq Polymerase Dye Terminator fluorescent sequencing kit (Perkin-Elmer) and analyzed using an ABI 373 automated sequencer and ABI Prism analysis software.

Evaluation of Ecnos, TGF-β, HO-1 And P22^{phox} Gene Products

ecNOS, TGF-βHO-1, p22^{phox} and β-actin gene expressions were quantified using a PCR-based densitometric semiquantitative analysis using NIH image. The ratio between ecNOS, TGFβ, HO-1 and p22^{phox} to β-actin PCR products were used as indexes of ecNOS, TGF-βHO-1 and p22^{phox} gene expressions.

NO Metabolites

NO metabolites plasma level was evaluated as nitrite/nitrate (NO_2^-/NO_3^-) by a colorimetric method as described by the manufacturer (Cayman Chemicals, USA). Briefly, in the first step nitrate was converted to nitrite by nitrate reductase. In a second step, the addition of Griess Reagents converted nitrite into a deep purple azo compound, revealed by absorbance measurement at 540 nm.

Hydroperoxides Plasma Level

The plasma level of hydroperoxides from cholesterol esters and from tryglycerides was determined according to Cawood et al [32] with minor modifications. Briefly, lipids were extracted by vortexing 1 ml of sample with 8 ml chloroform/methanol (2:1 v/v) for 2 minutes. After centrifugation, 5 ml of the chloroform layer were washed with 2 ml water in order to separate two phases. 3 ml of the chloroform phase were dried under nitrogen and dry extract was dissolved in 50µl exane. 20 µl of the exane phase were analysed by normal-phase HPLC using HI-bar LiChrosorb Si 60 column (Merck, Germany) with a mobile phase of exane/2-propanol (997:7 v/v). Isocratic HPLC analysis was performed on a Perkin Elmer apparatus comprising a Sigma 15 integrator, a series 10 pump and an LC-75 spectrophotometer with the detector set at 234nm. The flow rate was set at 2 ml/min. The data were expressed as integrated area units (i.a.u.).

Peroxynitrite Plasma Level

The plasma level of peroxynitrite was evaluated by HPLC as 3-nitrotyrosine following the procedure described by Kaur and Halliwell [33] with minor modofications. Briefly, 2 ml plasma were analyzed for nitrotyrosine after filtration through Centricon 10 filters (Amicon, USA) at 3000 g/min for 1h at 4°C. HPLC separation of 3-nitro-L-tyrosine was achieved using a 3-□C-18 column (150 x 4.6 mm; Supelco, USA), with a guard column C-18 (20 x 3.2 mm; Supelco, USA). The eluant was 0.5 mol/l KH2PO4-H3PO4 (pH 3.0), with 10% methanol (v/v) at a flow rate of 0.9 ml/min and injecting a volume of 50µl. 3-nitrotyrosine was detected by UV detector set at 274 and 220 nm. Using this procedure the detection limit was 0.1 µmol/l. The identification of the 3-nitrotyrosine peak was confirmed by tests of addition of 3-nitrotyrosine standard (Sigma, USA) and by the ratio of absorbance at 220 and 274 nm in comparison with samples of 3-nitrotyrosine standard. The CV was 6.3 and 3.3% at the concentrations of 2.0 and 10.0 μmol/l respectively (n=15). The mean recovery of 3-nitrotyrosine in the tests of addition was 97% (n=5). In all serum samples from healthy controls (n=20), 3-nitrotyrosine was under the detection limit.

Plasma Anti Oxidant Power (PAO)

Total Plasma AOP has been evaluated using a commercially available kit (Med.Dia, Italy). It is assessed by evaluating Cu^{+} derived by reduction of added Cu^{++}. Cu^{+} is detected by the formation of Cu^{+}-Bathocuproine complex. This complex is stable and has a typical light absorption at 480-490 nm.

The results are compared with a standard curve obtained from samples of known concentration of uric acid, used as a typical reducent [34].

Statistical Analysis

Data were evaluated on a Power Macintosh 5400/180 computer (Apple Computer, USA) using the Statview II statistical package (BrainPower Inc., USA). Data are expressed as Mean ± SD and they were analyzed using the analysis of variance (ANOVA) with Bonferroni's post-hoc test for multiple analysis. Student "t" test for paired and unpaired data were also used.

Values at a 5% level or less ($p<0.05$) were considered statistically significant.

Results

Evaluation of Nitric Oxide System in Kidney Transplanted Patients Treated With CsA or FK-506

Nitric oxide system was upregulated in kidney transplanted patients chronically treated both with CsA (n=10) and FK-506 (n=6) as demonstrated by the increase ecNOS mRNA production in transplanted patients compared to healthy subjects (0.92±0.09 for CsA, 0.96±0.04 for FK-506 and 0.37±0.05 d.u. in controls, $p<0.001$) (Fig. 1A).

This results were further supported by an increase of nitrates and nitrites (NO_2^-/NO_3^-) concentration in patients compared to controls (32.03±11.62 vs 10.37±4.86 μmol/l, $p<0.004$) (Fig.1B).

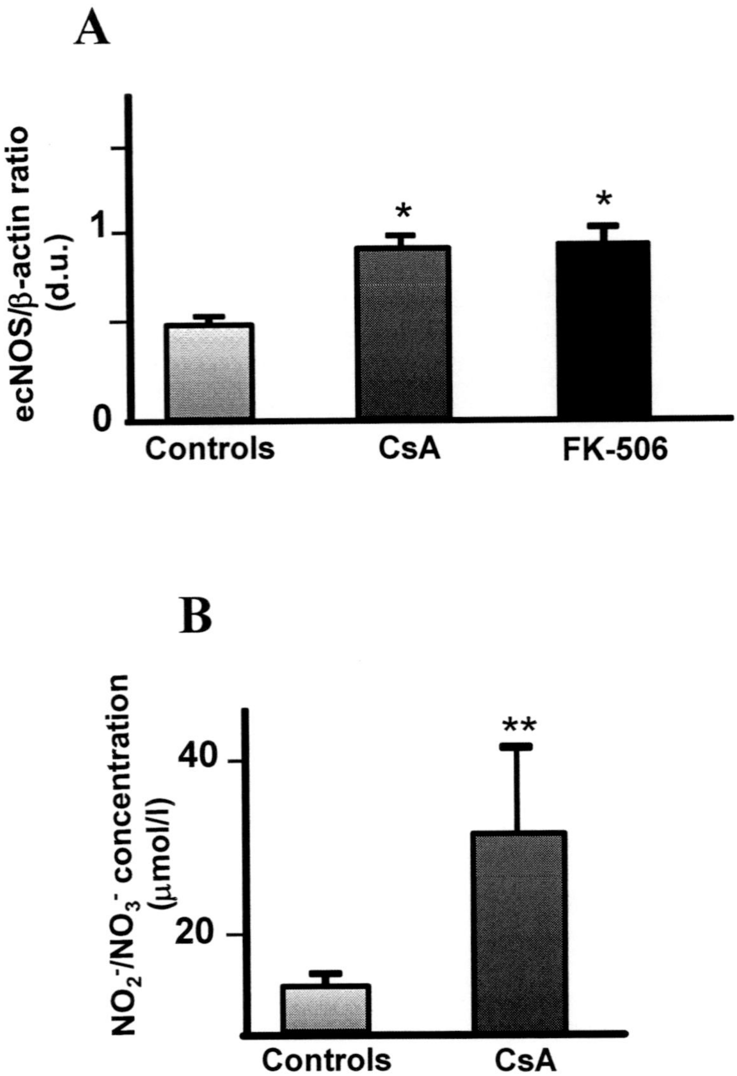

Figure 1.A. Densitometric analysis of ecNOS gene expression in mononuclear cells of healthy controls and of kidney transplanted patients with post-transplant hypertension under chronic CsA or FK-506 treatment. The expression of ecNOS was assessed by reverse transcription polymerase chain reaction (RT-PCR), using specific primers and is adjusted for the expression of the housekeeping gene β-actin, as reported in the method section. B. Plasma level of NO metabolites, in kidney transplanted patients chronically treated with CsA and in healthy subjects. *=p<0.001 vs controls; **=p<0.004 vs controls

Evaluation of Oxidative Stress Markers in Kidney Transplanted Patients Treated With Csa or FK-506

Oxidative status evaluation in kidney transplanted patients treated both with CsA (n=10) and with FK-506 (n=6) demonstrated a gene expression increase of oxidative stress related proteins in patients compared to controls.

p22^{phox} gene expression was, in fact, increased in transplanted patients than in healthy controls (0.90±0.05 d.u. for patients treated with CsA and 0.89±0.07 vs 0.53±0.07 u.d. for patients treated with FK-506, $p<0.001$) (Fig.2A).

TGF-β mRNA level was increased in patients treated with CsA or FK-506 compared to controls (0.83±0.05 for CsA, 0.84±0.05 for FK-506 and 0.75±0.03 d.u. for controls, $p<0.001$) (Fig.2B).

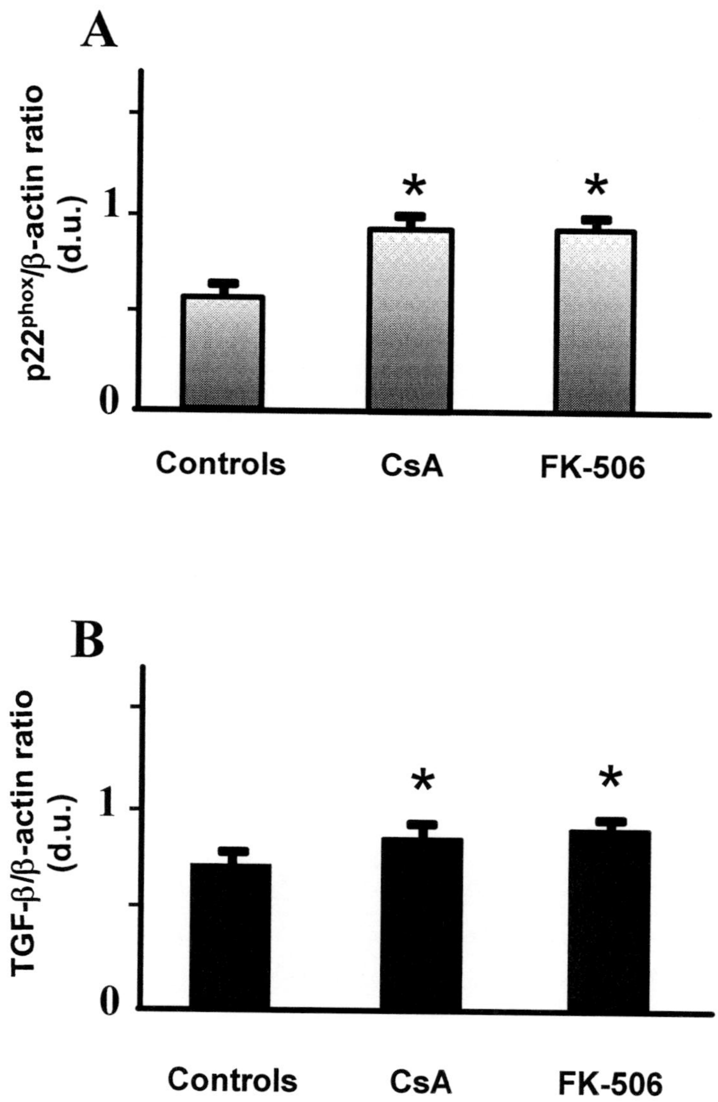

Figure 2. Densitometric analysis of p22^{phox} (A) and TGFβ B) gene expression in mononuclear cells of kidney transplanted patients with post-transplant hypertension under chronic CsA or FK-506 treatment and of healthy controls. The expression of p22^{phox} and TGFβ was assessed by reverse transcription polymerase chain reaction (RT-PCR), using specific primers and is adjusted for the expression of the housekeeping gene β-actin, as reported in the method section. *=$p<0.001$ vs controls

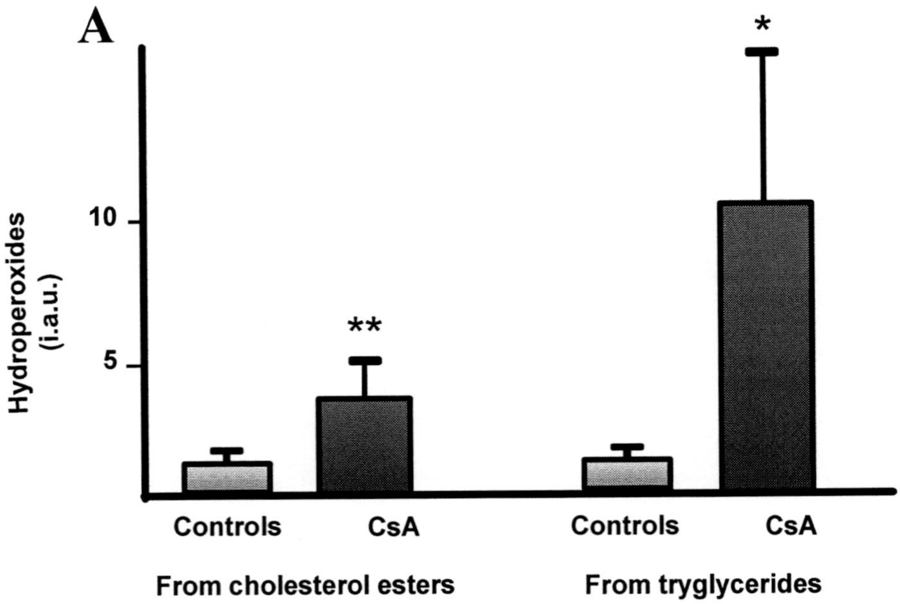

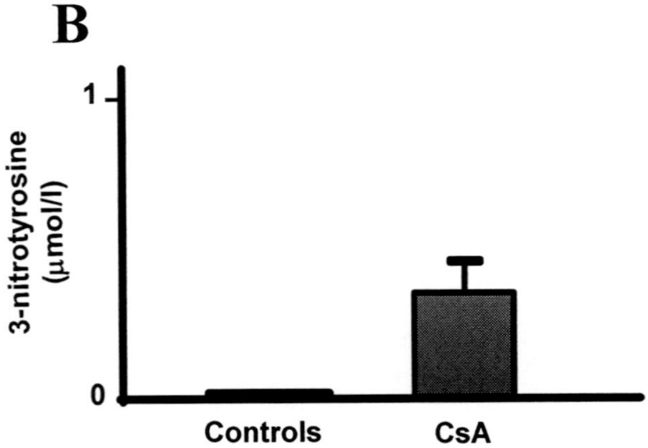

Figure 3. Plasma level of hydroperoxides (A) and peroxynitrites (B) in kidney transplanted patients chronically treated with CsA and healthy subjects (i.a.u. integrated area unit, n.d. not detectable). *=p<0.007, **=p<0.008

Plasma levels of hydroperoxides from cholesterol esters and from tryglycerides were higher in patients treated with CsA compared to controls (3.4±1.4 vs 1.3±0.6 i.a.u., $p<0.008$ for the former 10.6±6.4 vs 1.3±0.8 i.a.u., $p<0.007$ for the latter). Also, plasma level of peroxynitrites, products of the reaction between NO and ROS expressed as 3-nitrotyrosine concentration, was increased (0.32±0.11 μmol/l) in transplanted patients treated with CsA, while it was undetectable in healthy subjects (Fig.3A and B).

Evaluation of Treatment with ACE Inhibitor Ramipril

16 kidney transplanted patients chronically treated with CsA (n=10) and FK-506 (n=6) were treated for 2 months with ACE inhibitor Ramipril. Treatment with Ramipril significantly reduced blood pressure both for CsA (from 140±11/91±7 mmHg to 129±6/85±5 mmHg) and for FK-506 (from 138±7/92±7 mmHg to 127±10/82±6 mmHg) (Table 2)

Table 2. Sitting office blood pressure at baseline and after 2 months treatment with Ramipril in kidney transplanted patients with post-transplant hypertension chronically treated with either CsA and FK-506. *=p<0.01 versus baseline values.

	Ramipril	**mmHg**
CsA	Baseline	140±11 / 91±7
	2 months	129±6* / 85±5*
FK-506	Baseline	138±7 / 92±7
	2 months	127±10* / 82±6*

Treatment with Ramipril had no effect on nitric oxide system, which remained upregulated in transplanted patients treated both with CsA and with FK-506. ecNOS mRNA level, in fact, was: 0.92±0.09 at basal level vs 0.88±0.07 d.u. after 2 months of treatment with Ramipril for CsA and 0.96±0.04 vs 0.92±0.06 d.u. for FK-506 (Table 3).

Table 3. Effect of 2 months treatment with Ramipril on ecNOS gene expression in the kidney transplanted patients.

	Ramipril	**ecNOS/β-actin ratio (d.u.)**
CsA	Baseline	0.92±0.09
	2 months	0.88±0.07
FK-506	Baseline	0.96±0.04
	2 months	0.92±0.06

Treatment with Ramipril reduced $p22^{phox}$ gene expression both in transplanted patients treated with CsA (0.98±0.03 vs 0.83±0.05 d.u., p<0.03) and with FK-506 (0.94±0.2 vs 0.87±0.1 d.u., p<0.01) (Fig. 4A). Treatment with Ramipril reduced mRNA level of TGF-β (0.83±0.1 vs 0.74±0.1 d.u. both for CsA, p<0.01 and 0.83±0.05 vs 0.81±0.1 d.u. for FK-506, p<0.02) (Fig.4B). On the contrary, Ramipril was uneffective on HO-1 gene expression both in patients treated with CsA and in those treated with FK-506 (0.73±0.05 vs 0.73±0.03 d.u. for CsA and 0.72±0.05 vs 0.73±0.12 d. u. for FK-506) (Fig.4C).

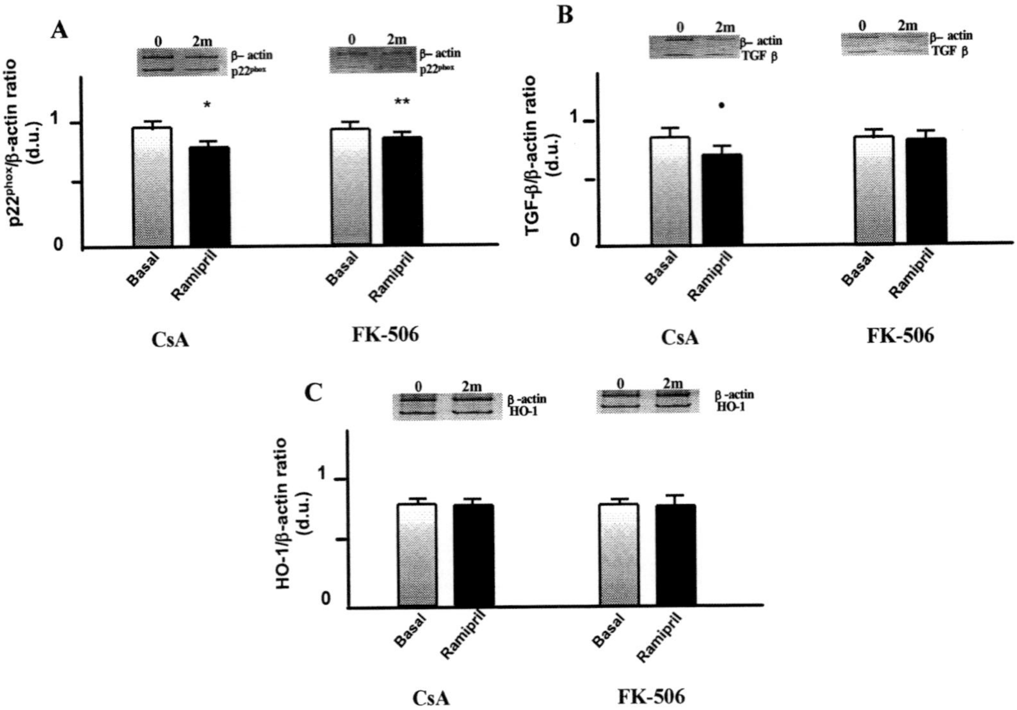

Figure 4. Densitometric analysis of p22^{phox} (A), TGFβ (B) and HO-1 (C) gene expression in mononuclear cells of kidney transplanted patients with post-transplant hypertension under chronic CsA treatment, at baseline and after 2 month treatment with Ramipril. The top side of each panel shows polyacrilamide gel from a representative patient. •: p=0.02; *=: p=0.01; **: p=0.004 vs. baseline

Evaluation of Treatment with the Dihydropiridine Calcium Channel Blocker Nifedipine

10 kidney transplanted patients chronically treated with CsA were treated for 4 months with the dihydropiridine calcium channel blocker Nifedipine. The calcium antagonist significantly reduced blood pressure already at the second months of treatment in CsA patients (from 149±7/97±4 mmHg to 135±6/87±4 mmHg and 131±3/85±3 mmHg after 2 and 4 months respectively) (Table 4).

Table 4. Sitting office blood pressure at baseline and after 2 and 4 months treatment with Nifedipine in kidney transplanted patients with post-transplant hypertension chronically treated with CsA. *=p<0.01 versus baseline values.

	Nifedipine	mmHg
CsA	Baseline	149±7 / 97±4
	2 months	135±6* / 87±4*
	4 months	131±3* / 85±3*

Treatment with Nifedipine also had no influence on upregulation of nitric oxide system observed in transplanted patients treated with calcineurin inhibitors. In fact, ecNOS mRNA level was: 1.88±0.2 at basal level vs 1.83±0.1 and 1.82±0.2 d.u. after 2 and 4 months respectively of treatment with Nifedipine (Fig.5A). NO metabolites, NO_2^-/NO_3^-, plasma concentration remained unchanged in patients treated with Nifedipine: 34.6±2.7 at basal level vs 34.8±2.1 and 35.6±2.7 μmol/l respectively at 2 and 4 months of treatment (Fig.5A).

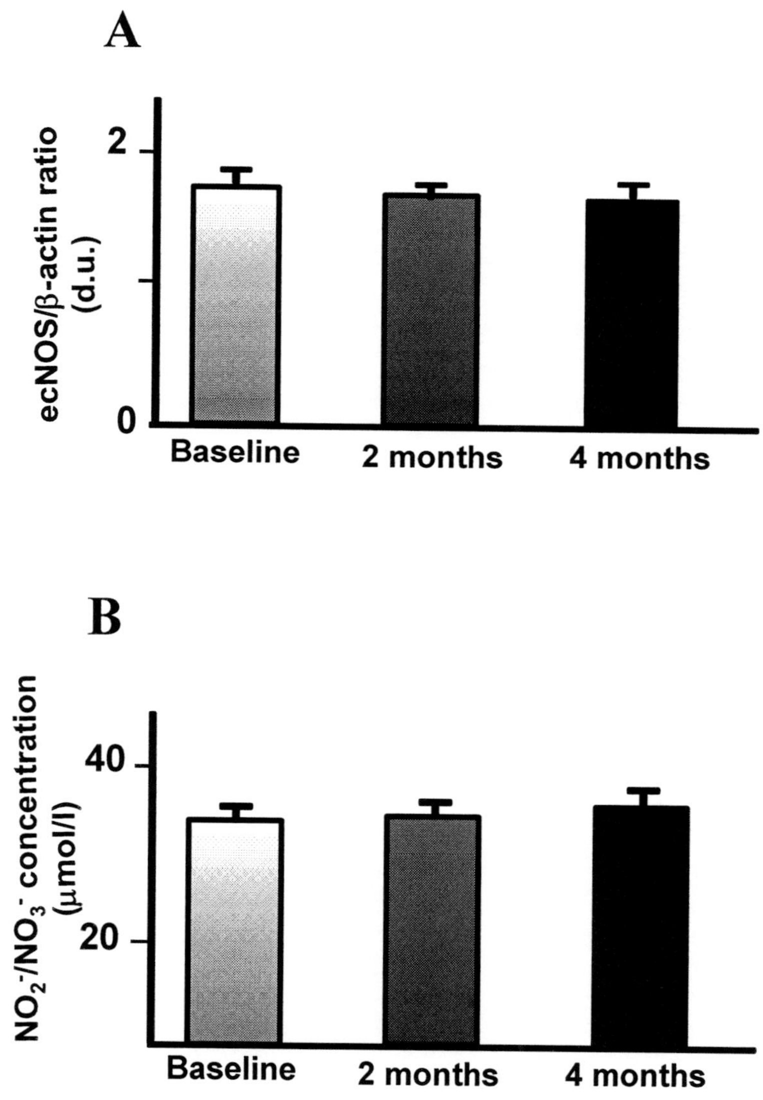

Figure 5. Mononuclear cell ecNOS gene expression as densitometric units (u.d.) (A) and plasma concentration of NO metabolites (B) in kidney transplanted patients with post-transplant hypertension induced by CsA, at basal level and after 2 and 4 months of treatment with Nifedipine.

Treatment with Nifedipine did not modify mRNA levels of TGFβ: 0.83±0.1 at basal level vs 0.83±0.03 e 0.81±0.1 d.u. and HO-1: 0.59±0.1 at basal level vs 0.56±0.1 and 0.59±0.1 d.u. after 2 and 4 months in patients chronically treated with CsA (Fig.6).

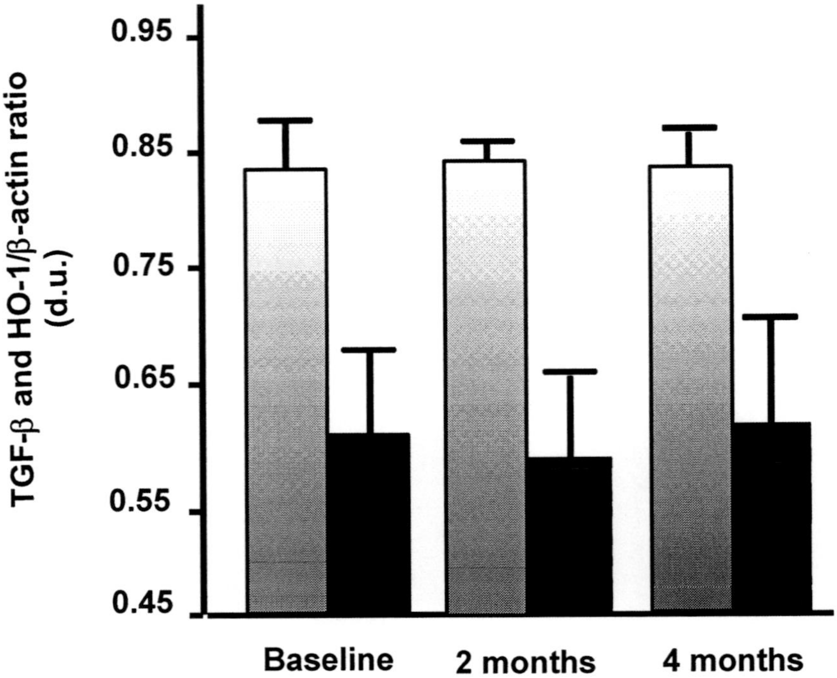

Figure 6. Effect of Nifedipine on TGFβ (white bars) and HO-1 (black bars) gene expression in kidney transplanted patients with post-transplant hypertension before and after 2 and 4 months of treatment.

Nifedipine was also neutral on 3-nitrotyrosine plasma concentration (0.6±0.1 vs 0.57±0.2 and 0.57±0.1 nmol/l) and plasma antioxidant power (1105±138.3 vs 1095±149.9 and 1110±137.0 μmol/l).

Evaluation of Treatment with the α1-β Blocker Carvedilol

15 kidney transplanted patients chronically treated with CsA were treated for 4 months with the α1-β blocker Carvedilol.

Treatment with Carvedilol reduced significantly blood pressure already at the second months of treatment in CsA patients (from 151±6/98±3 mmHg to 133±6/88±4 mmHg and 129±5/84±3 mmHg after 2 and 4 months respectively) (Tab.5).

Table 5. Sitting office blood pressure at baseline and after 2 and 4 months treatment with Carvedilol in kidney transplanted patients with post-transplant hypertension chronically treated with CsA. *=p<0.01 versus baseline values.

	Carvedilol	mmHg
CsA	Baseline	151±6 / 98±3
	2 months	133±6* / 88±4*
	4 months	129±5* / 84±3*

Treatment with Carvedilol left unchanged the upregulation of nitric oxide system observed in transplanted patients treated with calcineurin inhibitors. In fact, ecNOS gene expression was: 1.9±0.3 at basal level vs 1.85±0.2 e 1.82±0.3 after respectively 2 and 4 months of treatment with Carvedilol (Fig.7A). NO metabolites, NO_2^-/NO_3^-, plasma concentration remained unchanged in patients treated with Carvedilol: 34.6±4.3 at basal vs 33.4±4.5 and 32.9±5.0 µmol/l respectively 2 and 4 months of treatment (Fig.7B).

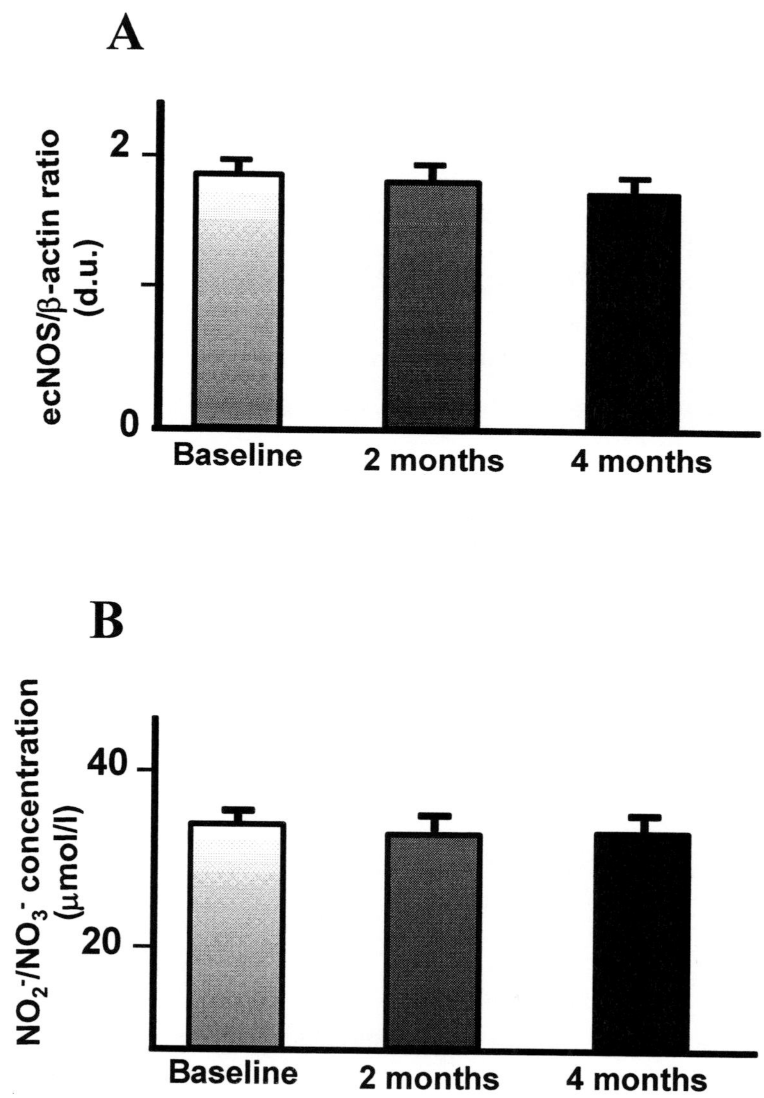

Figure 7. Mononuclear cell ecNOS gene expression as densitometric units (u.d.) (A) and plasma concentration of NO metabolites (B) in kidney transplanted patients with post-transplant hypertension induced by CsA, at basal level and after 2 and 4 months of treatment with Carvedilol.

Treatment with Carvedilol reduced significantly mRNA level of TGF-β: 0.86±0.06 at basal level vs 0.77±0.05 e 0.70±0.05 d.u. after 2 e 4 months respectively in patients chronically treated with CsA (Fig.8A). On the contrary, treatment with Carvedilol increased significantly HO-1 gene expression after 4 months (0.64±0.05 at basal level vs 0.66±0.03 e 0.72±0.03 d.u. after 2 and 4 months) (Fig.8A). Moreover, Carvedilol reduced significantly 3-nitrotyrosine plasma concentration, used as index of peroxynitrites plasma concentration (0.58±0.2 vs 0.4±0.2 and 0.29±0.1 nmol/l) (Fig.8B) and increased plasma antioxidant power (1075±164.96 vs 1260.7±254.33 and 1425.0±270.86 μmol/l) (Fig.8C).

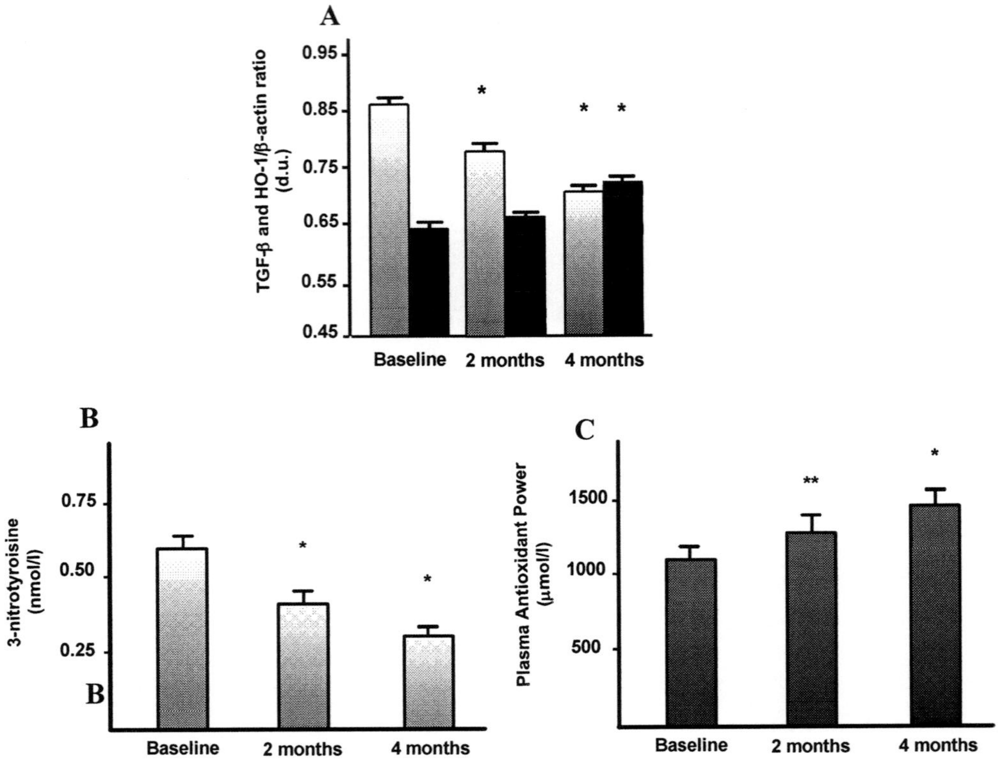

Figure 8. Effect of Carvedilol on TGFβ (white bars) and HO-1 (black bars) gene expression (A), on 3-nitrotyrosine plasma levels (B) and Plasma Antioxidant Power (C) in kidney transplanted patients with post-transplant hypertension before and after 2 and 4 months of treatment. *= p<0.001 and **=p<0.05 vs baseline

Discussion

Hypertension is a common occurrence during treatment with calcineurin inhibitors such as CsA and FK-506 [1,2]. The results of our study showing increased ecNOS gene expression and NOS enzimatic activity, have clearly confirmed that vasoconstriction and hypertension induced by calcineurin inhibitors cannot be a consequence of decreased levels of ecNOS as indicated by increased plasma level of NO metabolites in patients chronically treated with both CsA and FK-506. These evidences, therefore, confirm the association of treatment with CsA and FK-506 with an increased activity of ecNOS and NO production [6-8].

One possible mechanism which could link calcineurin inhibitors to nitric oxide system, involves an increased reactive oxygen species production induced by these drugs [7]. It is well known, in fact, that oxidative stress is a crucial pathophysiologic factor which contributes to the induction of post-transplant hypertension and its long term complications such as chronic rejection [2]. In fact, excessive production of superoxide anions reacting with NO, leads to formation of peroxynitrites therefore inducing a reduction of NO bioavailability and, therefore, of its vasodilatory and antiproliferative effects.

NO reduced bioavailability induced by CsA may explain vasocontriction in transplanted patients treated with CsA and FK-506 although ecNOS mRNA level is increased. This hypothesis is confirmed by the fact that, in patients treated both with CsA and FK-506, 3-nitrotyrosine plasma level, and indicator of peroxynitrite level, is increased while plasma 3-nitrotyrosine was undetectable in healthy subjects. In addition, the role of oxidative stress induced by calcineurin inhibitors through reduction of NO is further confirmed by increased plasma concentration of hydroperoxydes in transplanted patients treated with calcineurin inhibitors compared to healthy subjects.

The antioxidant activity of antihypertensive treatments is gaining always more recognition as strategy not only to reduce blood pressure but also to protect cardiovascular system from the damage induced by reactive oxygen species leading to long term complications of hypertension. The antioxidant activity of ACE inhibitors, dihydropiridine calcium channel blockers and α1-β blocker Carvedilol is well documented [12,13,16-21].

Angiotensin II (ATII) has been implicated both in cardiovascular remodeling and fibrogenesis as well as in oxidative stress, by stimulating cytokine and growth factors release [35]. ATII increases the activity of membrane bound multicomponent NADH/NADPH oxidase system, which is considered the major source of O_2^- in the vascular cells [35-37]. This enzyme includes $p22^{phox}$, a 22-kD α subunit of cytochrome b558, which functions as an integral subunit of the final electron transport from NADPH to heme to molecular oxygen in generating O_2^- [36]. It has been recently shown that $p22^{phox}$ expression and NADPH oxidase activity are elevated in rats infused with Ang II. Moreover, they are directly correlated with blood pressure and reduced by AT-1 receptor inhibition, suggesting that upregulation of $p22^{phox}$ may contribute to oxidase activation and the development of hypertension in Ang II dependent hypertension [38].

In our post-transplant hypertensive patients, treatment with the ACE inhibitor Ramipril reduced blood pressure and oxidative stress, induced by treatment with calcineurin inhibitors as indicated by the decline of $p22^{phox}$ gene expression. In addition, Ramipril reduced TGFβ gene expression, a key fibrogenic cytokine and effector signal of oxidative stress [39,40], and, therefore, it may be active in the prevention of the long term complications of the treatment with calcineurin inhibitors, such as fibrogenesis and remodeling [40]. It is known, in fact, that the effect of the oxidative stress induced by increased NADH/NADPH oxidase activity is in part mediated by TGFβ [35,40] and altered cell signalling mediated by oxidative stress may favor post-transplant nephropathy and chronic rejection also through other mechanisms such as activation of growth factors, mainly TGFβ [41]. Moreover, the treatment with Ramipril does not alter the nitric oxide system upregulation observed in postransplant patients treated with calcineurin inhibitors (ecNOS mRNA level does not change after 2 months of treatment both in patients treated with CsA and FK-506).

Carvedilol, another antihypertensive drug considered in our study, was effective in patients with post-transplant hypertension not only reducing blood pressure, but also markers of oxidative stress. Carvedilol, in fact, reduced plasma concentration of oxidative stress markers such as peroxynitrites and increased plasma antioxidant power. In addition, it left unchanged NO production induced by CsA compared with healthy controls as demonstrated by unchanged production of ecNOS mRNA and plasma concentration of NO metabolites. The reduction of 3-nitrotyrosine, an extimate of peroxynitrites, and the parallel increase of ecNOS mRNA and plasma NO_2^-/NO_3^- suggests that Carvedilol in patients with post-transplant hypertension was effective on NO bioavailability which, therefore, may contribute to decline in blood pressure seen in these patients. In addition to its effects on blood pressure and oxidative stress, Carvedilol treatment reduced also gene expression of TGF-β, fibrogenic cytokine, whose production is induced by CsA. This effect is confirmed by the antiproliferative action of Carvedilol demonstrated by studies on smooth muscle cells [29,42]. Carvedilol is a α1-β blocker antihypertensive drug with vasodilating properties secondary to α1 blocking activity [42]. In addition, Carvedilol in vitro and in vivo is a scavenger of ROS and has an antiproliferative effect in vascular smooth muscle cells [27-29]. The antioxidant properties of Carvedilol are greater than vitamin E [27] and animals treated with Carvedilol show delayed atherogenesis, reduced infarct size, and improved survival [29,30]. The antioxidant activity of Carvedilol is further suggested by the decline of TGFβ gene expression. In fact, in vitro, oxidative stress enhances TGFβ gene expression in renal fibroblasts and mesangial cells [39]. In vivo, rats placed on antioxidant deficient diets (i.e. free of selenium and vitamin E) demonstrate increased TGFβ expression, renal hypertrophy, proteinuria, tubulointerstitial thickening and loss of glomerular filtration rate, associated to increased lipid peroxidation of the renal membranes [39]. Stimulation of the ROS-induced TGFβ signalling system by immunosuppressive agents, such as CsA or FK 506, is one of their most important side effects that may lead to graft related long term complications such as fibrogenesis and chronic rejection [41]. The decline seen in TGFβ is, therefore, potentially significant in these patients to protect from chronic rejection, given its role as a major profibrotic cytokine [43].

Carvedilol was also able to increase HO-1 gene expression.

HO-1 is regulated by oxidative stress and by non oxidant mediators [44] and, relevant to our findings, it has been found that overexpression of TGFβ down regulates HO-1 [45]. It is therefore reasonable to hypothesize that the reduction of TGFβ seen in our patients treated with Carvedilol, may increase HO-1 gene expression. HO is a rate limiting enzyme that catalyzes the degradation of heme into biliverdin and CO [46]. Biliverdin is further metabolized to bilirubin, which is a very potent antioxidant [47]. There are three different isoforms of HO, HO-1, HO-2 and HO-3. HO-1 has a very low basal expression but it increases rapidly upon oxidative stress [48], while HO-2 and the recently identified isoenzyme HO-3 are constitutively expressed. Enhanced HO-1-mediated production of the vasodilator CO [46] may contribute to the decline in blood pressure seen upon Carvedilol treatment. Finally, HO-1 could induce long term antiinflammatory and antiproliferative effect [49]. In fact, studies in heart and liver transplant have shown that the expression of the HO-1 gene is functionally associated with graft survival [50] and with a better immunologic response, supporting a role for HO-1 in immune regulation [51-52]. Therefore, increased HO-1 may represent a powerful additive compensatory mechanism towards oxidative stress, hypertension and remodeling. Finally, the increased HO-1 gene expression seems a peculiar effect of this drug since it was not observed with the other drugs evaluated in our studies.

Some authors have investigated the antioxidative activity of calcium antagonists in patients with essential hypertension and shown that Nifedipine and Lacidipine improve endothelial function by restoring NO availability, an effect that was attributed to their antioxidant activity [22,23]. Furthermore, it has been suggested that nifedipine has an antioxidant activity by sparing endogenous scavengers [53]. Our study shows that in kidney transplanted patients with a very high basal NO production, Nifedipine does not further increase ecNOS mRNA expression and NO metabolites. However, the methods used in the studies showing an antioxidant activity of Nifedipine, were based essentially on the evaluation of circulating markers of oxidative stress and on the effect of this drug on endothelium dependent vasodilation induced by infusion of acetylcholine [22,23]. This approach leaves totally unexplored the effects of the drug upstream, that is to say on the production of reactive oxygen species such as an effect on NAD(P)H oxidase, the major O_2^- producing system, which should be evaluated to establish a role of Nifedipine on oxidative stress other than direct free radical scavenging activity.

Conclusions

The results of our studies confirm a role in the induction of endothelial dysfunction and post-transplanted hypertension by calcineurin inhibitors and demonstrate that treatment with either CsA and FK-506, induce comparable oxidative stress in kidney transplanted patients with post-transplant hypertension.

In addition our studies show that antihypertensive drugs such as Carvedilol and Ramipril potentiate protective mechanisms towards long term complications of chronic treatment with CsA or FK-506, such as endothelial dysfunction, fibrogenesis and chronic rejection, in which oxidative stress plays a pivotal role. These mechanisms include increased NO availability, TGF-β and $p22^{phox}$ reduction and HO-1 increase.

Finally, the demonstration that in kidney transplanted patients treated with calcineurin inhibitors and post-transplanted hypertension, treatment with different antihypertensive drugs inducing a comparable blood pressure reduction but different effect on oxidative stress related proteins could, give further information on the choice of the antihypertensive treatment for transplanted patients chronically treated with calcineurin inhibitors which take advantage from effects of these drugs beyond blood pressure reduction.

References

[1] First, M.R.; Neylan JF; Rocher LL et al. (1994). Hypertension after renal transplantation. *J Am Soc Nephrol*, *4* (Suppl), S30-S36.

[2] Zeier, M.; Mandelbaum, A.; Ritz, E. (1998). Hypertension in the transplanted patient. *Nephron, 80*, 257-268

[3] Pochet, J.M.; Pirson, Y.; Van Ypersele De Strihou C. (1989). Is post renal transplantation hypertension more severe or more frequent on cyclosporine than on conventional therapy? *Nephrol Dial Transplant, 4*, 507-510.

[4] Remuzzi, G. and Perico, N. (1998). Protecting single-kidney allografts from long-term functional deterioration. *J Am Soc Nephrol, 9*, 1321-1332.

[5] Paul, L.C. (1999). Chronic allograft nephropathy: an update. *Kidney Int, 56*, 783-793.

[6] Diederich, D., Skopec, J., Diederich, A. et al. (1994) Cyclosporine produces endothelial dysfunction by increased production of superoxide. *Hypertension, 23(part 2)*, 957-961.

[7] Lopez-Ongil, S., Hernandez-Perera, O., Navarro-Antolin, J., et al. (1998) Role of reactive oxigen species in the signalling cascade of Cyclosporine A-mediated up-regulation of cNOS in vascular endothelial cells. *Br J Pharmacol, 124*, 447-454.

[8] Navarro-Antolin, J., Hernandez-Perera, O., Lopez-Ongil, S., et al. (1998) CsA and FK506 up regulate cNOS expression: role of reactive oxigen species and AP-1. *Kidney Int Suppl, 68*, S20-S24.

[9] Griendling, K.K., Minieri, C.A., Ollerenshaw, J.D., Alexander, R.W. (1994) Angiotensin II stimulates NADH and NADPH oxidase activation in cultured vascular smooth muscle cells. *Circ Res, 74*, 1141-1148.

[10] Katusic, Z.S. (1996). Superoxide anion and endothelial regulation of arterial tone. *Free Radic Biol Med, 20*, 443-448.

[11] Stroes, E.S.G., Luscher, T.F., deGroot, F.G., et al. (1997). Cyclosporin A increases nitric oxide activity in vivo. *Hypertension, 29*, 570-575.

[12] Rajagopalan, S., Kurz, S., Munzel, T., et al. (1996). Angiotensin II-mediated hypertension in the rat increases vascular superoxide production via membrane NADH/NADPH oxidase activation. Contribution to alterations of vasomotor tone. *J Clin Invest, 97(8)*, 1916-1923.

[13] Warnholtz, A., Nickenig, G., Schulz, E., et al. (1999). Increased NADH-oxidase-mediated superoxide production in the early stages of atherosclerosis: evidence for involvement of the renin-angiotensin system. *Circulation, 99(15)*, 2027-2033.

[14] Hornig, B., Landmesser, U., Kohler, C., et al. (2001). Comparative effect of ace inhibition and angiotensin II type 1 receptor antagonism on bioavailability of nitric oxide in patients with coronary artery disease: role of superoxide dismutase. *Circulation, 103(6)*, 799-805.

[15] Hornig, B. (2002). Vitamins, antioxidants and endothelial function in coronary artery disease. *Cardiovasc Drugs Ther, 16(5)*, 401-409.

[16] Mak, I.T., Boheme, P., Weglicki, W.B. (1992) Antioxidant effect of calcium channel blockers against free radical injury in endothelial cells. Correlation of protection with preservation of glutatione levels. *Circ Res, 70*, 1099-1103.

[17] Diaz-Araya, G., Godoy, L., Naranjo, L. (1998). Antioxidant effects of 1,4-dihydropyridine and nitroso aryk derivatives on the Fe3+ /Ascorbate-stimulated lipid peroxidation in rat brain slices. *Gen Pharmac, 31*, 385-391.

[18] Mason, R.P., Mak, I.T., Trumbore, M.W., Mason, P.E. (1999). Antioxidant properties of calcium antagonists related to membrane biophysical interactions. *Am J Cardiol, 84*, 16L-22L.

[19] Sevanian, A., Shen, L., Ursini, F. (2000). Inhibition of LDL oxidation and oxidized LDL-induced cytotoxicity by dihydropyridine calcium antagonists. *Pharm Res, 17*, 999-1006.

[20] Mak, I.T., Zhang, J., Weglicki, W.B. (2000). Cytoprotective properties of nisoldipine and amlodipine against oxidative endothelial cell injury. *Ann N Y Acad Sci, 899*, 403-406.

[21] Mak, I.T., Zhang, J., Weglicki, W.B. (2002). Protective effects of dihydropyridine Ca-blockers against endothelial cell oxidative injury due to combined nitric oxide and superoxide. *Pharmacol Res, 45*, 27-33.

[22] Taddei, S., Virdis, A., Ghiadoni, L., et al. (2001). Effect of calcium antagonist or beta blockade treatment on nitric oxide-dependent vasodilation and oxidative stress in essential hypertensive patients. *J Hypertens, 19*, 1379-1386.

[23] Taddei, S., Virdis, A., Ghiadoni, L., et al. (2001) Restoration of nitric oxide availability after calcium antagonist treatment in essential hypertension. *Hypertension, 37*, 943-948.

[24] Nichols, A.J., Sulpizio, A.C., Ashton, D.J., et al. (1989). In vitro pharmacologic profile of the novel beta-adrenoceptor antagonist and vasodilator, carvedilol. *Pharmacology, 39(5),* 327-336.

[25] Ruffolo, R.R. Jr, Gellai, M., Hieble, J.P., et al. (1990). The pharmacology of carvedilol. *Eur J Clin Pharmacol, 38(S2),* S82-88.

[26] Nichols, A.J., Gellai, M., Ruffolo, R.R. Jr. (1991). Studies on the mechanism of arterial vasodilation produced by the novel antihypertensive agent, carvedilol. *Fundam Clin Pharmacol, 5(1),* 25-38.

[27] Yue, T., Cheng, H., Lysko, P.G., et al. (1992). Carvedilol, a new vasodilator and beta adrenoceptor antagonist, is an antioxidant and free radical scavenger. *J Pharmacol Exper Ther, 263*, 92-98.

[28] Moser, M. and Frishman, W. (1998). Results of therapy with carvedilol, a β-blocker vasodilator with antioxidant properties, in hypertensive patients. *Am J Hypertens, 11*, 15S-22S.

[29] Sung, C.P., Arleth, A.J., Ohlstein, E.H. (1993). Carvedilol inhibits vascular smooth muscle cell proliferation. *J Cardiovasc Pharmacol, 21*, 221-227.

[30] Hamburger, S.A., Barone, F.C., Feuerstein, G.Z. et al. (1991). Carvedilol (Kredex®) reduces infarct size in a canine model of acute myocardial infarction. *Pharmacology, 43*, 113-120.

[31] Rozen, S. Skaletsky. Primer3. 1998. Code available at: *http://www-genome.wi.mit.edu/genome_software/other/primer3.html.*

[32] Cawood, P., Wickinsens, D.G., Iverson, S.A. et al. (1983). The nature of diene conjugation in human serum bile and duodenal juice. *FEBS Lett, 162*, 239-243.

[33] Kaur, H., Halliwell, B. (1994). Evidence for nitric oxide-mediated oxidative damage in chronic inflammation. Nitrotyrosine in serum and synovial fluid from rheumatoid patients *FEBS Lett, 350*, 9-12.

[34] Bellomo, G., Perrugini, C., Bagnati, M., et al. (1998) Biochemical markers of lipoprotein oxidation. *Clin Chem Lab Med, 36*, A41.

[35] Griendling, K.K., Sorescu, D., Lassegue, B., et al. (2000). Modulation of protein kinase activity and gene expression by reactive oxygen species and their role in vascular physiology and pathophysiology. *Arterioscler Thromb Vasc Biol, 20(10),* 2175-2183.

[36] Griendling, K.K., Ushio-Fukai, M, Lassegue, B, et al. (1997). Angiotensin II signaling in vascular smooth muscle. *New Concepts Hypertension, 29(part2)*, 366-373.

[37] Pagano, P.J., Ito, Y., Tornheim, K., et al. (1995). A NADPH superoxide-genereting system in the rabbit aorta. *Am J Physiol, 268*, H2264-H2280.

[38] Fukui, T. Ishizaka, N., Rajagopalan, S., et al. (1997). P22phox mRNA expression and NADPH oxidase activity are increased in aortas from hypertensive rats. *Circ Res, 80*, 45-51.

[39] Nath, K.A., Grande, J., Croatt, A., et al. (1998). Redox regulation of renal DNA synthesis, transforming growth factor-beta-1 and collagen gene expression. *Kidney Int, 53*, 367-381.

[40] Silacci, P. and Hayoz, D. (1998). Oxidative stress as the triggering event for vascular remodelling. *Nephrol Dial Transplant, 13*, 1343-1346.

[41] Ader, J.L. and Rostaing, L. (1998). Cyclosporin nephrotoxicity: pathophysiology and comparison with FK-506. *Curr Op Nephrol Hyperten, 7*, 539-545.

[42] Bril, A., Slivjak, M., Di Martino, M.J., et al. (1992). Cardioprotective effects of carvedilol, a novel β-adrenoreceptor antagonist with vasodilating properties, in anaesthetized minipigs: comparison with propanolol. *Cardiovasc Res, 26*, 518-525.

[43] Basile, D.P. (1999). The transforming growth factor β system in kidney disease and repair: recent progress and future directions. *Curr Opin Nephrol Hypertens, 8*, 21-30.

[44] Immenschuh, S. and Ramadori, G. (2000). Gene regulation of Heme Oxygenase-1 as a therapeutic target. *Biochem Pharmacol, 60*, 1121-1128.

[45] Pellacani, A., Wiesel, P., Sharma, A., et al. (1998). Induction of heme oxygenase-1 during endotoxemia is downregulated by transforming growth factor-beta1. *Circ Res, 83*, 396-403.

[46] Maines, M.D. (1997). The hene oxygenase system: a regulator of second messenger gases. *Annu Rev Pharmacol Toxicol, 37*, 517-554.

[47] Stocker, R., Yamamoto, Y., McDonagh, A.F., et al. (1987). Bilirubin is an antioxidant of possible physiological importance. *Science, 235*, 1043-1046.

[48] Ewing, J.F. and Maines, M.D. (1993). Glutathione depletion induces heme oxygenase-1 (HSP32) mRNA and protein in rat brain. *J Neurochem, 60*, 1512-1519.

[49] Nath, K.A., Vercellotti, G.M., Grande, J.P., et al. (2001).Heme protein-induced chronic renal inflammation: suppressive effect of induced heme oxygenase-1. *Kidney Int, 59*, 106-117.

[50] DeBruyne, L.A. Magee, J.C., Buelow, R., et al. (2000). Gene transfer of immunomodulatory peptides correlates with heme oxygenase-1 induction and enhanced allograft survival. *Transplantation, 69*, 120-128.

[51] Willis, D., Moore, A.R., Frederick, R., et al. (1996). Heme oxygenase: a novel target for the modulation of the response. *Nat Med, 2*, 87-90.

[52] Woo, J., Iyer, S., Cornejo, M.C., et al. (1998). Stress protein-induced immunosuppression : inhibition of cellular immune effector functions following overexpression of haeme oxygenase (HSP32). *Transpl Immunol, 6*, 84-93.

[53] Tong, Y.C., Mak, I., Boehme, P. and Weglicki, W. (1992). Antioxidant effect of calcium channel blockers against free radical injury in endothelial cells. Correlation of protection with preservation of glutathione levels. *Circ Res, 70*, 1099-1103.

In: Trends in RNA Research
Editor: Peter A. McNamara, pp. 145-157

ISBN: 1-59454-506-5

Chapter VII

Correlation of Laboratory Test Results Supports the Quality Assurance of the AMPLICOR HCV Test in Detection of Hepatitis C Virus RNA

***Hayato Miyachi*[1], *Atsuko Masukawa*[2], *and Satomi Asai*[1]**
[1]Department of Laboratory Medicine, Tokai University School of Medicine, Bohseidai, Isehara, Japan
[2]Clinical Laboratory, Tokai University Hospital, Bohseidai, Isehara, Japan

Abstract

Detection of hepatitis C virus (HCV) RNA by AMPLICOR system (Roche Diagnostic Systems) has been improved in assay quality assurance through placement of new reagents and automated instruments as well as quality control methods. We correlated laboratory test results of consecutive 5395 serum samples received for routine detection of HCV by the qualitative AMPLICOR HCV test (AMPLICOR, version 1.0), and investigated causes of the inconsistency. After excluding patients under treatment with anti-viral agents, 519 and 866 of 1133 samples were positive for HCV RNA by the qualitative AMPLICOR HCV test and HCV antibodies by a third-generation enzyme immunoassay test, respectively. Titers of positive HCV antibodies were widely distributed with two groups: low (<40.0) and high (≧ 40.0), which showed distinctive positive rates for HCV RNA in 19/371 (5.1%) and 488/495 (98.6%), respectively ($p<0.0001$). Patterns of laboratory test results suggested false-negative results for HCV RNA in a total of 8 samples. Seven of them had high-titers of HCV antibodies. One sample with negative HCV antibodies and increased levels of aminotransferases was considered a possibility of acute hepatitis C infection. Four of the 8 samples were

[1] Corresponding author:Mailing address: Department of Laboratory Medicine, Tokai University School of Medicine, Bohseidai, Isehara, Kanagawa 259-1193 Japan. Phone: 81-463 (93)1121.Fax: 81-463(93)8607. E-mail: miyachi@is.icc.u-tokai.ac.jp.

suggested the presence of an inhibitor by a dilution study. Results for HCV RNA of another 3 samples turned to be positive, when assayed by a second version of the AMPLICOR HCV test (AMPLICOR, version 2.0), the reagent with more equivalent detection sensitivity of all HCV genotypes than the first version. All of the 3 samples were revealed to have genotype 2a. The last one sample was suggested a negative PCR result under the detection limit caused by RNA degradation. False-positive results were not observed. In summary, individual correlation of laboratory test results of each sample followed by further assessment in detection of HCV RNA by the qualitative AMPLICOR HCV test would support the quality assurance of each result as well as the system of assay.

Key words: Hepatitis C virus, RNA, PCR assay, Clinical validation, Quality control, Quality assurance

Introduction

Polymerase chain reaction (PCR) detection of hepatitis C virus (HCV) RNA in serum has been used for the diagnosis of HCV infection as well as the assessment of response to anti-viral therapy [1,2]. Quantitative measurements of HCV RNA have been used to see treatment response in the combination therapy with rivabirin and interferon alfa that depends on the pretreatment viral load [3,4]. Qualitative measurements of HCV RNA are presently the method of choice to confirm active infection and to assess virological clearance in response to the anti-viral therapy. Detection sensitivity of assay for HCV RNA by the qualitative AMPLICOR HCV test is 10- to 100-fold higher than those for quantitative tests.

One of the major advantages of nucleic acid amplification over conventional methods in diagnosis of infectious diseases is sensitive detection of agents directly from clinical specimens. However, major problems in PCR assays are false-positive results because of carry-over contamination of previously amplified products, and false-negative results because of amplification inhibitors present in clinical specimens [5, 6]. Automated systems have been developed for amplification and detection of nucleic acid sequences for HCV RNA using PCR [7, 8]. An example of this is the COBAS AMPLICOR (Roche Diagnostic Systems, Branchburg, NJ), which amplifies target nucleic acid sequences, captures the biotinylated and amplified products on oligonucleotide-coated paramagnetic microparticles, and detects the products with an avidin-horseradish peroxidase (HRP) conjugate system [9, 10]. In addition to labor savings, and containment of amplification and detection, this system has provided accuracy of assay results. In the COBAS AMPLICOR system, the use of dUTP and uracil-*N*-glycosylase (UNG) reduces the risk of carryover contamination [11]. However, it dose not avoid false-positive results due to cross contamination between samples.

With regards to total quality control, false-negative PCR results become more problematic. Serum may contain a variety of biological substances and therapeutic reagents that may interfere enzymatic amplification to cause false-negative PCR results [5,6, 12]. Biological substances include hemoglobin, bilirubin, hem and so on. Heparin is used as anticoagulant at blood sampling and has an inhibitory effect on amplification. In particular, the inhibitory effect of heparin on PCR has been problematic in hemodialysis patients

[13,14], who have a high prevalence of HCV infection [15]. For avoidance of false-negative PCR results, it is necessary to eliminate and monitor amplification inhibitors. The internal control has been introduced to the COBAS AMPLICOR to assure efficient amplification in each assay [16]. Coamplification and detection of the internal control are useful in estimating effects of inhibitors on amplification in each assay for detection of HCV RNA. Coextraction of the internal control also appears to be useful in estimating efficiency of the extraction process [16]. However, it can not detect other causes for false-negative PCR results such as RNA loss or degradation during the preanalytic phase, from sampling, transport to storage of samples. Despite improvement in assay quality assurance in AMPLICOR system for detection of HCV RNA through placement of new reagents as well as quality control methods, it is necessary to make further efforts to ensure the each assay result consistent with the clinical diseases. We correlated laboratory test results of consecutive 5395 serum samples received for routine detection of HCV by the qualitative AMPLICOR HCV test (AMPLICOR, version 1.0), and investigated causes of the inconsistency.

Materials and Methods

Clinical Specimens

Serum specimens used in this study were received for routine detection for HCV RNA from consecutive 5395 patients referred to Tokai University Hospital for liver diseases. All samples were separated from clots within 4 h of collection, divided into aliquots, and stored at -80°C until RNA were extracted.

RNA Extraction

RNA was isolated from 300 μL of serum by a manual method based on guanidinium thiocyanate lysis and isopropanol precipitation (SepaGene RV-R, Sanko Junyaku Co., Ltd. Tokyo, Japan). Briefly, 300 μL of guanidinium thiocyanate solution was added to the specimen, and the phases were then separated by adding 300 μl of sodium acetate and 600 μL of chloroform / agglutination solution followed by centrifugation for 5 min at 15000 rpm. RNA was precipitated from the upper phase using isopropanol and resuspended in 50 μL of a specimen diluent.

Amplicor HCV Test

HCV RNA was qualitatively measured by the AMPLICOR HCV Test, version 1.0, or COBAS AMPLICOR HCV, version 2.0 (Roche Diagnostic Systems, Branchburg, NJ), as previously described [9]. The second version of AMPLICOR HCV test (AMPLICOR, version 2.0) was developed with the intention of achieving more equivalent detection of all HCV genotypes [10], and placed into the COBAS system (COBAS AMPLICOR, version 2.0)

[17]. Optical density of each reaction was measured at 660 nm. In the AMPLICOR HCV Test, version 1.0, optical density readings for HCV-RNA of above 0.400 were considered positive. Optical density readings for HCV RNA of 0.200 to 0.400 were considered equivocal. In COBAS AMPLICOR HCV, version 2.0, optical density readings for HCV-RNA of above 1.00 were considered positive. Optical density readings for HCV RNA of 0.15 to 1.00 were considered equivocal. When needed, HCV RNA was quantitatively measured by the AMPLICOR MONITOR HCV test, version 2.0. A single amplification and detection was performed on each extract except as otherwise stated. One positive and two negative controls provided with the kit were run with each batch of patient specimens.

Sample Dilutions

Clinical specimens that were assayed for detection of HCV RNA by the AMPLICOR HCV test, version 1.0, was diluted by 10-fold in HCV seronegative normal human serum. Duplicate determinations were made on the diluted samples in the AMPLICOR HCV test, version 1.0.

In-House PCR for HCV RNA

HCV RNA in the serum was assayed using in-house nested PCR (lower detection limits were 10^2-10^3 copies /mL). Primers were targeted to 5' untranslated region of HCV. Sequences of primers used were; (sense) 5'ACTCCACCATAGATCACTCC3' and (antisense) 5'CATGGTGCACGGTCTACGAG3' for the first round PCR, and (sense) 5'AGGAACTACTGTCTTCACGC3' and (antisense) 5'CGGGGCACTCGCAAGCACCC3' for the second round of nested PCR.

The target sequences were amplified by PCR in a DNA thermal cycler (Perkin-Elmer 9600, Perkin-Elmer Cetus, Wellesley, MA, USA) for 38 cycles with a DNA thermal amplification reagents kit (Perkin-Elmer Cetus), which generated a product of 321 bp. Each reaction cycle included denaturation at 94°C for 1 min, primer annealing at 60 °C for 1 min, and primer extension at 72°C for 2 min. For the second round of PCR, products of the first round PCR were amplified for 22 cycles, which generated a product of 269 bp.

Genotypes of HCV

Genotypes of HCV were determined by a method based on the sequence variation of core region, as described by Okamoto [18]. Briefly, the core gene sequence of HCV RNA was amplified by PCR using a universal primer (sense) and a mixture of four type-specific primers (antisense); I (1a), II (1b), III (2a) and IV (2b), and the genotypes were determined by the size of amplified products specific to each of them.

Correlation of Laboratory Test Results

Laboratory test results of consecutive 5395 serum samples received for routine detection of HCV by the qualitative AMPLICOR HCV test (AMPLICOR, version 1.0) were correlated to see clinical validity of the test. Relationship was studied between the detection of HCV RNA and HCV antibodies assayed by a third-generation enzyme immunoassay (EIA) test (International Reagents Co. Kobe, Japan), core antibodies assayed by radioimmunoassay tests (Ortho Clinical Diagnostics, Raritan, New Jersey, USA), or recombinant immunoblot assay (RIBA) II tests (Ortho Clinical Diagnostics). When aminotransaminases were increased, laboratory data regarding causes for hepatitis were reviewed including surface antigen and core antibodies of hepatitis B virus assayed by EIA (Sysmex Co. Kobe, Japan).

Statistical Method

Statistical analysis was performed to estimate the differences between positive rates for HCV RNA in two groups with high and low titers of positive HCV antibodies using the chi-square test.

Results

HCV RNA Detection by the AMPLICOR HCV Test and Titers of HCV Antibodies

After excluding patients under treatment with anti-viral agents, 519 and 866 of 1133 samples were positive for HCV RNA by the qualitative AMPLICOR HCV test and HCV antibodies by a third-generation enzyme immunoassay test, respectively (Table 1). Titers of positive HCV antibodies were widely distributed with two groups: low (<40.0) and high (>40.0) which had distinctive positive rates for HCV RNA in 19/370 (5.1%) and 488/495 (98.6%), respectively ($p<0.0001$). (Table 2, Figure 1)

Table 1. Correlation of anti-HCV antibodies with detection of HCV RNA by the AMPLICOR HCV Test

Anti-HCV anitibodies	HCV RNA in samples		
	Positive	Negative	
Positive	507	359	866
Negative	12	255	267
Total	519	614	1133

HCV RNA was measured in singlet using the version 1.0 AMPLICOR HCV tests. Optical density readings for HCV RNA of above 0.400 were considered positive. Anti-HCV anitibodies were assayed by the third generation EIA test. Cut-off values of above 1.0 were considered positive.

Table 2. Correlation of anti-HCV antibodies titers with detection of HCV RNA by the AMPLICOR HCV Test

Anti-HCV anitibodies		HCV RNA in samples	
titers	Positive	Negative	
≥40	488	7	495
<40	19	352	371
Total	507	359	866

HCV RNA was measured in singlet using the version 1.0 AMPLICOR HCV tests. Optical density readings for HCV RNA of above 0.400 were considered positive. Anti-HCV anitibodies were assayed by the third generation EIA test. Cut-off values above 1.0 were considered positive.

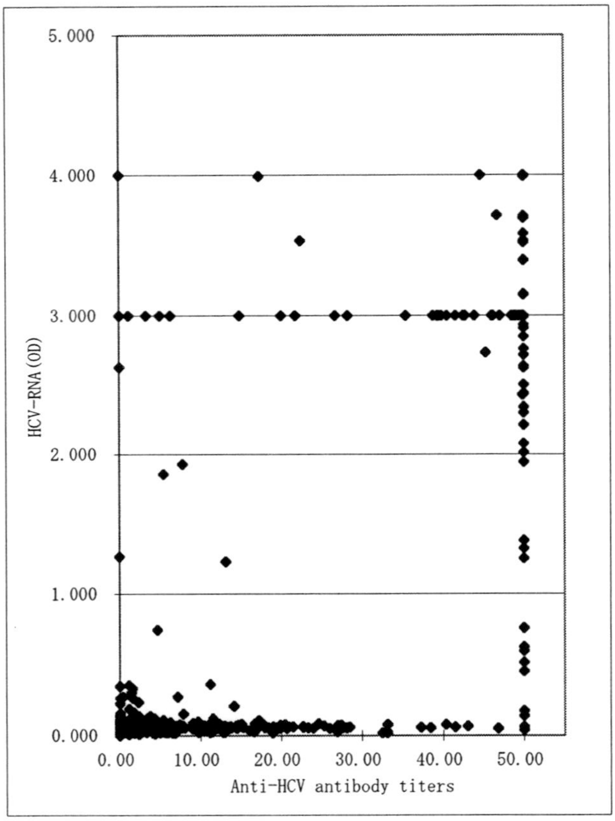

Figure 1. Correlation of anti-HCV antibodies titers with detection of HCV RNA by the AMPLICOR HCV Test HCV RNA was measured in singlet using the version 1.0 AMPLICOR HCV tests. Anti-HCV anitibodies were assayed by the third generation EIA test.

Correlation of Laboratory Test Results

When interpreting patterns of laboratory test results, a total of 8 samples were suggested to have false-negative results for HCV RNA as assayed by the AMPLICOR HCV test, version 1.0. Seven of them had high-titers of HCV antibodies. Other laboratory data also supported the presence of HCV RNA, including positive HCV core antibodies, and/or positive RIBA II assay for HCV. One sample with negative HCV antibodies and elevated levels of serum aspartate aminotransferase (68 U/L) and alanine aminotransferase (477 U/L) was considered a possibility of acute hepatitis C infection, because of a lack of evidence for other causes including surface antigen and core antibodies of hepatitis B virus.

Sample Dilution

The 8 samples with suggested false-negative results for HCV RNA were subjected to further investigations. When the serum was diluted by 10-fold and the assay was repeated, four of them tuned to be positive, suggesting the presence of an inhibitor (Table 2). Reviewing medical charts and interviewing physicians of these patients revealed the presence of heparin as an inhibitor in the serum, including those from patients with hemodialysis.

Genotyping and Quantitative Measurement of HCV RNA

When genotypes of HCV of the 4 samples that had no inhibition were assayed based on the sequence of the core region, they were revealed to be 2a (2 case), 1b+2a (1 cases), and 1b (1 case). When these samples were assayed by a second version of the AMPLICOR HCV test (AMPLICOR, version 2.0), all of the assay results turned to be positive except for 1 case (1b case) (Table 3). HCV RNA levels in 2 samples with genotype 2a quantitatively measured by the AMPLICOR HCV MONITOR Test, version 2.0, were 1.0 and <1.0 K copies/mL, respectively.

HCV RNA by in-House PCR

When the one serum sample that had no inhibition and a negative result by a second version of the AMPLICOR HCV test was assayed for HCV RNA by in-house PCR, the result was also negative.

Discussion

Quality control plays a vital role helping to ensure the reliability of test results of PCR detection for HCV RNA as well as others. Comparing the relationship between laboratory test results obtained from the same patient may sometimes identify analytical errors [19, 20]. Inconsistent results should prompt further investigations. To know consistency of the assay result of qualitative AMPLICOR HCV test with clinical diseases, we correlated laboratory test results of consecutive 5395 serum samples and causes of the discrepancy were investigated.

Table 3. Investigations for false negative PCR results for HCV RNA

Case	HCV RNA Version 1.0	HCV RNA Version 1.0 Diluted	HCV RNA Version 2.0	HCV RNA Levels (copies/ml)	HCV RNA In-house	Genotypes	
1	0.103(-)	>3.0(+)	ND	ND	ND	ND	
2	0.076(-)	>3.0(+)	ND	ND	ND	ND	
3	0.035(-)	>3.0(+)	ND	ND	ND	ND	
4	0.285(-)	>3.0(+)	ND	ND	ND	ND	
5	0.061(-)	0.018(-)	3.700 (+)		<1.K	(+)	2a
6	0.073(-)	0.058(-)	3.996(+)	ND	(+)	1b+2a	
7	0.257(-)	0.069(-)	3.998(+)	1.0K	(+)	2a	
8	0.057(-)	0.037(-)	0.032(-)	ND	(-)	1b	

Eight serum samples were assayed for detection of HCV RNA by the AMPLICOR HCV Test, version 1.0, or COBAS AMPLICOR HCV, version 2.0. Samples were also diluted by 10-fold and assay was repeated for detection of HCV RNA by the AMPLICOR HCV test, version 1.0. In the AMPLICOR HCV Test, version 1.0, optical density readings for HCV-RNA of above 0.400 and of 0.200 to 0.400 were considered positive and equivocal, respectively. In COBAS AMPLICOR HCV, version 2.0, optical density readings for HCV RNA of above 1.00 and of 0.15 to 1.00 were considered positive and equivocal, respectively. HCV RNA was quantitatively measured by the AMPLICOR MONITOR HCV test, version 2.0. Genotypes of HCV were determined by the sequence of core region.

ND, not determined.

Low-positive HCV EIA test results have been suggested as an important predictor of low likelihood of hepatitis C infection. Low- positive HCV antibodies frequently represent false-positive reactivity. Alternatively, low-positive HCV antibodies in coupling with negative HCV RNA indicate that individuals have been truly infected in the past, resolved the infection, and had a decrease in their titers of antibody. Actually in our study, a majority of persons with low-positive HCV titers were negative on HCV RNA. In contrast, high-titer HCV antibodies suggest the presence of HCV RNA in serum [1,2], if samples were taken before installation of anti-viral agents. For unusual patterns such as a combination of negative results for HCV RNA and high-titer HCV antibodies, the possibility of laboratory error should be considered. Seven samples with high-titer HCV antibodies were suggested to have false-negative results for HCV RNA as assayed by the first version of AMPLICOR HCV test. Other laboratory data also supported the presence of HCV RNA, including positive HCV core antibodies, and/or positive RIBA II assay for HCV [21]. One sample with negative HCV antibodies and increased levels of aminotransferases was considered a possibility of acute hepatitis by HCV infection [1,2], because of a lack of evidence for other causes including hepatitis B virus. These samples were subjected to further investigations for negative results for HCV RNA.

First of all, false-negative PCR results due to inhibitory effects on amplification were considered. When the serum was diluted by 10-fold and the assay was repeated, the PCR results in 4 samples turned to be positive, suggesting the presence of an inhibitor. This is supported by previous our report that 10-fold dilution of the serum sample was enough to eliminate the inhibitory effect on amplification. For avoidance of false-negative PCR results, it is necessary to eliminate and monitor amplification inhibitors by the internal control. Coamplification and detection of the internal control is useful in estimating the effects of inhibitors on amplification. However, the internal control is not routinely used in our laboratory, because its payment is not covered by health insurance in Japan. Reviewing medical charts and interviewing the physicians of patients revealed the presence of heparin as an inhibitor in the serum, including those from patients with hemodialysis. The inhibitory effect of heparin on PCR has been problematic in hemodialysis patients [13,14], who have a high prevalence of HCV infection [15]. To eliminate inhibitory substances in samples, it will be also necessary to educate physicians and nurses regarding proper sampling based on the limitations of the assay for detection of HCV RNA by PCR.

Recently there has been substantial progress in automation of the nucleic acid extraction from serum in addition to amplification and detection. In our previous study, we developed and evaluated a prototype automated specimen preparation instrument for specific capture of HCV RNA with probes and magnetic bead/fluid (B/F) separation [22, 23]. HCV RNA was isolated from serum by lysis of virus particles with a chaotropic agent, followed by hybridization of the RNA with biotinylated probes and capture of the hybridized RNA with streptavidin-coated paramagnetic particles. After washing of the hybrid-particle complexes to remove non-specifically bound materials, the particles were resuspended in a specimen diluent and were then ready for amplification and detection using a fully automated PCR system (COBAS AMPLICOR). The prototype automated RNA extraction system was shown to successfully eliminate the inhibitory effect of heparin from serum in the qualitative assay of HCV RNA, and was proved to be applicable to serum specimens from hemodialysis patients [22]. Based on the promising data that the prototype automated RNA extraction system had similar assay performance to the conventional manual method for the qualitative assay of HCV RNA, it has been further developed to a commercially available system (AmpliCap/GT-12; Roche Diagnostics). This system was shown to successfully eliminate the inhibitory effect of heparin as well as dextran sulfate from serum in the qualitative as well as the quantitative assay of HCV RNA. The extraction method used is theoretically suitable in eliminating any inhibitors in the serum, because it is based on the specific capture with probes and magnetic B/F separation. On the other hand, conventional manual methods such as that used in this study are not able to eliminate heparin and dextran sulfate, which are thought to bind nucleic acids [24]. Therapeutic reagents for infectious agents may have inhibitory effects [24, 25], and novel reagents will be constantly developed and introduced into the treatment of infectious diseases [26, 27]. It is increasingly important to accurately detect and measure viral loads in serum that may contain various therapeutic reagents. Therefore it is desirable to use an extraction method that can eliminate inhibitors as much as possible.

The significant differences in the efficiencies of detection of genotype 1 and non-genotype 1 isolates were observed in the first version of the AMPLICOR HCV MONITOR test (AMPLICOR, version 1.0) [28, 29], and recently a second version of the AMPLICOR HCV MONITOR Test (AMPLICOR, version 2.0) was developed with the intention of achieving more equivalent quantification of all HCV genotypes [30]. In the first version of AMPLICOR HCV MONITOR test (version 1.0), amplification in non-genotype 1 isolates has been compromised. This is also seen in the first version of the qualitative AMPLICOR HCV test. The significant differences in the efficiencies of detection of genotype 1 and non-genotype 1 isolates has been attributed to the genotype-specific differences in the strength of base pairings and in the secondary structure of the 5' noncoding region in which the primers anneal [31]. The difference was minimized in the second version of the AMPLICOR HCV Test (AMPLICOR, version 2.0) using reagents with optimized conditions for magnesium ion and with addition of dimethylsulfoxide. When the 3 samples with genotype 2a were assayed by the second version of the AMPLICOR HCV test, the assay results turned to be positive. As a cause of false-negative PCR results in these samples, 2a genotype in coupling with very low levels of HCV was suggested.

In the last one sample, the presence of an inhibitor and the influence of genotypes were unlikely for a cause for a false-negative PCR result. A negative result of under the detection limit due to RNA degradation was suggested, since in-house PCR was also negative for HCV RNA, while the genotype was successfully determined as 1b type. Unfortunately, there was insufficient volume for further testing of this sample. In such a case, sampling serum again for re-testing would be necessary to provide an accurate assay result.

Despite introduction of the reagents and methods for quality control, each assay result cannot be totally assured. Neither false-negative PCR results for HCV RNA due to RNA degradation nor false-positive results due to crossover contamination can be found. There are various applications of laboratory quality control procedures and their role in identifying laboratory errors [19,20]. An example is the use of patient-derived quality control procedures as a means of assessing laboratory performance. Early examples of these types of applications include use of Bull's algorithm, anion gap measurements, and delta checking. More recent applications include the correlation of laboratory test results. Optimization of quality control rules can vary depending on the application for which a test is to be used. In the present study, review and correlation of clinical and laboratory data followed by further assessment supported the quality assurance of each result, enabling us to ensure the test results consistent with clinical diseases as well as to monitor false-negative and -positive results.

There are possible limitations in the correlation of laboratory results to detect analytical errors. Decisions on performing additional testing were largely based on inconsistent data such as a combination of negative HCV RNA with high-titer HCV antibodies. However, a minority of patients with low-titer HCV antibodies also had HCV RNA. In such a group with low-titer HCV antibodies, there might have been also false-negative results for HCV RNA. To solve this problem, there has been improvement in assay quality assurance in AMPLICOR system for detection of HCV RNA through placement of new reagents such as the second version of AMPLICOR HCV test as well as quality control methods such as the internal control.

In summary, individual review and correlation of clinical and laboratory data of each sample followed by further assessment support the quality assurance of each result as well as the system of assay. False-negative results in detection of HCV by the AMPLICOR HCV test appeared to be attributed from the presence of inhibitor, genotype-specific difference in amplification efficiency or probably RNA degradation. It would be useful to introduce real-time review to confirm the assay result consistent with clinical diseases to provide reliable test results in routine diagnostic PCR.

Reference

[1] Lok, A.S.F., and Gunaratnam, N.T. 1997. Diagnosis of hepatitis C. Hepatology 26: 48S-56S.

[2] Carrithers, R.L.Jr., Marquardt, A., and Gretch, D.R. 2000. Diagnostic testing for hepatitis C. Semin Liver Dis 28: 159-171.

[3] Trimoulet, P., De Lédinghen, V., Foucher, J., Castéra, L., Fleury, H., and Couzigou, P. 2004. Predictive Value of Early HCV RNA Quantitation for Sustained Response in Nonresponders Receiving Daily Interferon and Ribavirin Therapy. J. Med. Virol. 72: 46-51.

[4] McHutchison, J.G., Poynard, T., Esteban-Mur, R., Davis, G.L., Goodman, Z.D., Harvey, J., Ling, M.-H., Garaud, J.J., Albrecht, J.K., Patel, K., and Dienstag J.L. 2002. Hepatic HCV RNA before and after treatment with interferon alone or combined with ribavirin. Hepatology. 35: 688-693.

[5] Rolfs, A., Schuller, I., Finckh, U., and Weber-Rolfs, I. 1992. PCR: Clinical diagnostics and research, p.51-67. Springer-Verlag, Berlin.

[6] Greenfield, L., and White, T.J. 1993. Sample preparation methods, p. 122-137. *In* D.H. Persing, T.F. Smith, F.C. Tenover, T.J. White (ed.), Diagnostic molecular microbiology. Principles and applications. American Society for Microbiology, Washington, D.C.

[7] DiDomenico, N., Link, H., Knobel, R., Caratsch, T., Weschler, W., Loewy, Z.G., and Rosenstraus, M. 1996. COBAS AMPLICOR™: fully automated RNA and DNA amplification and detection system for routine diagnostic PCR. Clin Chem 42: 1915-1923.

[8] Jungkind, D., DiRenzo, S., Beavis, K.G., and Silverman, N.S. 1996. Evaluation of automated COBAS AMPLICOR PCR system for detection of several infectious agents and its impact on laboratory management. J. Clin. Microbiol. 34: 2778-2783.

[9] Poljak, M., Seme, K., and Koren S. 1997. Evaluation of the automated COBAS AMPLICOR hepatitis C virus PCR system. J. Clin. Microbiol. 35: 2983-2784.

[10] Eralis, M., Ashwood, E.R., and Hillyard, D.R. 2000. Performance characteristics of the COBAS AMPLICOR hepatitis C virus MONITOR test, version 2.0. Am. J. Clin. Pathol. 114: 180-187.

[11] Longo, M. C., Berninger, M. S., and Hartley, J. L.. 1990. Use of uracil DNA glycosylase to control carry-over contamination in polymerase chain reactions. Gene. 93: 125-128.

[12] Masukawa, A., Miyachi, H., Ohshima, T., and Ando, Y. 1999. Inhibitory effects of therapeutic reagents on the PCR detection of hepatitis C virus in serum and their elimination by RNA extraction method. Jpn. J. Clin. Pathol. 47: 949-955. [Japanease with English abstract]

[13] Willems, M., Moshage, H., Nevens, F., Fevery, J., and Yap, S.H. 1993. Plasma collected from heparinized blood is not suitable for HCV-RNA detection by conventional RT-PCR assay. J. Virol. Methods. 42: 127-130.

[14] Beutler, E., Gelbart, T., and Kuhl, W. 1990. Interference of heparin with the polymerase chain reaction. BioTechniques. 9: 166.

[15] Silini, E., Bono, F., Cerino, A., Piazza, V., Solcia, E., and Mondelli, M.U. 1993. Virological features of hepatitis C virus infection in hemodialysis patients. J. Clin. Microbiol. 31: 2913-2917.

[16] Miyachi, H., Masukawa, A., Ohshima, T., Fusegawa, H., Hirose, T., Impraim, C., and Ando, Y. 1998. Monitoring of inhibitors of enzymatic amplification in polymerase chain reaction and evaluation of efficacy of RNA extraction for the detection of hepatitis C virus using the internal control. Clin. Chem. Lab. Med. 36: 571-575.

[17] Krajden, M., Ziermann, R., Khan, A., Mak, A., Leung, K., Hendricks, D., and Comanor, L. 2002. Qualitative detection of hepatitis C virus RNA: Comparison of analytical sensitivity, clinical performance, and workflow of the Cobas Amplicor HCV test version 2.0 and the HCV RNA transcription-mediated amplification qualitative assay. J. Clin. Microbiol. 40: 2903-2907.

[18] Okamoto, H., Sugiyama, Y., Okada, S., Kurai, K., Akahane, Y., Sugai, Y., Tanaka, T., Sato, K., Tsuda, F., Miyakawa, Y., and Mayumi, M. 1992. Typing hepatitis C virus by polymerase chain reaction with type-specific primers: Application to clinical surveys and tracing infectious sources. J. Gen. Virol. 73: 673-679.

[19] Kazmierczak, S.C. 2003. Laboratory quality control: Using patient data to assess analytical performance. Clin. Chem. Lab. Med. 41: 617-627.

[20] Ng, R.H. Quality performance in the physician's office. 1987. Med. Clin. North. Am. 71: 677-890.

[21] Dufour D.R., Talastas M., Fernandez M.D.A., Harris B., Strader D.B. and Seeff L.B. 2003. Low-positive anti-hepatitis C virus enzyme immunoassay results: An important predictor of low likelihood of hepatitis C infection. Clin. Chem. 49: 479-486.

[22] Miyachi, H., Masukawa, A., Ohshima, T., Hirose, T., Impraim, C., and Ando, Y. 2000. Automated specific capture of hepatitis C virus RNA with probes and paramagnetic particle separation. J. Clin. Microbiol. 38: 18-21.

[23] Miyachi, H., Masukawa, A., Asai, S., Miura, T., Tamatsukuri, S., Hirose, T., and Ando, Y. 2003. Quantitative assay of hepatitis C virus RNA using an automated extraction system for specific capture with probes and paramagnetic particle separation. J. Clin. Microbiol. 41: 572-575.

[24] Demeke, T., and Adams, R.P. 1992. The effects of plant polysaccharides and buffer additives on PCR. BioTechniques 12: 332-334.

[25] Yedidag, E.N., Koffron, A.J., Mueller, K.H., Kaplan, B., Kaufman, D.B., Fryer, J.P., Stuart, F.P., and Abecassis. M. 1996. Acyclovir triphosphate inhibits the diagnostic polymerase chain reaction for cytomagalovirus. Transplantation 62: 238-242.

[26] Squires, K.E. 2001. An introduction to nucleoside and nucleotide analogues. Antiviral Ther. 6: Suppl.1-14.

[27] McHutchinson J.G., Gordon, S.C., Schiff, E.R., Shiffman, M.L., Lee, W.M., Rustgi, V.K., Goodman, Z.D., Ling, M.-H., Cort, S., and Albrecht, J.K. 1998. Interferon alfa-

2b alone or in combination with ribavirin as initial treatment for chronic hepatitis C. N. Engl. J. Med. 339, 1485-1492.

[28] Hawkins, A., Davidson, F., and Simmonds, P. 1997. Comparison of plasma virus loads among individuals infected with hepatitis C virus (HCV) genotypes 1, 2, and 3 by quntiplex HCV RNA assay versions 1 and 2, Roche Monitor assay, and an in-house limiting dilution method. J. Clin. Microbiol. 35: 187-192.

[29] Mellor, J., Hawkins, A. and Simmonds, P. 1999. Genotype dependence of hepatitis C virus load measurement in commercially available quantitative assays. J. Clin. Microbiol. 37: 2525-2532.

[30] Yu, M.-L., Chuang, W.-L., Dai, C.-Y., Chen, S.-C., Lin, Z.-Y., Hsieh, M.-Y., Wang, L.-Y., and Chang, W.-Y. 2000. Clinical evaluation of the automated COBAS AMPLICOR HCV MONITOR test version 2.0 for quantifying serum hepatitis C virus RNA and comparison to the quantiplex HCV version 2.0 test. J. Clin. Microbiol. 38: 2933-2939.

[31] Smith, D.B., Mellor, J., Jarvis, L.M., Davidson, F., Kolbeg, J., Urdea, M., Yap, P.L., Simmonds, P., Conradie, J.D., Neill, A.G.S., Dusheiko, G.M., Kew, M.C., Crookes, R., Koshy, A., Lin, C.K., Lai, C., Murray-Lyon I.M., El-Guneid, A., Gunaid, A.A. 1995. Variation of the hepatitis C virus 5' non-coding region: Implications for secondary structure, virus detection and typing. J. Gen. Virol. 76: 1749-1761.

In: Trends in RNA Research
Editor: Peter A. McNamara, pp. 159-189

ISBN: 1-59454-506-5

Chapter VIII

The Primary and Secondary Structures of HIV-1 Genomic RNA Region Encompassing DIS, SD and ψ Hairpins: *In Silico* Study

Margarita I. Zarudnaya, Iryna M. Kolomiets, Andriy L. Potyahaylo and Dmytro M. Hovorun
Molecular Biophysics Department, Institute of Molecular Biology and Genetics, National Academy of Sciences of Ukraine, 150, vul. Zabolotnoho, Kyiv, 03143, Ukraine

Abstract

The 5′-untranslated region of the human immunodeficiency virus type 1 (HIV-1) genomic RNA contains several stem-loops that play important roles in the viral replication cycle. Amongst these are the dimerization initiation site (DIS), the hairpin containing the major splice donor site (SD) and the packaging signal (ψ), all three hairpins are sequentially located in the region of approximately 100 nt length preceding GAG start codon. In this chapter, we analyzed the primary and the secondary structures of this region in all HIV-1 isolates for which this region was submitted in NCBI database, 350 isolates in total. The secondary structures were predicted by the mfold program (Zuker, 2003). We found that 94% of viral isolates (groups M and N) contain three hairpins (DIS, SD and ψ) in the region preceding GAG start codon. In the rest of HIV-1 isolates (group O) this region has four hairpins. Two hairpins are DIS- and ψ-like hairpins, a hairpin located between them being different from SD-like hairpin. The additional fourth hairpin is located immediately upstream of GAG start codon. Such structure of the region preceding GAG start codon has first been reported by Harrison et al. (1998). We determined the most frequently occurring secondary structures of DIS, SD and ψ hairpins in HIV-1 isolates of M group, and compared our phylogenetic findings with literature data on high-resolution structures for these hairpins. In particular, we found that 73% of HIV-1 group M isolates surveyed contain DIS hairpins with the AA[GCGCGC]A (Lai variant), AG[GUGCAC]A (Mal variant) or A[GUGCAC] apical loops. The last apical loop is typical for HIV-1 isolates (subtype C) from Botswana, South Africa and India. Contrary to the well-known structures of DIS hairpins in Lai and Mal isolates, the structure of the DIS hairpin with the A[GUGCAC] apical loop is not well studied yet. We observed that HIV-1 genomic RNAs with this loop often contain SD and ψ hairpins with the specific base changes (A^5 to G^5 in SD and U^1 to A^1 in ψ hairpins). This leads to alterations in the hairpins secondary structures. The stem of the SD hairpin in these isolates has no A-U-A base triple platform and the region corresponding to ψ hairpin is folded mainly in two alternative structures, including one with a truncated stem. Phylogenetic analysis data provide evidence that DIS, SD and ψ motifs are essential for HIV-1 replication cycle.

Introduction

The recent studies on structure of the 5'-terminal region of HIV-1 genomic RNA showed that this region adopts two mutually exclusive conformations, so called LDI (long-distance interaction) and BMH (branched multiple hairpins) ([1]-[6]). BMH structure consists of three domains - TAR, polyA and the domain closed by U5-AUG duplex (see Fig. 1 adapted from ([5])). The last domain includes the hairpin containing the primer binding site (PBS) and primer activation signal (PAS) ([7]), DIS hairpin, or stem-loop 1 (SL1) ([8], [9]), essential for initiation of dimerization of HIV-1 genomic RNAs, SD hairpin, or stem-loop 2 (SL2) ([10]), containing the major splice donor site that is required for the generation of subgenomic mRNAs, and also ψ hairpin, or stem-loop 3 (SL3) ([11], [12]), which is the major determinant for specific packaging.

In this chapter, we examine only the region encompassing three sequentially located DIS, SD and ψ hairpins, two of them (DIS and SD) are absent in LDI structure (Fig. 1). The specific structural features of these hairpins are believed to be essential for their functioning, but the secondary structures of these hairpins are different in diagrams of the 5'-terminal region of HIV-1 genomic RNA reported in literature, for example ([5], [13], [14]). The high-resolution structures for some DIS, SD and ψ constructs have been elucidated by crystal and NMR investigations ([14]-[29]). To elucidate which secondary structures of DIS, SD and ψ hairpins are the most frequent in HIV-1 isolates, we analysed all HIV-1 isolates in NCBI database for which the region of approximately 100 nt length preceding the GAG start codon was reported. In sum, 350 isolates from different world regions have been surveyed. Our detailed phylogenetic analysis showed that the secondary structures of DIS, SD and ψ hairpins in most isolates surveyed are similar to those shown in Fig. 1 (BMH structure) or moderately altered. Some structural features of these hairpins appeared to be virus subtype and/or country specific.

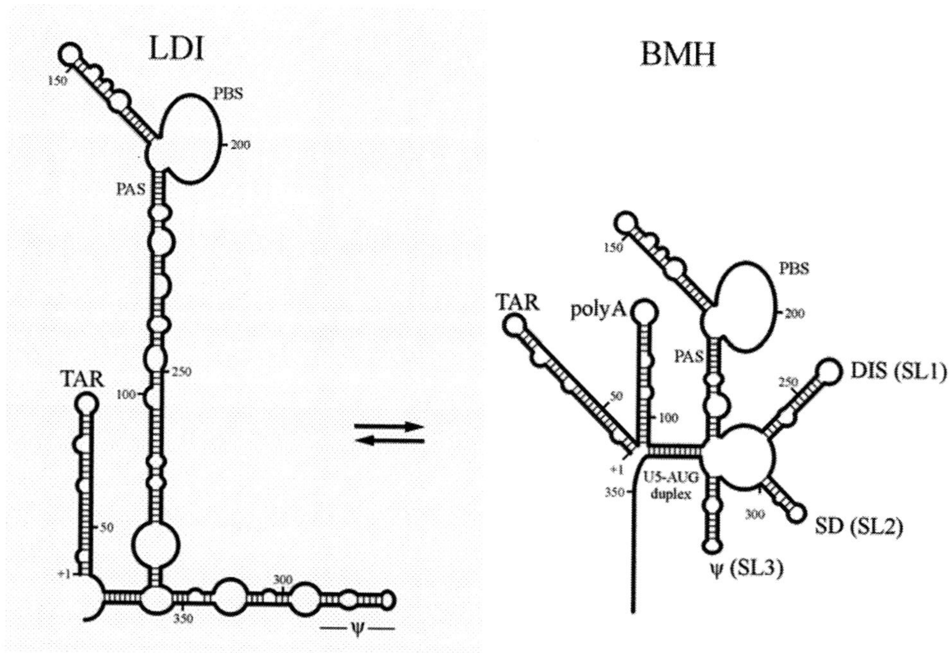

Fig. 1. LDI and BMH conformations of the 5'-terminal region of HIV-1_{Lai} RNA (adapted from ([5])). TAR – *trans*-activation response element, polyA – domain containing the polyadenylation signal, PBS – primer binding site, PAS – primer activation signal, DIS – dimerization initiation site, SD – major splice donor site, ψ - packaging signal.

Major and Minor Groups of HIV-1 Isolates Classified by the Structure of the Region Preceding GAG Start Codon

We found that the sequence corresponding to the DIS hairpin in isolates Lai or Mal is present in genomic RNAs of all HIV-1 isolates studied and located approximately 100 nt upstream of GAG start codon. In genomic RNA of each isolate, we analysed the primary and secondary structures of the region stretching from the beginning of DIS motif up to the nucleotide at the penultimate position of GAG start codon. The secondary structures were predicted using the mfold program by Zuker ([30]). Following the analysis of secondary structures we divided all the isolates surveyed into two groups: major and minor. The major group includes the isolates where the region preceding GAG start codon consists of three well-known hairpin structures: DIS, SD and ψ. As an example, the secondary structure of this region in genomic RNA of isolate Lai is shown in Fig. 2A, the structure being identical to that reported by Abbink and Berkhout ([5]). In some isolates of the major group, the secondary structures of DIS, SD and ψ hairpins are altered due to infrequent base changes.

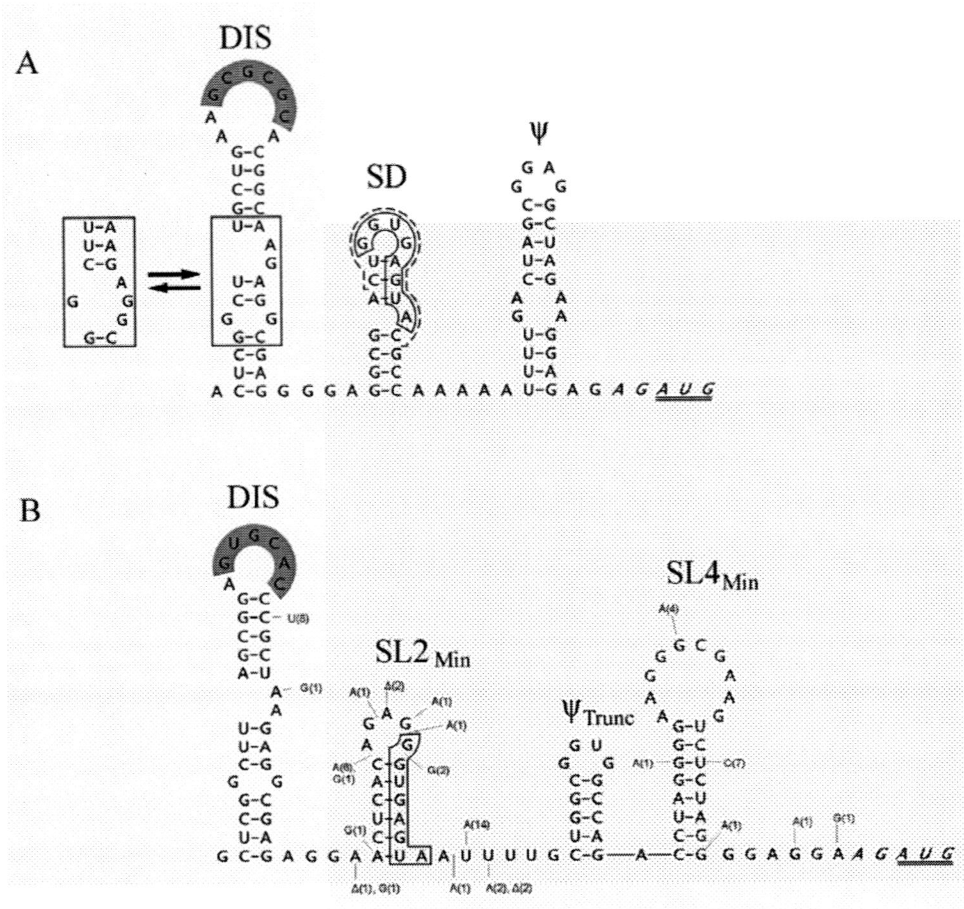

Fig. 2. The secondary structures of the region preceding GAG start codon in genomic RNA of HIV-1 isolates of the major and minor groups predicted by mfold program. The palindromes in the DIS hairpin are shadowed. The identical sequences in SD and $SL2_{Min}$ hairpins are outlined. The sequences in italics were not included in folding. The GAG start codon is underlined. (A) The major group isolate Lai. Two forms of the DIS stem defect are boxed. MSD in the SD hairpin is indicated by dotted lines. (B) The minor group isolate ANT70C. Base changes in the minor group isolates compared with the isolate ANT70C are indicated and the numbers of times a given nucleotide is found at a given position are bracketed; Δ - deletion.

HIV-1 is classified into three groups termed M, N and O ([31]). All 329 isolates from our major group appeared to belong to HIV-1 groups M or N, and all 21 isolates from our minor group - to group O. The minor group isolates contain DIS- and ψ-like hairpins (Fig. 2B) with numerous base changes relative to the isolates of group M (Fig. 2A). They also have a hairpin located between the DIS- and ψ-like hairpins, but this structure is different from SD-like motif. Another specific feature of the minor group isolates is the presence of an additional stem-loop structure ($SL4_{Min}$) between ψ-like hairpin and GAG start codon. Fig. 2B shows the secondary structure of the region preceding GAG start codon in the isolate ANT70C, first reported by Harrison ([32]) and presents the collected data on 20 isolates of the minor group. The 21st isolate is hypermutated, nevertheless it contains all the four hairpins shown in Fig. 2B, their structures being somewhat altered. Base changes (substitutions, deletions or insertions) occur mainly in the single-stranded portions of the structure shown in Fig. 2B.

Mfold program ([30]) predicts that none of these base changes leads to alterations of the secondary structure of the region preceding GAG start codon in the minor group isolates. The structure also remains similar to that in genomic RNA of the isolate ANT70C (or alters moderately) when mutations occur in the double-stranded regions, since all these mutations lead to base pair substitution. The only exception is the mutation that breaks the bottom C-G base pair in the hairpin $SL4_{Min}$ in the isolate 99USTWLA. This mutation leads only to one base-pair shortening of the stem. Thus, the secondary structures of the region preceding GAG start codon are quite similar in all the isolates of the minor group surveyed. But, as it was mentioned above, the situation is rather different for isolates of the major group, probably because of their numerical superiority.

Only for two isolates from group N (CM-YBF106 strain and isolate with the accession number AJ006022), the 100 nt region preceding GAG start codon is submitted to NCBI database. The most favorable secondary structures of this region in group N isolates are shown in Fig. 3.

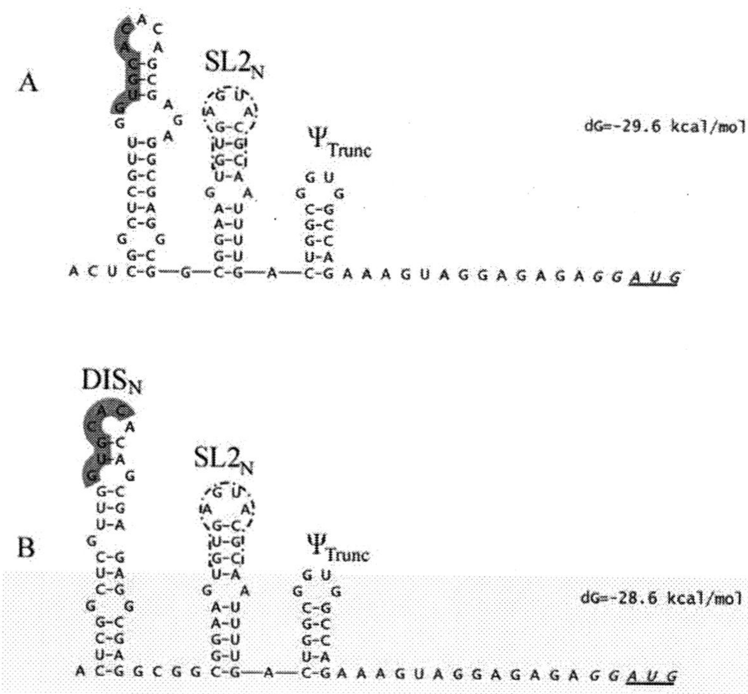

Fig. 3. The most favorable secondary structures of the region preceding GAG start codon in genomic RNA of group N isolates predicted by mfold program. Both the strain CM_YBF106 and the isolate with the accession no. AJ271370 have the identical structures (A) and (B). The palindromes in the DIS hairpin and in the sequence corresponding to this hairpin are shadowed. The sequence in $SL2_N$ indicated by dash-and-dot line is also present in the SD hairpin of Lai isolate. The sequences in italics were not included in folding. The GAG start codon is underlined.

Obviously, the structures are different from those both in HIV-1 group O and M isolates (Fig. 2). Contrary to isolates of the group O (Fig. 2B), genomic RNAs of the group N isolates are folded into three (not four) stem-loops in the region studied. On the other hand, similar to group O isolates, they have no SD-like hairpin and have a truncated ψ-like hairpin identical to that in the group O isolates (Fig. 2B). DIS hairpins and stem-loops located between DIS- and ψ-like hairpins in isolates of O and N groups are not similar (Figs 2B and 3).

Folding of HIV-1 Genomic RNA Region Preceding GAG Start Codon in Isolates from Group M

Mfold program ([30]) predicts different numbers of possible secondary structures of RNA fragment depending on percent suboptimality used. In this study, all predictions on RNA secondary structures were performed with a 15% suboptimality range. Figs 4 and 5 demonstrate three most favorable structures of the regions spanning DIS, SD and ψ hairpins in two typical HIV-1 isolates. The isolate with the accession no. NC_001802 (reference sequence) (Fig. 4) differs from the isolate Lai (Fig. 2A) only in one position (C instead A) in the linker between DIS and SD hairpins. As being discussed below, the GGGA linker occurs much more rare than GGGC. The most favorable structure of the isolate with the accession no. NC_001802 (Fig. 4A) contains DIS, SD and ψ hairpins, while SD-like hairpin is absent in the second structure (Fig. 4B) and DIS-like hairpin is absent in the third structure (Fig. 4C). Mfold program predicts a fewer number of structures for isolate Lai than for isolate with the accession no. NC_001802 (five and seven, respectively) but three of the most favorable structures are similar for both isolates. The free energy of the first structure is about 4 kcal/mole less than those of the second and third structures. Therefore, only the first structure consisting of three common hairpins (Fig. 4A) is supposed to form in this genomic RNA. This remains valid for many isolates of group M, the minimal difference between the free energies (dG) of the common structure and structures which do not contain DIS, SD or ψ motifs is mostly within the 2-5 kcal/mole interval. So, for the region preceding GAG start codon in HIV-1 genomic RNA, the structure containing these motifs is generally more favorable than alternative structures.

The most favorable secondary structures of the region preceding GAG start codon in one of Botswanian HIV-1 isolates (C-96BW04.09) is depicted in Fig. 5. The isolates from this country comprise the most numerous group among isolates studied (22% of group M isolates). As many as 92% of Botswanian isolates, including the isolate C-96BW04.09, have a base change U to A in the 5'-terminal nucleotide of the region corresponding to ψ hairpin that leads to alterations in its structure. Mfold program predicts two favorable structures of the region encompassing DIS, SD and ψ hairpins in genomic RNA of the isolate C-96BW04.09, which are very close in free energy (Fig. 5A and 5B). Evidently, the interconversion between these structures takes place. The first structure (Fig. 5A) contains a truncated ψ hairpin and another (Fig. 5B) has a ψ hairpin shortened only by a single base pair in comparison with that in isolate Lai. Besides, in many cases the third ψ-like structure is formed with an altered bottom segment of the stem (Fig. 5C). The base change U to A in the ψ hairpin does not affect the structures of the DIS and SD hairpins (Fig. 5).

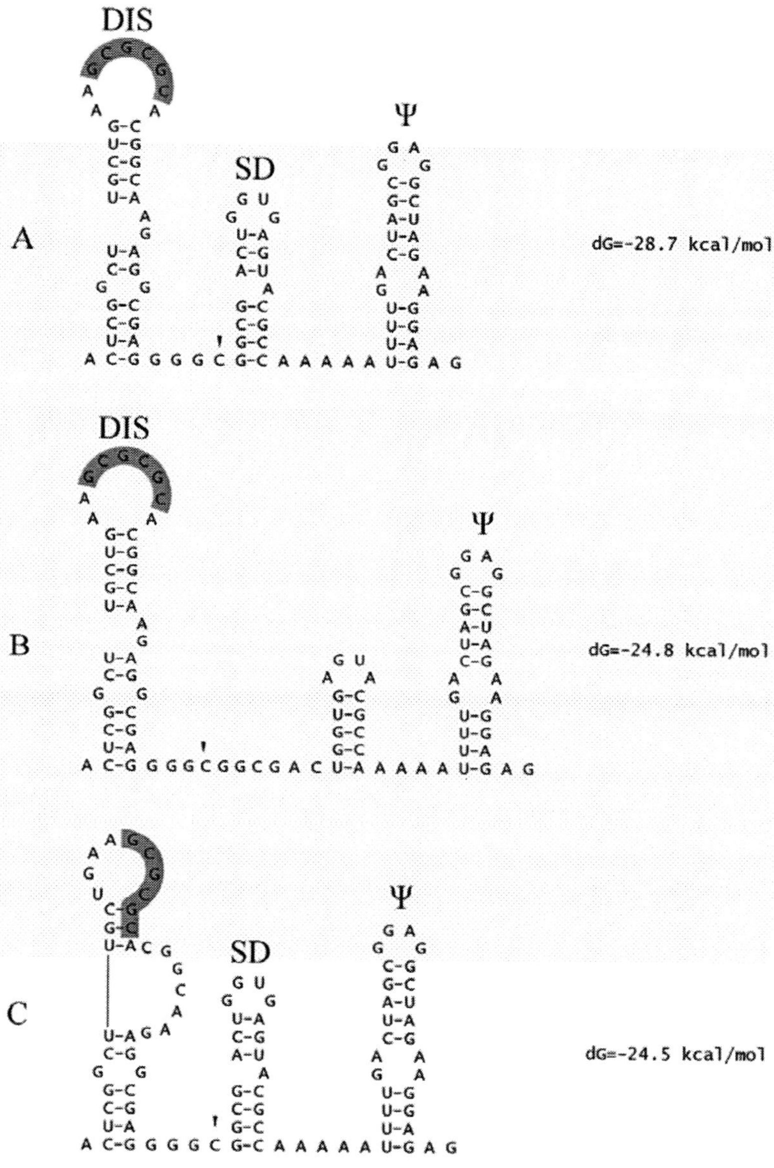

Fig. 4. The most favorable secondary structures of the region preceding GAG start codon in genomic RNA of group M isolate with the accession number NC_001802 predicted by mfold program. The palindromes in the DIS hairpin and in the sequence corresponding to this hairpin are shadowed. The single base change in comparison with Lai isolate is indicated by arrow.

Fig. 6A gives an example of when the lowest energy structure of the region preceding GAG start codon has neither DIS nor SD hairpins. The first suboptimal structure does not contain the SD hairpin (Fig. 6B), and only the second suboptimal structure contains all common hairpins, DIS, SD and ψ. (Fig. 6C). In the isolates in which DIS, SD and ψ hairpins are formed in the suboptimal structures, the dimerization and other processes of HIV-1 replication cycle are likely to be inhibited to a certain extent.

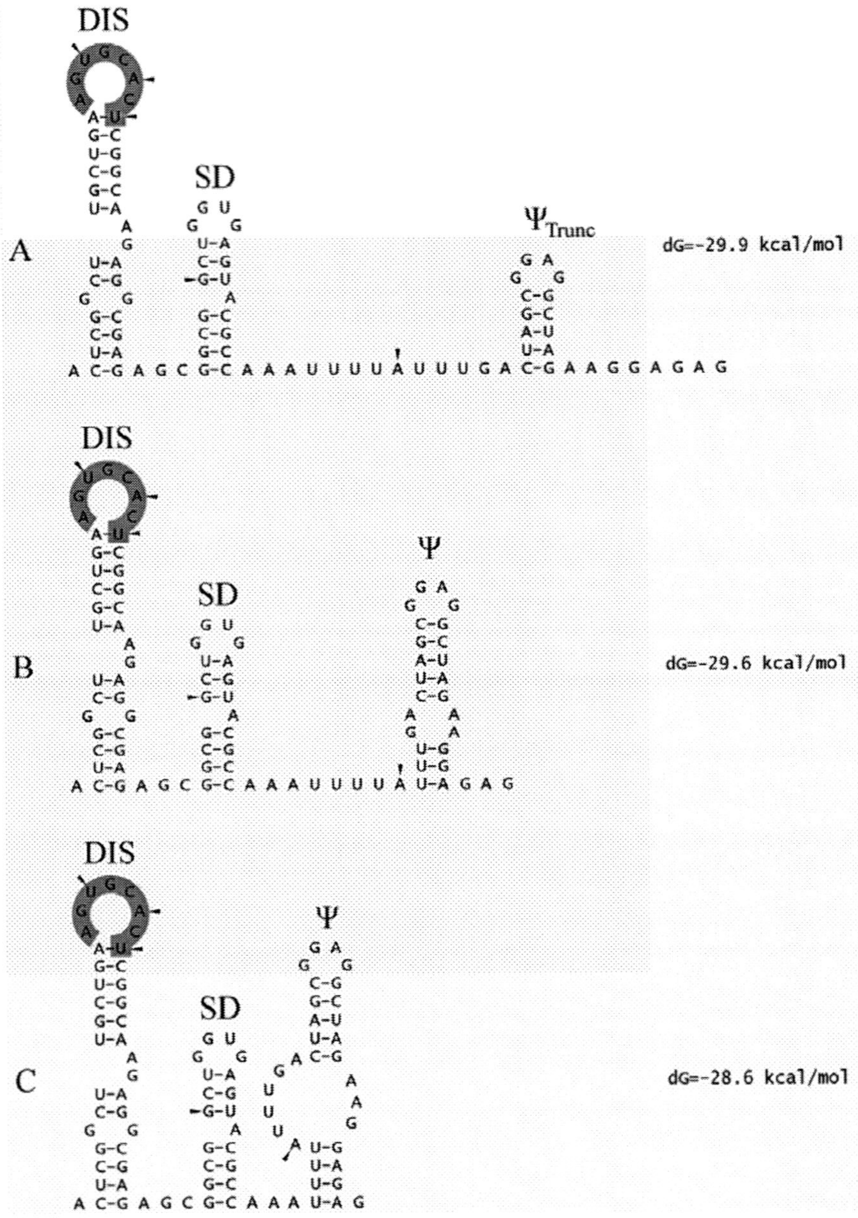

Fig. 5. The most favorable secondary structures of the region preceding GAG start codon in genomic RNA of Botswanian isolate C-96BW04.09 predicted by mfold program. The palindrome in the DIS hairpin is shadowed. Base changes in the DIS, SD and ψ-like hairpins (or the sequence corresponding to the ψ hairpin) compared with Lai isolate are indicated by arrows.

From 327 isolates of group M, 16% of isolates have the base changes in the region preceding GAG start codon, which lead to alternative structures that are more favorable or comparable in free energy with common structures. However, only in 5% of group M isolates the base changes leading to alternative structures were observed at least twice, in this case a base variation (≥ 0,6%) is probably higher than the rate of sequencing and data base errors (0,5%) ([23]).

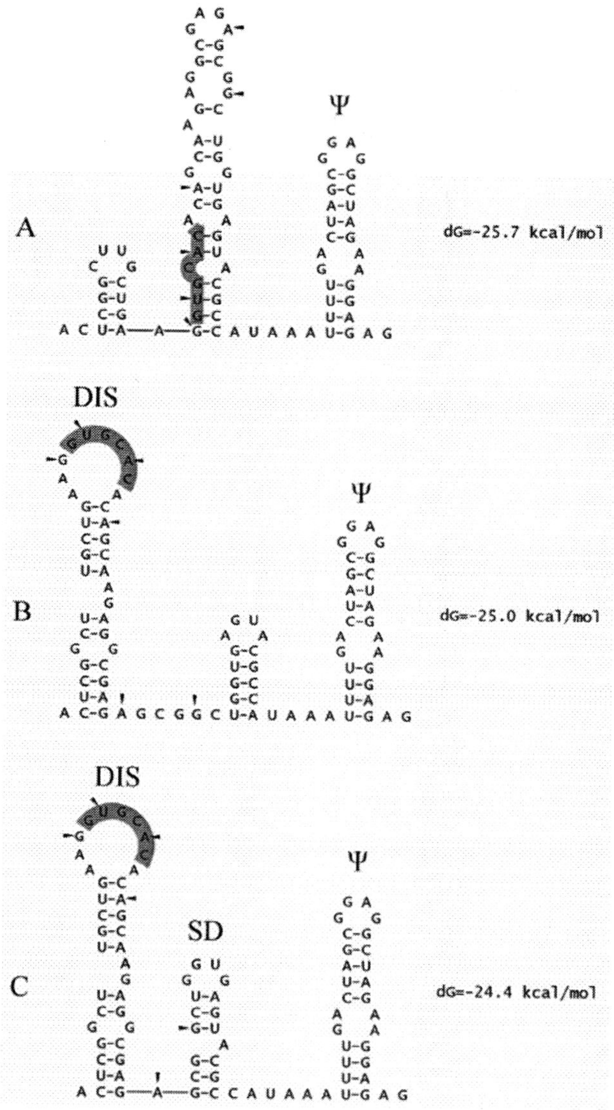

Fig. 6. The most favorable secondary structures of the region preceding GAG start codon in genomic RNA of three Spanish isolates predicted by mfold program. These are X421, X475 , and X605. The palindromes in the DIS hairpin or in the sequence corresponding to this hairpin are shadowed. Base changes in DIS and SD hairpins (or in the sequences corresponding to these hairpins) compared with Lai isolate are indicated by arrows.

The Secondary Structure of HIV-1 DIS Hairpins

The apical loop of the DIS hairpin contains self-complementary sequence (Figs 2, 4-7), which is required for dimerization of genomic RNA. According to the kissing-loop model ([8], [9]), two genomic RNAs initially interact through base pairing between these self-complementary sequences forming so-called loose dimer ([33]) (Fig. 7A and 7B). And then the DIS hairpin stems isomerize to form the extended inter-strand duplex (stable or tight dimer) (Fig. 7C). The extended duplex is believed to be formed by nucleocapsid (NC) protein during virion maturation ([34]). The other two specific features of the DIS hairpin are purines in the apical loop, which flank the palindrome, and a stem defect that has two alternative forms (Fig. 2A). These forms differ in free energy by only 0,1 kcal/mol ([23]).

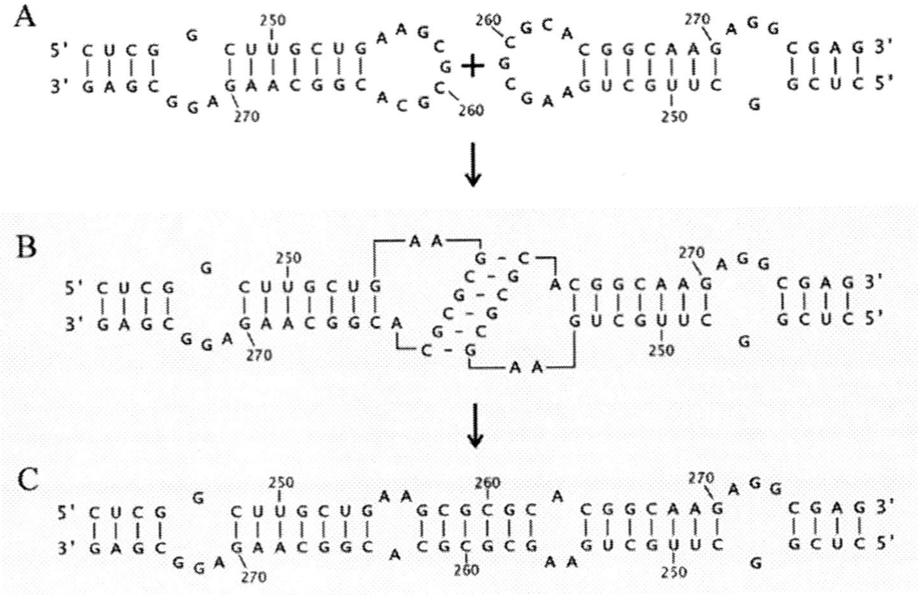

Fig. 7. Dimerization of HIV-1_{Lai} DIS hairpin. (A) DIS hairpins; (B) Loose dimer; (C) Extended duplex.

The nucleotide sequences of the regions corresponding to DIS hairpins are diverse in genomic RNAs of group M isolates. However, in most cases base changes lead only to subtle or moderate alterations in the hairpin structure (see for example Figs 5, 6B, and 8A-C). In particular, in most of the Botswanian isolates, contrary to Lai isolate, the nucleotide flanking the palindrome at the 3′-end is U-residue (but not a purine). This residue forms an additional base pair, which extends the stem and shortens the apical loop (Fig. 5).

Fig. 9 demonstrates sequence variations in the DIS, SD and ψ stem-loops of group M isolates in comparison with Lai isolate. Base changes in DIS hairpin stem occur rather seldom except for the substitutions of G^{23} (Fig. 9A). In many viral isolates this base is replaced by A^{23} forming U^{11}-A^{23} base pair. Among other base changes, substitutions A^{27} for G^{27} in the bulge and G^{35} for A^{35} in the bottom pair of DIS hairpin stem are the most frequent events.

Our findings on sequence variation in DIS hairpin are qualitatively consistent with the data of Yuan et al. ([23]) who surveyed 627 sequences of DIS hairpins from GenBank. However, by our evaluation the frequency of base changes in positions 27 or 35 is approximately twice as high as reported by Yuan et al. ([23]).

In mostly cases the change A^{27} to G^{27} results in substitution of the AG bulge by the GG bulge, while substitution G^{35} for A^{35} leads not only to destruction of the bottom base pair in the DIS hairpin stem but also to destruction of other base pairs and formation of truncated DIS-like hairpins (for example, Fig. 8D). Only in one out of 6 viral isolates with such mutation the DIS-like hairpin shortened by a single base pair (Fig. 8C) is formed along with the other forms.

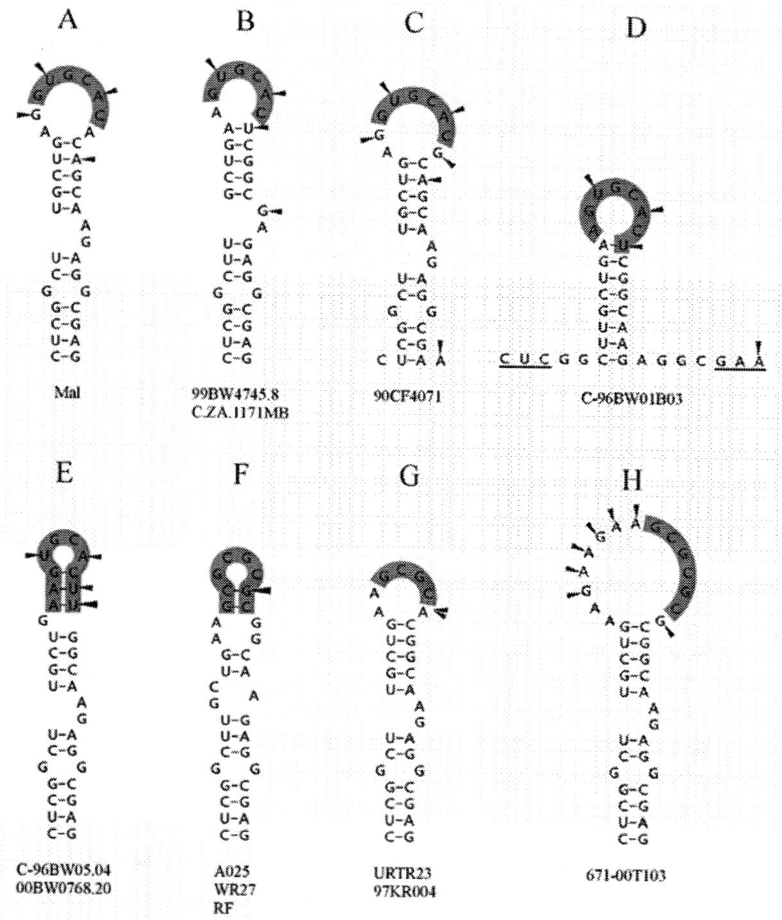

Fig. 8. Patterns of DIS-like hairpins folding in genomic RNA of HIV-1 group M isolates predicted by mfold program. The accession numbers of the isolates are given in brackets after the isolate name. Base changes (in comparison with Lai isolate) are indicated by arrows. The palindromes are shadowed. The underlined nucleotides in pattern (D) are base-paired with other nucleotides of the region preceding GAG start codon.

Other positions in DIS hairpin stem vary at almost three times, if any (Fig. 9A). In particular, the change C^{22} to U^{22} in DIS hairpin stems of isolates C-96BW05.04 and 00BW0768.20 results in crowning of the stems by tetraloop (Fig. 8E). In these isolates six-base palindrome appeared not to be exposed in the apical loop of the DIS hairpin. But during "breathing" of the upper double helical segment of the hairpin this palindrome may be temporally exposed.

The hairpins with tetraloop crowning the DIS hairpin stem are possibly able to form kissing-loop complex but with less efficiency than common DIS hairpins. Involvement of six-base palindrome in a secondary structure also occurs in some DIS hairpins with base changes in the apical loop (for example Fig. 8F). It is worth noting that the four-base palindromes are still exposed in the apical loops in the hairpins shown in Fig. 8E and 8F. Possibly the palindromes of such length are able to form a kissing-loop complex as it was proposed for dimerization of Moloney murine virus genomic RNA ([35]). DIS-like hairpins, which have no six-base palindrome but contain four-base selfcomplementary sequence (for example, the isolates URTR23 and 97KR004, Fig. 8G) may also dimerize through such short palindromes. Using the original experimental system to investigate the dimerization process, Windbichter et al. ([36]) have recently demonstrated that RNAs, which contain the sequences with strongly reduced complementarity in the apical loops, are still capable to dimerize although with greatly diminished efficiency.

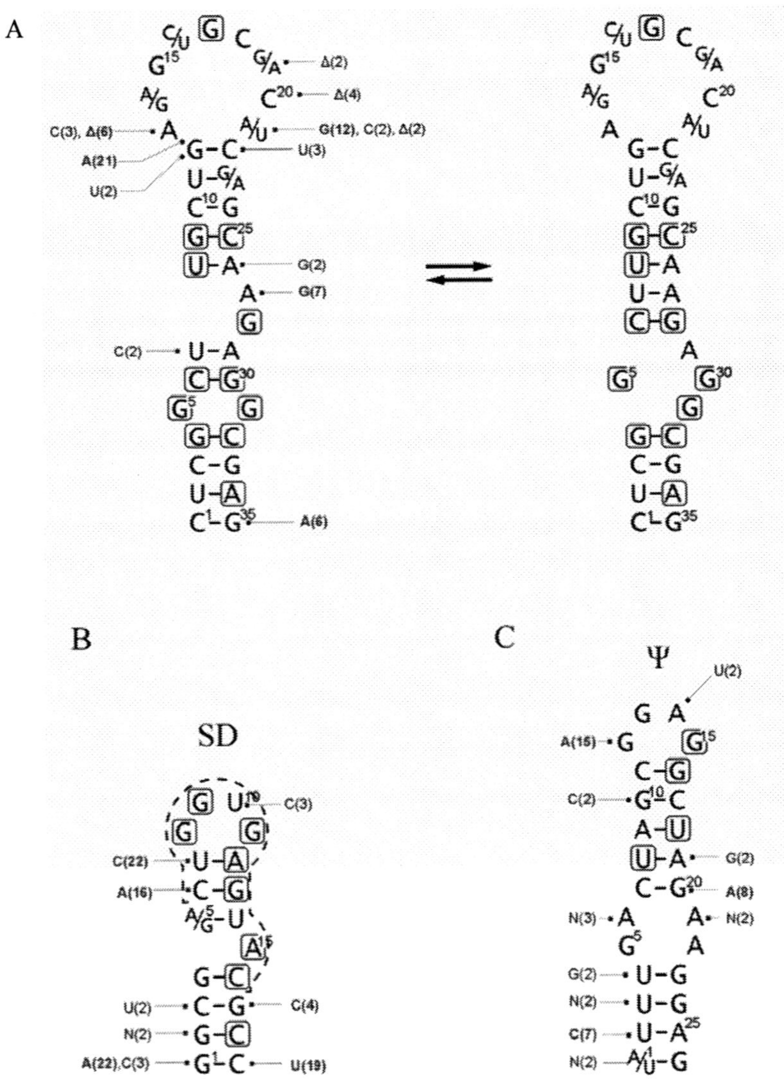

Fig. 9. Variable bases in DIS, SD and ψ hairpins in genomic RNA of HIV-1 group M isolates in comparison with Lai isolate. Only those changes are given that occur at least twice among isolates surveyed. The number of times a given nucleotide is found at a given position is indicated in brackets. N indicates the cases when a position is changed more than once but by different bases. Bases in boxes are absolutely conserved. (A) DIS hairpin with different forms of the stem defect; (B) SD hairpin, MSD is indicated by dotted line; (C) ψ hairpin.

In group N isolates, neither six-base nor four-base palindromes are exposed in the apical loop of DIS-hairpin (Fig. 3). Probably dimerization of genomic RNAs in these isolates occurs during breathing of the upper part of the DIS-like hairpins and is less efficient than in isolates with exposed palindromes.

The only alteration in the primary structure of DIS hairpin, which is not illustrated in Fig. 9A, is a large insertion (AAGAAG) in its apical loop. This insertion is present in genomic RNAs of all seven isolates from Netherlands recently submitted to NCBI. Mfold program predicts an unaltered structure of the DIS hairpin stem in these isolates, while the apical loop is greatly enlarged (Fig. 8H).

In sum, 19% of group M isolates surveyed have DIS-like hairpins which evidently function with lower efficiency (if any) because of great alterations in their secondary structure, absence of six-base palindrome or/and preferable folding of the region preceding GAG start codon into the structure that does not contain this hairpin. In 8% of group M isolates, base changes leading to inhibition of dimerization efficiency occur at least twice; that is probably higher than the rate of sequencing and data base errors. So our phylogenetic data entirely agree with the current opinion of many authors that though DIS hairpin is rather essential for dimerization of HIV-1 genomic RNA, it is not a necessary element (reviewed in ([37]) and ([38])). It is generally thought that the association of genomic RNAs through DIS hairpins mainly initiates the dimerization, while in mature virions the genomic RNAs are also held together through other locations or even through other locations solely. This opinion is mainly supported by the facts, that *in vivo* dimer-monomer ratio does not greatly decrease in mutated viruses containing disrupted or deleted DIS hairpin and the dimers extracted from the mutants exhibit thermal stability comparable with that from wild type HIV-1 ([37]).

The Apical Loop of DIS Hairpin in Group M Isolates

Phylogenetic data on structure of DIS-like hairpins in HIV-1 isolates from group M are summarized in Table 1, where we listed those hairpins which were observed at least twice among the viral isolates surveyed. As seen from the table, only three six-base palindromes are found in the apical loops of DIS hairpins: GUGCAC, GCGCGC or GUGCGC. The isolates with DIS-like hairpins occurring once (not listed in Table 1) contained either the same palindromes or the sequences differing from these palindromes in a single position, for example, the GUGCAA sequence (the isolate C.96BW06.H51).

Laughrea et al. ([40]) considered that the GUGCGC palindrome is found only in 2% of viral isolates merely because it is a transition sequence linking GCGCGC to GUGCAC via single-point mutations. Thus, formation of the kissing-loop complex between HIV-1 genomic RNAs occurs practically through either GUGCAC or GCGCGC self-complementary sequences. Such conservation of the palindromes is rather surprising because model RNAs with some other palindromes can effectively dimerize both in vitro and in vivo, for example RNAs with GGGCCC and GGUACC self-complementary sequences ([40]). The latter palindrome is contained in the apical loop of DIS hairpin in HIV-2 RNA ([41]).

Table 1. DIS hairpins in genomic RNA of HIV-1 group M isolates

No	DIS hairpin		The number of isolates with the hairpin	Subtype (the number of isolates of each subtype is given in brackets)
	Apical loop	Stem		
1	A[GUGCAC]	As in Lai except for an additional A-U at the top of the stem (Fig. 5)	96	C(89), IR*(7)
2	–//–	As in hairpin 1 except for the bulge AG substituted for GG	3	C(3)
3	–//–	See Fig. 8B	2	C(2)
4	–//–	As in Fig. 8D, truncated	2	C(2)
5	AA[GCGCGC]A	As in Lai	54	B(35), NI**(6), B/F(4), D(3), A(2), D/F(2), others (2)
6	–//–	As in Lai except for U^{11}-G^{23} substituted by U^{11}-A^{23}	17	IR(8), B(7), D(2)
7	AG[GUGCAC]A	–//–	46	IR(31), A(8), E(3), G(2), NI(2)
8	–//–	As in Lai	3	A(2), IR(1)
9	AA[GUGCAC]A	–//–	13	IR(4), F(4), C(3), others (2)
10	–//–	As in hairpin 6	4	H(2), IR(1), NI(1)
11	AAG[GUGCAC]A	–//–	9	B/G(6), others (3)
12	AAGAAGAA[GCG CGC]G	As in Lai	7	NI(7)
13	AA[GUGCAC]	As in hairpin 1	6	C(5), IR(1)
14	AG[GUGCAC]G	As in hairpin 6	5	IR(3), G(2)
15	AA[GUGCGC]A	As in Lai	5	IR(3), B(2)
16	[GCGC]	See Fig. 8G	3	B(2), IR(1)
17	CA[GCGCGC]A	As in Lai	2	IR(2)
18	A[GCGCGC]A	–//–	2	IR(1), D(1)
19	–//–	As in hairpin 6	2	IR(2)
20	AA[GCGC]A	As in Lai	2	IR(2)
21	A[GCGC]AC	As in hairpin 1	2	C(2)
22	[UGCA]	See Fig. 8E	2	C(2)
23	Others		39	-

* IR – intersubtype recombinant

** NI – subtype is not indicated

Such base changes can be expected from sequencing and database errors ([23]).

DIS hairpins with four-base palindrome or with tetraloop in the apical region have been discussed in the previous section. The apical loop of DIS hairpin 21 (A[GCGC]AC) is formally presented in Table 1 as one containing four-base palindrome. Most likely, the GCGCAC sequence in this loop resulted from a single-point mutation, either in G*U*GCAC or GCGC*G*C palindromes (mutated bases are underlined). The first way is more probable since isolates with DIS hairpin 21 belong to the same HIV-1 subtype C like those with A[GUGCAC] DIS hairpin (see the last column in Table 1). Model RNAs with the GCGCAC sequence in DIS hairpin apical loop did not form homodimers in vitro but efficiently formed heterodimers with RNAs with the native, rarely occurred, GUGCGC palindrome ([8]). We suppose HIV-1 viruses with the GCGCAC sequence in the apical loop of the DIS hairpin can survive due to heterodimerization of their genomic RNAs with mutant RNAs having GUGCGC palindrome.

Many viral isolates surveyed in this work have DIS hairpins with eight-base palindrome and several isolates contain the longer self-complementary sequences. But in all these cases, the whole palindrome is not exposed in the apical loop (see for example Figs 5 and 8D-F).

We observed that several DIS hairpins preferably occur in genomic RNAs of certain HIV-1 subtype, i. e., these hairpins are subtype specific (Table 1). For example, 93% isolates with the A[GUGCAC] apical loop in the DIS hairpin belong to HIV-1 subtype C (Table 1, lines 1-4, and the isolates omitted in Table 1), while 67% isolates with the AAG[GUGCAC]A loop in the DIS hairpin belong to subtype B/G (Table 1, line 11). But in other cases, identical DIS hairpins are found in isolates belonging to different virus subtypes (for example, Table 1, line 9).

As it was already mentioned, the second specific feature of the HIV-1 DIS hairpin is purine residues flanking the palindrome in the apical loop. In vitro experiments ([42]) demonstrated that a single adenine residue can be removed from either side of the palindrome with no effect on dimerization in RNA constructs containing DIS hairpin identical to that in the isolate Lai (Fig. 2A). However, deletion of all three adenine residues abolished dimer formation. As seen in Table 1, one to three purines flank the palindrome at the 5′ side in the apical loops of all DIS hairpins which are not crowned by tetraloops except for two DIS hairpins, the palindrome being flanked by eight purines in the first case (Table 1, line 12) and by the CA dinucleotide in the second (Table 1, line 17). At the 3′ side all the palindromes are flanked by one purine residue or not flanked at all (the apical loop in the DIS hairpin 21 is considered to contain a mutated palindrome).

In most of viral isolates that are not presented in Table 1, the palindromes are flanked by purines likewise in DIS hairpins described above. So, the purine flanking the palindrome at the 5′ side was found in an overwhelming majority of the isolates surveyed. According to the models based on X-ray crystallography, NMR spectroscopy or/and dynamic studies this purine (A^{14}/G^{14}, Fig. 9A) plays a special role in the formation both of kissing-loop complex and extended duplex. In particular, Pattabiraman et al. ([43]) suggested that the intramolecular tertiary interactions of this purine and 3′ stacking of the bases in the apical loop of the DIS hairpin keep the loop almost in the productive A-RNA form. This enables the loop to form a stable kissing stem with another loop. The authors also suggested that conversion of the kissing-loop complex to a stable extended duplex occurs without unwinding the stems by a strand exchange mechanism and linker nucleotide A^{14}/G^{14} is involved in the cleavage reaction. A little earlier, a strand exchange model for the extended duplex formation has been proposed by Ennifar et al. ([21]). By this model, the purine A^{14}/G^{14} is also involved in the cleavage reaction, but the authors proposed another mechanism of kissing-loop-to- duplex transition.

It should be noted that in vitro experiments unambiguously demonstrated that the formation of the extended duplex occurs via melting of the DIS stems but not via a strand exchange ([19], [36], [44]). However, all these experiments were fulfilled with short RNA constructs and a conversion mechanism for long viral RNAs may be different ([21], [43]).

Our phylogenetic data evidence that many isolates (for example Table 1, lines 1-4) have the seven-base apical loop where the single purine flanks the palindrome (at the 5′ side).

Ennifar et al. ([21]) supposed that the crystal structures of the RNA dimerization initiation sites for Lai and Mal isolates are also relevant to RNAs with seven-base apical loop in DIS hairpin. By their opinion, supressing A^{13} and A^{21} (Fig. 9A) would not result in any change in the loop-loop interaction or in the duplex structure.

In some models of kissing-loop complex ([45]-[47]), the first and the last nucleotides in nine-base apical loop form a non-canonical base pair, whereas in other models they do not ([15], [21], [48]). According to our phylogenetic data in rare cases one of the terminal nucleotides in the apical loop of DIS hairpin is represented by C residue (for example, Table 1, line 17). Lodmell et al. suppose that this residue is also capable of forming base pairs closing the apical loop of DIS hairpin so called sheared A·C and C·A base pairs ([49]).

The length of the apical loop in DIS hairpins of the major group isolates appeared not to exceed 10 nucleotides (Table 1 and other isolates not shown in this table) except for Netherlands isolates (Table 1, line 12). Only 13 isolates have four purine residues flanking the palindrome (for example, Table 1, line 11). According to Pattabiraman et al. ([43]), this number exceeds the optimal number of the linker nucleotides in the apical loop of DIS hairpin. The authors supposed that in case of more than three linker nucleotides with six or seven nucleotides forming the kissing stem the conformational flexibility of the loop distorts the conformation of the loop from A-RNA-like conformation required for the formation of a stable kissing stem. Possibly the genomic RNAs of the isolates with four purine residues flanking the palindrome (Table 1, line 11) dimerize with less efficiency than those containing 1-3 linker residues in the apical loop of DIS hairpin. The isolates with four purine residues flanking the palindrome are mainly from Spain.

The apical loops of DIS hairpins in Netherlands isolates appeared to be unexpectedly long (Fig. 8H). It would be interesting to learn how such a length can affect the dimerization process provided that the six-base insertion in these loops is not generated by error.

As we have already mentioned, the structures of DIS hairpins of some HIV-1 isolates are country specific. We surveyed, on the whole, 350 isolates from 46 countries. However in many cases only 1-3 HIV-1 genomes from each country, which contain the 100 nt sequence preceding GAG start codon, are presented in NCBI database. Moreover, country origin of some isolates is not specified. In Table 2, we present the apical loops of the DIS hairpin in genomic RNAs of HIV-1 isolates from several countries provided that a number of the isolates exceeds ten. As particularly seen from this table, the A[GUGCAC] apical loop crowns the stem of DIS hairpin in most isolates from Botswana, South Africa and India. On the contrary, DIS hairpins of viral isolates from Democratic Republic of Congo (DRC) and Cameroon are not homogeneous. Such diversity can be explained by the longest time of HIV-1 epidemics and therefore the widest variety of HIV-1 strains in these countries ([50]).

Although SD and ψ hairpins are not so diverse as DIS hairpins, we also observed a certain country specificity of their structures. Most isolates from Botswana (92%), South Africa (86%) and India (100%), in whole 111 isolates, have U^1 to A^1 change in the ψ hairpin (Fig. 5). This base change was also observed in 18 isolates of group M from other countries, 14 isolates containing the DIS hairpin of Botswanian type, i. e., with A[GUGCAC] sequence in the apical loop. The most frequent alteration in the SD hairpin is substitution A^5 to G^5, which was observed in 67% of Botswanian, 47% of South African and 7% of Indian isolates.

About 70% of all isolates with A^5 to G^5 change in SD hairpin are originated from Botswana and South Africa.

Stem Defect of DIS Hairpin

Our phylogenetic findings and Yuan's data ([23]) evidence a high conservation of the nucleotides forming the asymmetrical (1x3) internal loop in DIS hairpin stem (Fig. 9A, right) and base pairs flanking this loop. The other form of the defect (Fig. 9A, left) includes the bulge and the symmetrical (1x1) internal loop. Nucleotides of 1x1 internal loop and the base pairs flanking this loop are also well conserved. On the other hand, as it was already mentioned above, the bulge sequence is not highly conserved, the secondary structure of the bulge remaining unaltered.

As many as 19 isolates from group M have the DIS hairpin with truncated stem or the stem defect different from that in the isolate Lai, and only in 8 from these 19 isolates base changes leading to these alterations occurred more than once. So the secondary structure of the stem defect is rather conserved, especially that of the 1x1 loop.

Mechanism of the DIS stem defect functioning in dimerization process is not completely elucidated at present. Shen et al. ([51]) showed that mutations in the region of the defect impacted slightly in vitro dimerization of some partial HIV-1 RNAs with length of several hundreds nucleotides but inhibited dimerization of the whole genomic RNAs in vivo. The authors suggested that the lower part of DIS hairpin stimulates dimerization only in RNAs longer then the partial transcripts and/or in the presence of unidentified viral and cellular factors. In particular, it was shown that nucleocapsid protein binds specifically to 1x3 loop although its affinity to this site is not so high as to ψ and SD hairpins ([52]). Intermediated affinity of NC to 1x3 internal loop can be caused by low accessibility of loop guanosine residues for binding to NC, since according to NMR studies they are partially ordered and G^5 (Fig. 9A, right) forms a non-canonical base pair with A^{29} ([23], [24]). Binding of NC to the intermediated affinity sites is favorable during virion maturation when a large excess of the protein is produced. As it was mentioned above, at this stage of the viral replication cycle nucleocapsid protein is supposed to promote conversion of kissing loop complex into the extended dimer.

Takahashi et al. ([53]) and Yuan et al. ([23]) suggest that the internal 1x3 loop of the DIS stem may play a direct role in the NC-assisted conversion of loose kissing-loop dimer into tight dimer. Takahashi et al. ([53]) suppose that the presence of the internal loop is important in adjusting the stability of the DIS stem to make the conformational conversion possible. The authors demonstrated that NC did not convert DIS construct with highly stable stem from the kissing-loop dimer to the extended duplex. Yuan et al. ([23]) proposed a mechanism for the formation of the extended duplex involving branch migration. In their model the presence of NC bound at the DIS stem defect may accelerate the rate of the dimer transformation.

Table 2. Country specificity of apical loops in DIS hairpins of HIV-1 isolates

Country	The number of	Apical loop	Percentage of the isolates with the apical
Botswana	72	A[GUGCAC]	89
		AA[GUGCAC]	6
		AA[GUGCAC]A	4
		others	1
South Africa	36	A[GUGCAC]	81
		AA[GUGCAC]	6
		A[GCGCAC]	6
		others	7
USA	26	AA[GCGCGC]A	77
		A[GUGCAC] (group O)	8
		others	15
Cameroon	23	A[GUGCAC] (group O)	65
		AG[GUGCAC]A	13
		A[GCGCGC]A	9
		others	13
Spain	15	AAG[GUGCAC]A	53
		AA[GCGCGC]A	47
India	14	A[GUGCAC]	100
Democratic Republic of Congo	12	AA[GCGCGC]A	42
		AG[GUGCAC]A	25
		AA[GUGCAC]A	25
		others	8
Uganda	12	AG[GUGCAC]A	75
		AA[GCGCGC]A	25
Thailand	11	AG[GUGCAC]A	82
		AA[GCGCGC]A	18

[*] With the identical loop or the loop with one base change

SD Hairpin

HIV-1 major splicing sites (MSD) somewhat vary in genomic RNAs of HIV-1 groups M, N and O and located in the hairpins with quite different secondary structures (Figs 2 and 3). In group M isolates, MSD (CUG/GUGAGUAC, ([54])) encompasses the upper part of SD hairpin (Fig. 2A). The main features of this hairpin are the GGUG tetraloop crowning its stem and a bulge in the stem. As recent studies showed the stem has, in fact, a more complex structure ([25]). A^{15} residue does not form a predicted bulge and the residues A^5 and U^{14} do not form a Watson-Crick base pair as it is shown, for example, in Fig. 9B. But, according to the NMR study ([25]), these residues form a novel A^5-U^{14}-A^{15} base triple platform.

However, in 28% of group M isolates the SD hairpin has A^5 to G^5 base change and therefore A-U-A triple platform is not formed in such hairpin. Interestingly, about 70% of the isolates with A^5 to G^5 base change in the SD hairpin contain the DIS hairpin with the A[GUGCAC] apical loop. In regard to HIV-1 subtype, we found that about 80% of isolates with A^5 for G^5 substitution in the SD hairpin belong to subtype C.

Remarkably, among six isolates with AA[GCGCGC]A apical loop in the DIS hairpin (variant Lai), which have A^5 to G^5 base change in SD hairpin, four isolates with the code WCIPR were derived from the same patient who underwent a transition from long-term non-progressive to rapidly progressive infection. Fang et al. ([55]) cloned and sequenced the HIV-1 RNA genomes derived from this patient in 1985 (before the transition) and in 1990 (during the transition in disease states) and submitted five sequences corresponding to each year to NCBI. In all viruses derived in 1985, the region encompassing DIS, SD and ψ hairpins was identical to that in isolate with the accession no. NC_001802. But in four out of five viruses derived in 1990 the SD hairpin had A^5 to G^5 base change, while DIS and ψ hairpins were not altered. Two of these four isolates had additional base change in SD hairpin (G^{17} to C^{17}). Under the double mutation (in positions 5 and 17), the stem of SD hairpin appeared to be greatly altered (Fig. 10A), the apical loop remaining the same. Single base change (A^5 to G^5) in the SD hairpin also had an effect on SD hairpin folding in isolates with the code WCIPR. The SD hairpin was split into two small hairpins, the first including a part of linker (ggc) (Fig. 10B). The combination of G^5 in SD hairpin and GGGC linker adjacent to SD hairpin resulted in splitting in all genomic RNAs with this combination (7 isolates, including those with the code WCIPR). However, conversion of the SD hairpin into two hairpins is not particularly favorable, the free energy of a two-hairpin structure being 1,2 kcal/mol higher than that of a common structure.

Furthermore, base change A^5 to G^5 was badly compatible with the DIS hairpins identical to those in the isolate Mal (Table 1, line 7). The combination of G^5 in SD hairpin and DIS hairpin of Mal variant always led to coexistence between common and alternative structures of the region preceding GAG start codon in HIV-1 genomic RNA (for example, Fig. 11). Although either SD hairpin with A^5 to G^5 base change or DIS hairpin of Mal variant frequently occur among isolates surveyed, their combination is very rare (four among 327 group M isolates) possibly due to alternative structure formation.

A^5 to G^5 is the most frequent base change in the SD hairpin of HIV-1 group M isolates. Other frequent changes are substitution of the C^6U^7 dinucleotide for A^6C^7 and substitution of the G^1-C^{19} base pair for A^1-U^{19}. The first substitution led to the alteration of the hairpin secondary structure, the apical loop being elongated by one base, and 1x1 internal loop being formed in the stem instead of the bulge (Fig. 10C). The SD hairpin with such structure was found in 14 isolates from group M, particularly in 6 out of 8 Tanzanian isolates. Almost all isolates with C^6U^7 to A^6C^7 substitution have the DIS hairpin identical to that in the isolate Lai or altered in a single base. It is worth noting, variable nucleotides 6 and 7 are located at the beginning of the major splicing site, the rest of the positions in MSD are highly conserved (Fig. 9B).

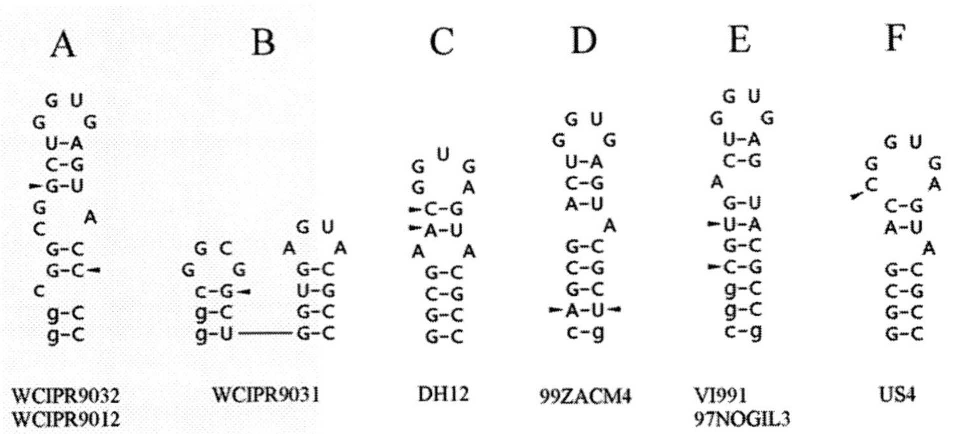

Fig. 10. Patterns of SD-like hairpins folding in genomic RNA of HIV-1 group M isolates predicted by mfold program. Base changes (in comparison with Lai isolate) are indicated by arrows. Nucleotides commonly located in the linker regions are given in lower-case letters.

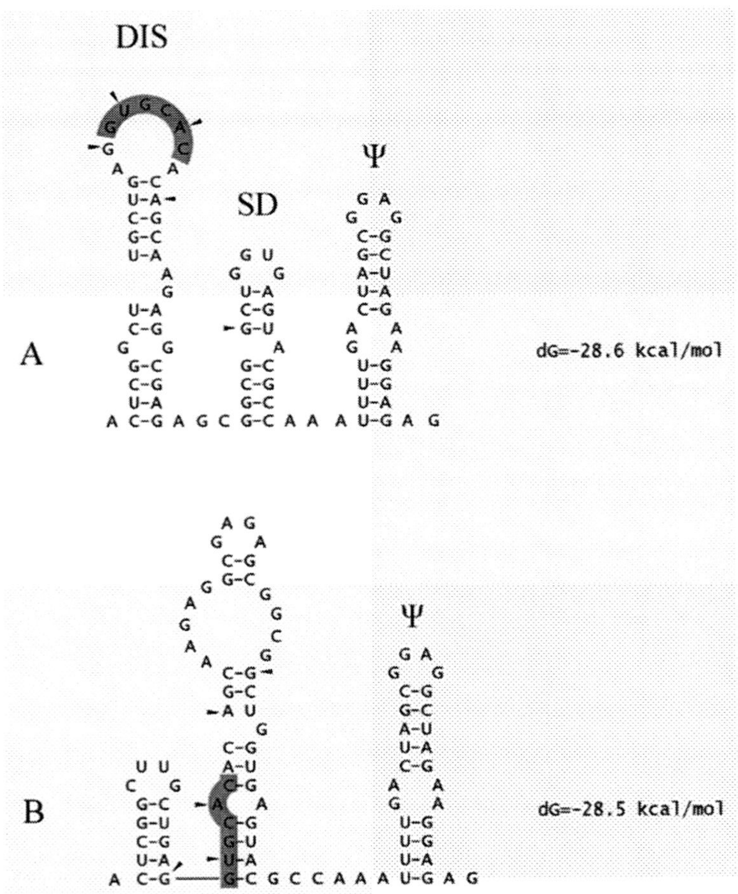

Fig. 11. The most favorable secondary structures of the region preceding GAG start codon in genomic RNA of the isolate 99TH.R2399 predicted by mfold program. The palindromes in the DIS hairpin and in the sequence corresponding to this hairpin are shadowed. Base changes in DIS and SD hairpins or in the sequences corresponding to these hairpins compared with Lai isolate are indicated by arrows.

The substitution of the G^1-C^{19} base pair for A^1-U^{19} not altering the SD hairpin structure was found in 17 isolates from group M. Besides, in about 60% of these isolates the SD hairpin was elongated by one base pair formed by adjacent linker nucleotides (for example, Fig. 10D). SD hairpin elongation due to flanking nucleotides was also found in other isolates, 9% of group M isolates have an elongated SD hairpin.

Fig. 10E and 10F demonstrates other SD-like hairpins with great structural alterations in the stem or the apical loop, which occur at least twice among group M isolates surveyed, that is probably higher than the rate of sequencing and data base errors. In sum, about 20% of group M isolates have base changes in the region preceding GAG start codon which led to great alterations in the SD hairpin structure, folding of this region into an alternative structure without the SD hairpin or coexistence between common and alternative structures. In 11% of these isolates, such base changes occur more than once. As reported above, corresponding percentages for the DIS hairpin are slightly lower (19% and 8%). Moreover, considering that group O and N isolates have no SD-like hairpins one may conclude that the SD hairpin is less conserved than the DIS hairpin.

SD and ψ hairpins represent the most preferable sites for nucleocapsid protein binding in the 5′-terminal region of HIV-1 genomic RNA ([52]). These two hairpins have similar apical loops (GGUG tetraloop in SD and GGAG tetraloop in ψ) and different stems. The structures of these tetraloops significantly differ from those of the GNRA class. G residue located at the 5′ end of each tetraloop is stacked above the last base pair of the respective stem and the remaining tetraloop residue are disordered to some extent and available for potential interaction with either RNA or protein compound ([25], [29], [56]). Nucleocapsid protein interacts similarly with the apical loops of SD and ψ hairpins but differently with the stems of these hairpins ([26]). As reported by Amarasinghe et al. ([26]), NC interacts with A-U-A base triple platform of the SD hairpin stem. The authors suppose that NC binds in an adaptive manner to SD and ψ hairpins via different subsets of inter and intra-molecular interactions. Possibly this protein also adapts to SD-like hairpins with altered structure (Fig. 10A, C, E, and F) and to frequently occurring hairpins which have no potential to form A^5-U^6-A^7 base triple platform because of A^5 to G^5 change in the hairpin stem (Fig. 5).

ψ Hairpin

The ψ hairpin is considered to be the major determinant for the specific package because its linkage to heterologous RNAs is sufficient to direct their recognition and packaging into virus-like particles ([11]). However, ψ hairpin alone is not sufficient to target RNA into HIV-1 virions, it functions together with other signal elements ([13], [57]). Some of these elements are located in DIS, ψ and stem-loop 4 (SL4) ([13]) and are very likely in SD ([26]). SL4 is not formed in the secondary structure of the 5' end of HIV-1 genomic RNA, as presented by Abbink and Berkhout ([5]). We consider this structure as the most realistic since an alternative structure with the elongated DIS hairpin and SL4 proposed by Zeffman et al. ([29]) and Harrison et al. ([32]) cannot be formed in those numerous isolates which do not contain GC sequence at the 3' end of the linker between DIS and SD hairpins (Table 3). Therefore, SL4 is not being discussed here. The other packaging signal elements are located upstream of DIS hairpin: in TAR, poly(A) and PBS hairpins ([13], [57]). Unlike DIS, SD, and ψ, these elements bind to NC domain of the GAG protein with low affinity and contribute to packaging through a mechanism distinct from these hairpins.

Table 3. Linkers between DIS and SD hairpins in genomic RNA of group M isolates*

No	Linker	Percentage of the isolates with the linker
1	AGC	47
2	GGGC	31
3	AGG	8
4	GGG	2
5	GGGU	2
6	GGGAC	2
7	AGA	1
8	Absent	1
9	AGGC	1
10	GGGA	1
11	Others	4

* Linkers are formally presented as the stretches between the sequences corresponding to DIS and SD hairpins. In fact, they can be shortened due to involvement in elongation of the SD hairpin or elongated due to base changes in the SD hairpin leading to alterations of its structure.

In the secondary structure of the 5′-terminal region of HIV-1 genomic RNA presented by Abbink and Berkhout ([5]) the stem defect of ψ hairpin represents 2x2 symmetrical internal loop (Figs 2A and 9C). Based on Damgaard et al. data ([58]), G^5 residue and residues 23 and 24 adjacent to the loop (Fig. 9C) are somewhat protected from RNAse T1 attack when NC protein binds to HIV-1 RNA 5′ leader region. These results provide evidence for specific interacting between NC and the stem defect of ψ hairpin.

Damgaard et al. ([58]), however, depicted ψ hairpin as a short stem-loop structure without an internal loop, which is identical to that in Fig. 5A. Such truncated ψ hairpin is also depicted in many works ([13], [26], [39], and others). The high-resolution structure of nucleocapsid protein bound to ψ hairpin was also determined for truncated variant (nt 7-20, Fig. 9C) elongated by one A-U and two G-C base pairs, which did not contain an internal loop ([27]). On the contrary, ψ element is presented as elongated hairpin with 2x1 asymmetrical internal loop by Zeffman et al. ([29]) and Harrison et al. ([32]). This structure is similar to ψ-like hairpin shown in Fig. 12A except that U^2 residue was not substituted for C. However, as it was mentioned above, the region preceding GAG start codon in genomic RNAs of many HIV-1 isolates have no the potential to fold into the structure proposed in works ([29]) and ([32]). In this structure the nucleotide corresponding to G^{26} of ψ hairpin structure (Fig. 9C) is involved in base pairing with the sequence located upstream of DIS hairpin that prevents formation of 2x2 internal loop. According to our data, the formation of 2x1 internal loop in ψ-like hairpins shown in Figs 12A and 12B was conditioned by base changes in their stems.

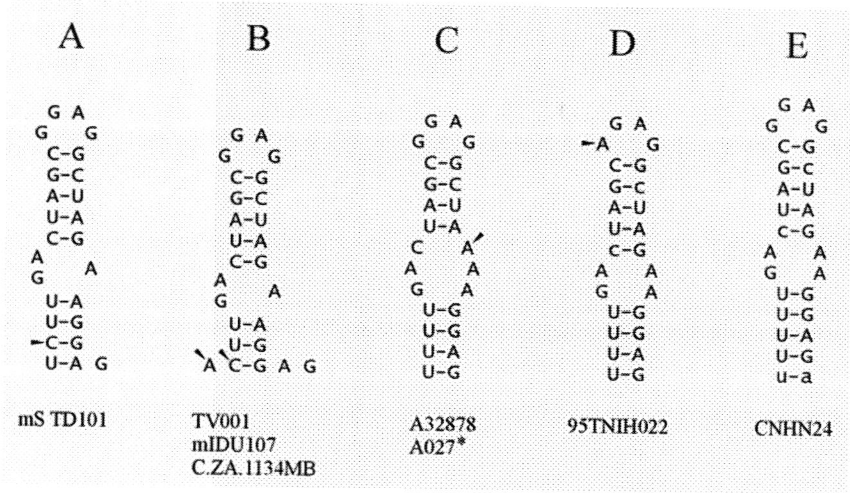

Fig. 12. Patterns of ψ-like hairpins folding in genomic RNA of HIV-1 group M isolates predicted by mfold program. Base changes (in comparison with Lai isolate) are indicated by arrows. Nucleotides commonly located in the linker regions are given in small letters. * The isolate AO27 has the AGAG apical loop in ψ-like hairpin.

Zeffman et al. ([29]) studied structure of the elongated ψ hairpin by NMR spectroscopy. Although the authors reported ψ hairpin structure with 2x1 internal loop, the structure with 2x2 internal loop (Fig. 9C) was possibly formed in the RNA construct under their study, since their NMR data do not contradict the formation of 2x2 loop. The authors showed that each of the purine residues on the 3′-side of the internal loop (nt 20-23, Fig. 9C) is stacked on a neighboring nucleotide and the opposite side of the loop (nt 3-6) is in exchange between different conformations. Besides, they demonstrated an absence of interstrand interactions stabilising loop.

According to our phylogenetic data, 91% of group M isolates surveyed can form the ψ hairpin with 2x2 internal loop as shown in Fig. 9C. In the rest of isolates ψ-like hairpins with altered internal loops, bulges or truncated stems are formed due by mutations. Base changes in ψ hairpin occurring at least twice among isolates studied are shown in Fig. 9C. Positions 1, 2 and 20 in the hairpin stem appeared to be most variable. As it was already noted above U^1 to A^1 base change leads to coexistence of hairpins with truncated stem and the stem shortened only by one base pair (Figs 5A and 5B). About 35% of group M isolates have such ψ hairpins. Base change U^2 to C^2 and G^{20} to A^{20} result in shortening (Fig. 12A and 12B) or elongation (Fig. 12C) of the internal loop, respectively.

The sequence of the apical loop of the ψ hairpin was found to be highly conserved except that 5% of group M isolates had A residue in position 12 (Fig. 9C). This led to substitution of the GGAG apical loop for AGAG, the stem remaining unaltered (Fig. 12D). Our data on base changes in the apical loop of the ψ hairpin agree with findings of Paoletti et al. ([59]) who surveyed 489 HIV-1 sequences deposited in GenBank. By the authors opinion, rather frequent changes G^{12} to A^{12} can be explained by the fact that A-residue in position 12, like G residue, stacks upon the base-paired stem, stabilizing the loop, and does not make specific H-bonding contacts with the protein. Paoletti et al. ([59]) have measured the affinities of different ψ-like constructs for NC protein and showed that G^{15} in G^{13}-N^{14}-G^{15} sequence of the apical loop (Fig. 9C) is the most important nucleotide for high affinity binding. They also showed that NC binding with the wild type GGAG apical loop is not the most stable, G^{13}-U^{14}-G^{15} and G^{13}-G^{14}-G^{15} bind more strongly.

According to our phylogenetic data, A^{14} to U^{14} base change was found only in two isolates from group M. One of them (strain 98ZA528) along with A^{14} to U^{14} base change had a frequently occurring A^5 to G^5 and U^1 to A^1 base changes in SD and ψ hairpins, respectively, and contained frequently occurring the AGC linker between DIS and SD hairpins. Such combinations led to preferable folding of the region preceding GAG start codon into a structure that had neither SD nor ψ hairpin and possibly was inhibitory for replication of this isolate. So, the GGUG tetraloop crowing ψ hairpin is not compatible with the presence of the AGC linker, A^5 to G^5 (in SD) and U^1 to A^1 (in ψ) base changes. Although the GGUG tetraloop is a very rare event in ψ hairpins of group M isolates, all isolates from groups O and N have Ψ-like hairpin with GGUG apical loop.

Like SD hairpins, ψ hairpins are often elongated by one base pair formed by the linker residues flanking the hairpin (for example, see Fig. 12E). As many as 26% of group M isolates have elongated ψ hairpins.

According to our phylogenetic findings ψ hairpin appeared to be the most conserved among hairpins analysed. Only 11% of group M isolates (that is approximately two times less than in case of DIS and SD hairpins) have base changes in the region preceding GAG start codon leading to coexistence between common and alternative structures or great alterations in the ψ hairpin secondary structure. In 3% of group M isolates such base changes occur at least twice, that is approximately three and four times less than in the case of DIS and SD hairpins, respectively.

Linker Sequences

Linker sequences between DIS and SD hairpins are short (not more than 6 nt in length) and rather conserved. As many as 86% of the major group isolates have the AGC, GGGC or AGG sequences in this region (Table 3). The rest of the isolates (except for three ones without linker) also contain one or more G residues in DIS-SD linker region. Contrary to the linker between DIS and SD hairpins, the linker sequence located between SD and ψ hairpins is more heterogeneous in length (nt 1-11) and in sequence, which mainly consists of A and U residues (Table 4). In most cases, diversity of linker sequences does not greatly affect the secondary structure of the region preceding GAG start codon. The main alteration is elongation of SD or/and ψ hairpins (see previous sections) due to base pairing between nucleotides adjacent to these hairpins.

Only in 9 isolates linker nucleotides affected the structure of the region encompassing DIS, SD and ψ hairpins. In two of them, the presence of G residues adjacent to the 5' ends of DIS and SD hairpins led to coexistence between common and alternative structures. It is worth noting that though G residue adjacent to SD hairpin at the 5' end is observed rather frequently in group M isolates (Table 3), only four isolates have this nucleotide in the position adjacent to DIS hairpin at the 5' end, A residue commonly flanking DIS in this position. In the other four isolates, DIS-SD linkers contained the GAG sequence, which formed double helix with the CUC sequence located at the 5' end of the DIS hairpin. This also led to coexistance between common and alternative structures. In the rest of the three cases, base changes in linker sequences affecting the structure of the region encompassing DIS, SD and ψ hairpins occurred only once among all isolates surveyed.

Table 4. Linkers between SD and ψ hairpins in genomic RNAs of the major group isolates*

No	Linker	Percentage of isolates with the linker	No	Linker	Percentage of isolates with the linker
1	AAAUUUU	14	15	GAAAU	2
2	AAAAA	9	16	AUU	2
3	AAUUUUU	9	17	AUUUUU	1
4	AAA	7	18	AAAUUUUU	1
5	AAUUUU	5	19	AAAAAUU	1
6	AAAAU	4	20	AAAAAAUU	1
7	AAAA	3	21	AAAUUUUA	1
8	AAU	3	22	AAAUU	1
9	AUAAA	2	23	AAAAAA	1
10	GAAUUUU	2	24	GAAA	1
11	AAAAAAU	2	25	GAAAA	1
12	AAAAAU	2	26	AAAAUUU	1
13	UAAAA	2	27	Others	20
14	AAAAUUUU	2			

* Linkers are formally presented as the stretches between the sequences corresponding to SD and ψ hairpins. In fact, they can be shortened due to involvement in elongation of SD or/and ψ hairpins or elongated due to base changes in SD or/and ψ hairpins leading to alterations of their structures.

According to Abbink and Berkhout ([5]), the linker region between the ψ hairpin and U5-AUG duplex is very short in group M isolates; it represents only two nucleotides (AG). The AG AGAUGGGUGCG sequence consisting of the AG linker and the downstream strand of U5-AUG duplex is highly conserved. Each residue in this sequence is changed not more than three times (if any). Phylogenetic analysis showed that the U5-AUG duplex is highly conserved in HIV-1 isolates from groups O and N. Unfortunately, the sequences corresponding to the upstream strand of this duplex are not available in NCBI database for most of group M isolates surveyed. But it is also highly conserved in 16 isolates for which this sequence was submitted.

Various views on the secondary structure of the hairpin adjacent to the DIS hairpin at the 5'-end are currently known, the linker region between the DIS hairpin and the hairpin containing PBS varies in different works. In particular, there is no linker at all between these hairpins in some diagrams of the 5'-terminal region of HIV-1 genomic RNA (for example ([51])). According to recent data of Damgaard et al. ([60]), the linker is the GA dinucleotide, while it is GCAGGA in Abbink and Berkhout work ([5]). The hexamer *GCAGGA* (and therefore the GA dinucleotide) appeared to be rather conserved in group M isolates. G^1, C^2

and G^5 residues of the hexamer (underlined) are absolutely conserved, while the others are changed not more than 5 times. However, ten isolates have one base insertion between A^3 and G^4 residues.

Conclusion

Phylogenetic analysis of the region of HIV-1 genomic RNAs preceding GAG start codon, which is folded into three well-known hairpins (DIS, SD and ψ) in group M isolates, showed that some base changes in this region can lead to great alterations in the secondary structure of these hairpins, folding of the region preceding GAG start codon into the structure without some of these hairpins or coexistence between common and alternative structures. However, in almost all these cases it is possible to suggest that the hairpins still function, but less efficiently. Moreover, in many cases mutations leading to alternation in the structure of HIV-1 genomic RNA region encompassing DIS, SD and ψ hairpins were observed only once among HIV-1 isolates surveyed and most likely were caused by sequence and database errors. This provides evidence that formation of full DIS, SD and ψ hairpins, especially of ψ hairpin, is essential for HIV-1 replication cycle.

References

[1] Huthoff, H; Berkhout, B. Two alternating structures of the HIV-1 leader RNA. *RNA,* 2001, 7, 143-157.

[2] Huthoff, H; Berkhout, B. Mutations in the TAR hairpin affect the equilibrium between alternative conformations of the HIV-1 leader RNA. *Nucleic Acids Res,* 2001, 29, 2594-2600.

[3] Huthoff, H; Berkhout, B. Multiple secondary structure rearrangements during HIV-1 RNA dimerization. *Biochemistry,* 2002, 41, 10439-10445.

[4] Berkhout, B; Ooms, M; Beerens, N; Huthoff, H; Southern, E; Verhoef, K. *In vitro* evidence that the untranslated leader of the HIV-1 genome is an RNA checkpoint that regulates multiple functions through conformational changes. *J Biol Chem*, 2002, 277, 19967-19975.

[5] Abbink, TEM; Berkhout, B. A novel long distance base-pairing interaction in human immunodeficiency virus type 1 RNA occludes the Gag start codon. *J Biol Chem,* 2003, 278, 11601-11611.

[6] Ooms, M; Verhoef, K; Southern E; Huthoff H; Berkhout B. Probing alternative foldings of the HIV-1 leader RNA by antisense oligonucleotide scanning arrays. *Nucleic Acids Res,* 2004, 32, 819-827.

[7] Beerens, N; Groot, F; Berkhout, B. Initiation of HIV-1 reverse transcription is regulated by a primer activation signal. *J Biol Chem*, 2001, 276, 31247-31256.

[8] Skripkin, E; Paillart, J-C; Marquet, R; Ehresmann, B; Ehresmann, C. Identification of the primary site of the human immunodeficiency virus type 1 RNA dimerization *in vitro. Proc. Natl. Acad. Sci. USA*, 1994, 91, 4945-4949.

[9] Laughrea, M; Jetté, L. A 19-nucleotide sequence upstream of the 5′ major splice donor is part of the dimerization domain of human immunodeficiency virus 1 genomic RNA. *Biochemistry,* 1994, 33, 13464-13474.

[10] Sakaguchi, K; Zambrano, N; Baldwin, ET; Shapiro BA; Erickson, JW; Omichinski, JG; Clore, GM; Gronenborn, AM; Appella, E. Identification of a binding site for the human immunodeficiency virus type 1 nucleocapsid protein. *Proc. Natl. Acad. Sci USA*, 1993, 90, 5219-5223.

[11] Hayashi, T; Shioda, T; Iwakura, Y; Shibuta, H. RNA packaging signal of human immunodeficiency virus type 1. *Virology,* 1992, 188, 590-599

[12] Hayashi, T; Ueno, Y; Okamoto, T. Elucidation of a conserved RNA stem-loop structure in the packaging signal of human immunodeficiency virus type 1. *FEBS Lett,* 1993, 327, 213-218.

[13] Clever, JL; Miranda, DJr; Parslow, TG. RNA structure and packaging signals in the 5′ leader region of the human immunodeficiency virus type 1 genome. *J. Virol.*, 2002, 76, 12381-12387.

[14] Greatorex, J; Gallego, J; Varani, G; Lever, A. Structure and stability of wild-type and mutant RNA internal loops from the SL-1 domain of the HIV-1 packaging signal. *J. Mol. Biol.,* 2002, 322, 543-557.

[15] Mujeeb, A; Clever, JL; Billeci, TM; James, TL; Parslow, TG. Structure of the dimer initiation complex of HIV-1 genomic RNA. *Nat. Struct. Biol.,* 1998, 5, 432-436.

[16] Girard, F; Barbault, F; Gouyette, C; Huynh-Dinh, T; Paoletti, J; Lancelot, G. Dimer initiation sequence of HIV-1 Lai genomic RNA: NMR solution structure of the extended duplex. *J. Biomol. Struct. Dyn.*, 1999, 16, 1145-1157.

[17] Mujeeb, A; Parslow, TG; Zarrinpar, A; Das, C; James, TL. NMR structure of the mature dimer initiation complex of HIV-1 genomic RNA. *FEBS Lett.,* 1999, 458, 387-392.

[18] Ennifar, E; Yusupov, M; Walter, P; Marquet, R; Ehresmann, B; Ehresmann, C; Dumas, P. The crystal structure of the dimerization initiation site of genomic HIV-1 RNA reveals an extended duplex with two adenine bulges. *Structure Fold. Des.* 1999, 7, 1439-1449.

[19] Takahashi, K-i; Baba, S; Hayashi, Y; Koyanagi, Y; Yamamoto, N; Takaku, H; Kawai, G. NMR analysis of intra- and inter-molecular stems in the dimerization initiation site of the HIV-1 genome. *J. Biochem.,* 2000, 127, 681-686.

[20] Theilleux-Delalande, V; Girard, F; Huynh-Dinh, T; Lancelot, G; Paoletti, J. The HIV-1 (Lai) RNA dimerization. Thermodynamic parameters associated with the transition from the kissing complex to the extended dimer. *Eur J. Biochem.,* 2000, 267, 2711-2719.

[21] Ennifar, E; Walter, P; Ehresmann, B; Ehresmann, C; Dumas, P. Crystal structures of coaxially stacked kissing complexes of the HIV-1 RNA dimerization initiation site. *Nat. Struct. Biol.*, 2001, 8, 1064-1068.

[22] Kieken, F; Arnoult, E; Barbault, F; Paquet, F; Huynh-Dinh, T; Paoletti, J; Genest, D; Lancelot, G. HIV-1_{Lai} genomic RNA: combined used of NMR and molecular dynamics stimulation for studying the structure and internal dynamics of a mutated SL1 hairpin. *Eur Bioph J.,* 2002, 31, 521-531.

[23] Yuan, Y; Kerwood, DJ; Paoletti, AC; Shubsda, MF; Borer, PN. Stem of SL1 RNA in HIV-1: structure and nucleocapsid protein binding for a 1 x 3 internal loop. *Biochemistry,* 2003, 42, 5259-5269.

[24] Lawrence, DC; Stover, CC; Noznitsky, J; Wu, Z; Summers, MF. Structure of the intact stem and bulge of HIV-1 ψ-RNA stem-loop SL1. *J. Mol. Biol.,* 2003, 326, 529-542.

[25] Amarasinghe, GK; De Guzman, RN; Turner, RB; Summers, MF. NMR structure of stem-loop SL2 of the HIV-1 ψ RNA packaging signal reveals a novel A-U-A base-triple platform. *J. Mol. Biol.*, 2000, 299, 145-156.

[26] Amarasinghe, GK; De Guzman, RN; Turner RB; Chancellor, KJ; Wu, ZR; Summers, MF. NMR structure of the HIV-1 nucleocapsid protein bound to stem-loop SL2 of the ψ-RNA packaging signal. Implications for genome recognition. *J. Mol. Biol.*, 2000, 301, 491-511.

[27] De Guzman, RN; Wu, ZR; Stalling, CC; Pappalardo, L; Borer, PN; Summers, MF. Structure of the HIV-1 nucleocapsid protein bound to the SL3 ψ-RNA recognition element, *Science,* 1998, 279, 384-388.

[28] Pappalardo, L; Kerwood, DJ; Pelczer, I; Borer, PN. Three-dimensional folding of an RNA hairpin required for packaging HIV-1. *J. Mol. Biol,.* 1998, 282, 801-818.

[29] Zeffman, A; Hassard, S; Varani, G; Lever, A. The major HIV-1 packaging signal is an extended bulged stem loop whose structure is altered on interaction with the Gag polyprotein. *J. Mol. Biol.,* 2000, 297, 877-893.

[30] Zuker, M. Mfold web server for nucleic acid folding and hybridization prediction. *Nucleic Acids Res.,* 2003, 31, 3406-3415. (http://www.bioinfo.rpi.edu/applications /mfold).

[31] Robertson, DL; Anderson, JP; Bradac, JA, et al. HIV-1 nomenclature proposal. In: Kuiken, CL; Foley, B; Hahn, B; Korber, B; McCutchan, F; Marx, PA; Mellors, JW; Mullins, JI; Sodroski, J; Wolinksy S. editors. *Human Retroviruses and AIDS 1999: A compilation and Analysis of Nucleic Acid and Amino Acid Sequences.* Los Alamos, NM: Theoretical Biology and Biophysics Group, Los Alamos National Laboratory; 1999; 492-505.

[32] Harrison, GP; Miele, G; Hunter, E; Lever, AML. Functional analysis of the core human immunodeficiency virus type 1 packaging signal in a permissive cell line. *J. Virol.* 1998, 72, 5886-5896.

[33] Laughrea, M; Jetté, L. Kissing-loop model of HIV-1 genome dimerization: HIV-1 RNAs can assume alternative dimeric forms, and all sequences upstream or downstream of hairpin 248-271 are dispensable for dimer formation. *Biochemistry,* 1996, 35, 1589-1598.

[34] Muriaux, D; De Rocquigny, H; Roques, B-P; Paoletti, J. NCp7 activates HIV-1_{Lai} RNA dimerization by converting a transient loop-loop complex into a stable dimer. *J. Biol. Chem.,* 1996, 271, 33686-33692.

[35] Ly, H; Parslow, TG. Bipartite signal for genomic RNA dimerization in Moloney murine leukemia virus. *J. Virol.*, 2002, 76, 3135-3144.

[36] Windbichler, N; Werner, M; Schroeder, R. Kissing complex-mediated dimerisation of HIV-1 RNA: coupling extended duplex formation to ribozyme cleavage. *Nucleic Acids Res.,* 2003, 31, 6419-6427.

[37] Greatorex, J; Lever, A. Retroviral RNA dimer linkage. *J. Gen. Virol.,* 1998, 79, 2877-2882.

[38] Zarudnaya, MI; Potyahaylo, AL; Hovorun, DM. Structural model of dimerization region of human immunodeficiency virus type 1 genome. *Biopolimery i Kletka,* 2003, 19, 37-42 (in Russian).

[39] Laughrea, M; Jette, L. HIV-1 genome dimerization: kissing-loop hairpin dictates whether nucleotides downstream of the 5′ splice junction contribute to loose and tight dimerization of human immunodeficiency virus RNA. *Biochemistry,* 1997, 36, 9501-9508.

[40] Laughrea, M; Shen, N; Jette, L; Wainberg, MA. Variant effects of non-native kissing-loop hairpin palindromes on HIV replication and HIV RNA dimerization: role of stem-loop B in HIV replication and HIV RNA dimerization. *Biochemistry,* 1999, 38, 226-234.

[41] Dirac, AMG; Huthoff, H; Kjems, J; Berkhout, B. The dimer initiation site hairpin mediates dimerization of the human immunodeficiency virus, type 2 RNA genome. *J. Biol. Chem.*, 2001, 276, 32345-32352.

[42] Clever, JL; Wong, ML; Parslow, TG. Requirements for kissing-loop-mediated dimerization of human immunodeficiency virus RNA. *J. Virol.,* 1996, 70, 5902-5908.

[43] Pattabiraman, N; Martinez, HM; Shapiro, BA. Molecular modeling and dynamics studies of HIV-1 kissing loop structures. *J. Biomol. Struct. Dyn.,* 2002, 20, 397-411.

[44] Rist, MJ; Marino, JP. Mechanism of nucleocapsid protein catalyzed structural isomerization of the dimerization initiation site of HIV-1. *Biochemistry,* 2002, 41, 14762-14770.

[45] Paillart, JC; Westhof, E; Ehresmann, C; Ehresmann, B; Marquet, R. Non-canonical interactions in a kissing loop complex: the dimerization initiation site of HIV-1 genomic RNA. J. *Mol. Biol.,* 1997, 270, 36-49.

[46] Dardel, F; Marquet, R; Ehresmann, C; Ehresmann, B; Blanquet, S. Solution studies of the dimerization initiation site of HIV-1 genomic RNA. *Nucleic Acids Res.,* 1998, 26, 3567-3571.

[47] Jossinet, F; Paillart, JC; Westhof, E; Hermann, T; Skripkin, E; Lodmell, JS; Ehresmann, C; Ehresmann, B; Marquet, R. Dimerization of HIV-1 genomic RNA of sybtypes A and B: RNA loop structure and magnesium binding. *RNA,* 1999, 5, 1222-1234.

[48] Réblová, K; Špačková, N; Šponer, JE; Koča, J; Šponer, J. Molecular dynamics simulations of RNA kissing-loop motifs reveal structural dynamics and formation of cation-binding pockets. *Nucleic Acids Res.,* 2003, 31, 6942-6952.

[49] Lodmell, JS; Ehresmann, C; Ehresmann, B; Marquet, R. Structure and dimerization of HIV-1 kissing loop aptamers. *J. Mol. Biol.,* 2001, 311, 475-490.

[50] Gisselquist D. Emergence of the HIV type 1 epidemic in the twentieth century: comparing hypotheses to evidence. *AIDS Res. Hum. Retroviruses,* 2003, 19, 1071-1078.

[51] Shen, N; Jetté, L; Wainberg, MA; Laughrea, M. Role of the stem B, loop B, and nucleotides next to the primer binding site and the kissing-loop domain in human

immunodeficiency virus type 1 replication and genomic-RNA dimerization. *J. Virol.* 2001, 75, 10543-10549.

[52] Shubsda, MF; Paoletti, AC; Hudson, BS; Borer, PN. Affinities of packaging domain loops in HIV-1 RNA for the nucleocapsid protein. *Biochemistry,* 2002, 41, 5276-5282.

[53] Takahashi, K-i; Baba, S; Chattopadhyay, P; Koyanagi, Y; Yamamoto, N; Takaku, H; Kawai, G. Structural requirement for the two-step dimerization of human immunodeficiency virus type 1 genome. *RNA,* 2000, 6, 96-102.

[54] Ashe, MP; Pearson, LH; Proudfoot, NJ. The HIV-1 5′ LTR poly(A) site is inactivated by U1 snRNP interaction with the downstream major splice donor site. *EMBO J.,* 1997, 16, 5752-5763.

[55] Fang, G; Burger, H; Chappey, C; Rowland-Jones, S; Visosky, A; Chen, C-H; Moran, T; Townsend, L; Murray, M; Weiser, B. Analysis of transition from long-term nonprogressive to progressive infection identifies sequences that may attenuate HIV type 1. *AIDS Res. Hum. Retroviruses,* 2001, 17, 1395-1404.

[56] Yu, E; Fabris, D. Direct probing of RNA structures and RNA-protein interactions in the HIV-1 packaging signal by chemical modification and electrospray ionization Fourier transform mass spectrometry. *J Mol. Biol.,* 2003, 211-223.

[57] Sakuragi, J-I; Ueda, S; Iwamoto, A; Shioda, T. Possible role of dimerization in human immunodeficiendy virus type 1 genome RNA packaging. *J Virol.,* 2003, 77, 4060-4069.

[58] Damgaard, CK; Dyhr-Mikkelsen, H; Kjems, J. Mapping the RNA binding sites for human immunodeficiency virus type-1 GAG and NC proteins within the complete HIV-1 and -2 untranslated leader regions. *Nucleic Acids Res.,* 1998, 26, 3667-3676.

[59] Paoletti, AC; Shubsda, MF; Hudson, BS; Borer, PN. Affinities of the nucleocapsid protein for variants of SL3 RNA in HIV-1. *Biochemistry,* 2002, 41, 15423-15428.

[60] Damgaard, CK; Andersen, ES; Knudsen, B; Gorodkin, J; Kjems, J. RNA interactions in the 5′region of the HIV-1 genome. *J. Mol. Biol.*, (2004) 336, 369-379.

Appendix

Names and accession numbers of all the isolates being mentioned in the text are listed below.

00BW0768.20 (accession no. AF443089); 90CF4071 (accession no. AF197341); 95TNIH022 (accession no. AB032740); 97KR004 (accession no. AF286239); 97NOGIL3 (accession no. AJ237565); 98ZA528 (accession no. AY158535); 99BW4745.8 (accession no. AF443085); 99TH.R2399 (accession no. AF530576); 99USTWLA (accession no. AY169814); 99ZACM4 (accession no. AF411964); 671-00T103 (accession no. AY423386);

A025 (accession no. AF408626); A027 (accession no. AF332867); A32878 (accession no. AF408632); ANT70C (accession no. L20587); C-96BW01B03 (accession no. AF110959); C-96BW04.09 (accession no. AF110965); C-96BW05.04 (accession no. AF110968); C.96BW06.H51 (accession no. AF290027); CM-YBF106 (accession no. AJ271370); CNHN24 (accession no. AY180905); C.ZA.1134MB (accession no. AY463217); C.ZA.1171MB (accession no. AY463232); DH12 (accession no. AF069140); Lai (accession no. K02013); Mal (accession no. X04415); mIDU107 (accession no. AB097868); mS TD101 (accession no. AB097870); RF (accession no. M17451); TV001 (accession no. AY162223); URTR23 (accession no. AF385934); US4 (accession no. AY173955); VI991 (accession no. AF190127); WCIPR9012 (accession no. U69590); WCIPR9031 (accession no. U69592); WCIPR9032 (accession no. U69593); WR27 (accession no. AF286365); X421 (accession no. AF423757), X475 (accession no. AF423758); X605 (accession no. AF450096).

In: Trends in RNA Research
Editor: Peter A. McNamara, pp. 191-223

ISBN: 1-59454-506-5

Chapter IX

Genetic Diversity and Pathogenic Relevance of Sodium Hydrogen Exchanger

Islam Khan

Department of Biochemistry, Faculty of Medicine, Kuwait University, Kuwait PO Box 24923.

Abstract

Sodium hydrogen exchanger [NHE] plays an important role in the regulation of intracellular pHi and net uptake of NaCl and HCO_3^- from intestine and renal tubules. Several pharmacologically and biochemically distinct NHE isoforms have been identified, and classified broadly as amiloride sensitive or resistant ones. Sardet et al cloned an amiloride sensitive human NHE-1 isoform for the first time in 1989. Using this cDNA as a probe, NHE-1 isoform was cloned from several other species including chinese hamster, mouse, rabbit, rat and swine. So far nine distinct NHE genes [NHE-1–NHE-9], and an alternatively spliced variant of NHE-1 have been reported. These isoforms show considerable heterogeneity in their tissue/cell expression and membrane localization, their sensitivity to pharmacological agents, kinetic properties, and regulation by hormonal and physical stimuli. Although the primary amino acid sequence of the human NHE isoforms shares only limited [21-44%] identity, they show a common secondary structure consisting of 10-13 conserved trans membrane domains in their N-terminus, and a regulatory C-terminus located in the cytoplasm. Most of the NHE isoforms are located in the plasma membrane, but the NHE-5, NHE-6 and NHE-7 are reported to be present intracellularly. Analysis also predicts strong probability of a mitochondrial target sequence in the NHE-1 and NHE-3 which are normally located on the plasma membrane. Among the known isoforms, NHE-1 and NHE-3 are the most extensively studied isoforms. We and others have shown that NHE-1 is localized on basolateral domain, and NHE-3 on the apical domain of epithelial cells. Their distinct location on the epithelial cell suggests that the main function of NHE-1 is to regulate pHi, while that of NHE-3 is to regulate absorption of NaCl. Alterations in the NHE-1 and

NHE-3 activity and expression of mRNA and protein have been linked to pathogenesis of several diseases including colitis, essential hypertension, congenital secretory diarrhea, gastroesophageal reflux disease, diabetes, and tissue damage in ischemia/reperfusion. This chapter documents recent developments pertaining to the molecular diversity and the regulation of NHE isoforms in pathogenic conditions. A comprehensive analysis based physical properties, predictions on the primary and secondary structures, topology, and functional relevance of NHE in disease pathogenesis will also be presented.

1. Molecular and Structural Diversity

Na/H exchanger [NHE] is an integral membrane protein which takes up extra-cellular Na^+ and extrudes intra-cellular H^+ in an electrically neutral manner across the plasma membrane. Biochemical and pharmacological studies indicated several isoforms of this protein. Subsequent cloning studies have identified a family of multiple genes encoding several distinct NHE isoforms in humans and other species [Counillon and Pouyssegur, 1993; Hildebrandt et al, 1991; Orlowski et al, 1992; Reilly et al, 1991; Sardet et al. 1988, Sardet et al 1989; Shugrue et al, 1999; Takaichi et al, 1992; Tse et al, 1991; Yun et al, 1995,]. These isoforms differ in their kinetic behavior, sensitivity toward amiloride, cell/tissue selective expression, response to various growth factors, hormones and cation concentration [Attaphitaya et al, 1999; Brant et al 1995; Brett et al, 2002; Goyal et al, 2003; Lifton et al, 1990; Numata and Orlowski, 2001, Orlowski and Grinstein, 2004; Tse et al, 1993]. Cloning era for eukaryotic cDNAs, in general, started in early 1980s and continued until late 1990. Due to lack of cDNA probe, cloning of NHE cDNA was accomplished indirectly using a complementation technique [Sardet et al, 1989]. Later on using the cDNA probe from this study [Sardet et al, 1989], NHE-1 was cloned from chinese hamster, mouse, rabbit, rat, and swine by several other groups [Counillon and Pouyssegur, 1993; Fliegel et al, 1991; Hildebrandt et al, 1991; Malakooti et al, 1999; Orlowski et al, 1992; Reilly et al, 1991; Shugrue et al, 1999; Tse et al, 1991]. Subsequently, low stringency screening of cDNA libraries using probes generated from these studies has facilitated cloning of several other NHE isoforms. So far 8 NHE isoforms have been reported in literature in humans and other species [Attaphitaya et al, 1999; Brant et al 1995; Brett et al, 2002; Dibrov and Fliegel, 1998; Goyal et al, 2003; Lifton et al, 1990; Numata and Orlowski, 2001; Orlowski and Grinstein, 2004; Sardet et al. 1988, 1989; Takaichi et al, 1992; Tse et al, 1993; Yun et al, 1995]. Recently, a new alternatively spliced variant of NHE-1 lacking in amiloride binding site has been reported in human reticulocytes and erythrocytes [Zerbini et al, 2003]. Thus molecular and functional diversity of NHE is further increased. These isoforms show different chromosomal, cellular and sub cellular locations in the humans and rats [Table 1]. For example, human NHE-1, 2, 3, 4, 5, 6, 7 and 8 genes are located on 1p36.1-p35, 2q11.2, 5p15.3, 2q12.1, 16q22.1, Xq26.3, X and 20q13.13 chromosomes respectively [Table 1]. The rats NHE-1, 2, 3, 4 and 5 genes are located on 5q36, 9q21, 1p11, 9q21, and 19q11 chromosomes, whereas mouse NHE-6 and NHE-8 genes are present on XA5, and 2H3 chromosomes [Table 1].

Table 1. Chromosomal, cellular and subcellular localization of indicated NHE isoforms in the species shown

Isoforms	Human Chromosomal	Rat (Mouse) Chromosomal	Cellular	Subcellular
NHE-1	1p36.1-p35	5q36	Ubiquitous (epithelial)	PM (Basolateral)
NHE-2	2q11.2	9q21	Epithelial	PM (Apical)
NHE-3	5p15.3	1p11	Epithelial	PM (Apical)
NHE-4	2q12.1	9q21	Renal/hippocampal	PM (Apical?)
NHE-5	16q22.1	19q11	Neuronal	PM
NHE-6	Xq26.3	XA5 (mouse)	Ubiquitous	Mitochondrial
NHE-7	X	Not known	Vacuolar, endosomal	PM
NHE-8	20q13.13	2H3 (mouse)	Ubiquitous	PM

PM= Plasma membrane

Among these isoforms, NHE-1 and NHE-3 are the most extensively studied ones. NHE-1 is ubiquitous in nature and is present at basolateral domain in the epithelial cells [Khan et al, 2001], whereas NHE-3 is present on the apical domain of epithelial cells [Bookstein et al, 1994; Khan and Cheng, 2001; Khan et al, 2001; Khan, 2002; Maher et al, 1997]. NHE-1 is amiloride sensitive, whereas as NHE-2 and NHE-3 are relatively resistant to amiloride. NHE-2 and NHE-3 are the major epithelial isoforms [Chow, 1999; Hoogerwerf et al, 1996]. NHE-2 is primarily located at the plasma membrane, whereas NHE-3 under certain conditions is also sequestered by the endosome, an intracellular recycling source. Both NHE-2 and NHE-3 are located on the apical domain of epithelial cells [Chow, 1999]. Deletion of 731-777 aa region from rat NHE-2 led to mis targeting, suggesing that this region cosnsiting of a 45 aa long cytoplamsic domain is critical for apical targeting of the rat NHE-2 [Chow, 1999]. The aa in the C-terminal domain contains two proline-rich sequences, Pro-1 and Pro-2 that resembles to SH-3 (Src homology 3) binding domains [Chow, 1999]. The Pro-1 domain is essential for appropriate apical targeting of NHE-2 [Chow et al, 1999]. NHE-3 is highly epithelial cell selective protein and can be used as a marker of these cells. NHE-2 usually colocalizes with NHE-3 on the apical domain in epithelial cells [Dudja et al, 1996; Tse et al, 1993]. Presence of the both NHE-2 and NHE-3 isoforms on the apical domain of an epithelial cell in the intestine suggests their main function in absorption of NaCl, but they are believed to be regulated differently under the normal or pathophysiological state [Bookstein et al, 1994].

NHE-4 isoform is extremely resistant to amiloride and ethylisopropylamiloride inhibition. It shows tissue distribution in renal inner medullary collecting tubules and hippocampus [Bookstein et al, 1996]. Unlike other isoforms, NHE-4 plays a unique physiological role as a K^+/H^+, Na^+/H^+ and Li^+/H^+ exchanger [Chambrey et al, 1997]. Based on an abundant expression of NHE-4 mRNA in the parietal and chief cells, it is also suggested that this isoform is involved in volume regulation during acid secretion [Rossmann et al, 2001]. Rat and mouse NHE-4 proteins are 90% identical, whereas they are 52-60% identical to human NHE-4 protein sequence [not shown].

NHE-5 isoform which is relatively insensitive to the amiloride analogue has been cloned from rat and human [Attaphitaya et al, 1999; Baird et al, 1999]. This isoform has an open reading frame for 896 amino acids [Mr 99 kD] and is predicted to contain 10 transmembrane domains [Table 2]. Rat and human NHE-5 aa sequence shares 95% identity [not shown]. A high level expression of NHE-5 mRNA in dentate gyrus, and a lower level in the hippocampus and cerebral cortex in brain suggests its specialized role in neuronal tissues [Attaphitaya et al, 1999; Baird et al, 1999]. Inhibition of rat brain NHE-5 activity by hypertonic cell shrinkage is mediated by activation of PKA and PKC [Attaphitaya et al, 2001]. However, its activity remains unaffected by hyposmotic challenge [Attaphitaya et al, 2001]. NHE-5 is localized in the recycling endosomal pathway, and is regulated by phosphatidylinositol 3'-kinase and by the state of F-actin assembly [Szaszi et al, 2002]. A polymorphic dinucleotide (GT/CA)n repeat contained in the NHE-5 has been linked to D16S421 marker [Klanke et al, 1995]. Multipoint linkage analysis identified NHE-5 between D16S421 and D16S512 markers. This NHE-5 polymorphic marker can be used to study its involvement in the pathologic conditions [Klanke et al, 1995].

Table 2. Physical properties of the indicated NHE isoforms in different species as predicted using expasy.org (Protparam)

Isoform	aa	% identity	pI	MW	Instability index	Half-life (hrs)	TM
Human NHE-1	815	100	6.74	90.8	54.63	30	12
Mouse NHE-1	820	92	6.65	91.5	53.57	30	13
Rabbit NHE-1	816	95	6.61	90.7	55.77	30	13
Rat NHE-1	820	93	6.71	91.6	52.42	30	13
CH NHE-1	822	92	7.13	91.0	55.47	30	12
Pig NHE-1	818	95	6.60	91.3	54.91	20	12
Bovine NHE-1	817	94	7.13	91	55.47	30	13
D NHE-1	1099	08	6.31	121.8	47.04	30	13
Human NHE-2	812	44	9.2	91.5	46.13	30	13
Human NHE-3	834	34	7.32	92.9	40.02	30	12
Rat NHE-4	717	23	9.16	81.5	43.24	30	11
Human NHE-5	896	33	7.33	99.0	50.44	30	10
Human NHE-6	669	22	6.03	74.2	44.42	30	12
Human NHE-7	725	21	5.97	80.1	41.25	30	12
Human NHE-8	595	25	6.07	65.4	41.84	30	11

Note: aa= amino acid, pI=isoelectric pH, MW=molecular weight, TM=trans membrane domains, CH = Chinese hamster, D = Drosophila.

NHE-6, a ubiquitous isoform, was cloned by Numata et al, 1998. Co-localization of a green fluorescent chimera protein has proved its mitochondrial location [Numata et al, 1998]. However, this protein was also shown to target to the ER membrane which is sorted to organeller membranes derived from the ER, which argued against its mitochondrial location [Miyazaki et al, 2001]. This was further supported by a report that showed its location in endosome [Brett et al, 2002]. However, prediction of a mitochondrial target sequence [Table 3] in the NHE-6 isoform [Claros and Vincens, 1996] supports its mitochondrial location. It is generally believed that mitochondrial proteins carrying a specific target sequence do not go through ER-golgi modification and trafficking. For ER-golgi modification and trafficking, a specific signal peptide is a minimal requirement. Interestingly, analysis [Nielson et al, 1997] also predicted a signal peptide sequence in NHE-6 isoform [Table 4] showing a potential to undergo posttranslational modification and trafficking pathways as well. In addition, computer analysis [Claros and Vincens, 1996] predicted a mitochondrial target sequence in NHE-1 and NHE-3 isoforms as well, indicating their localization in mitochondria [Table 3]. However, experimental proofs to support these predictions and their functional relevance are lacking at present. Nonetheless, NHE-6 isoform is believed to play important role in mitochondrial integrity and lysosomal biogenesis. Blast analysis of human NHE-6 and mouse NHE-6 protein sequences revealed 96% identity [not shown].

Table 3. Predicted mitochondrial target sequence with targeting probability (parentheses).

Human Isoforms	Predicted Mitochondrial Signal Sequence
NHE-1	MVLRSGICGLSPHRIFPSLLVVVALVGLLPVLRS (0.8599)
NHE-2	MEPLGNWRSLRAPLPPMLLLLLLQVAGPVGALAETLLNAPRA (0.0842)
NHE-3	MWGLGARGPDRGLLLALALGGLARA (0.588)
NHE-4	Not predicted
NHE-5	Not predicted
NHE-6	MARRGWRRAPLRRG (0.9963)
NHE-7	Not predicted (0.024)
NHE-8	Not predicted (0.0035)

A novel isoform, NHE-7 was cloned by Numata and Orlowski, 2001 from human cells. This is ubiquitously expressed and is localized predominantly to the trans-Golgi network [Numata and Orlowski, 2001]. This is a relatively amiloride insensitive isoform that mediates influx of Na^+ or K^+ in exchange for H^+, thus displaying unique pharmacological properties. NHE-7 may play an important role in maintaining cation homeostasis of this important organelle, and in the secretory and endocytic pathways of mammalian cells [Numata and Orlowski, 2001].

Recently, a new isoform designated as NHE-8 has been cloned from human and mouse kidney which has an open reading frame for 595 and 576 amino acids [Table 2] [Goyal et al, 2003; Nagase et al, 1999)]. This isoform shows 25% amino acid identity with human NHE-1 sequence [Table 2]. However, the mouse and human NHE-8 protein isoforms are 96% identical to each other [Not shown]. This is expressed ubiquitously in mouse tissues, and is localized on the apical domain in epithelial cells. It is expressed in both cortex and medulla in the proximal tubule, indicating its contribution to apical membrane ion transport [Goyal et al, 2003].

Table 4. Predicted signal peptide sequence in indicated human NHE isoforms

Isoforms	Signal Peptide Sequence
NHE-1	MVLRSGICGLSPHRIFPSLLVVVALVGLLPVLRSH
NHE-2	MEPLGNWRSLRAPLPPMLLLLLLQVAGPVGAL
NHE-3	MWGLGARGPDRGLLLALALGGLARAG
NHE-4	MGPAMLRAFSSWKWLLLLMVLTCLEAS
NHE-5	MLRAALSLLALPLAGAA
NHE-6	MARRGWRRAPLRRGVGSSPRARRLMRPLWLLLAVGVFDWAGAS
NHE-7	MEPGDAARPGSGRATGAPPPRLLLLPLLLGWGLRVAAAASASSSGAAAE
NHE-8	Not predicted

1.1 Differential Expression

We have focused in our studies on the molecular regulation of NHEs both in the rats and humans. To address this issue we have utilized animal models of colitis and also human inflammatory bowel disease biopsy samples. Earlier we demonstrated that NHE-1, 2, 3, 4 mRNAs are expressed in rat intestine [Khan et al, 1998]. We have also reported that NHE-1 is ubiquitously present in rat tissues with maximum expression in the stomach and colon [Fig.1]. NHE-3 is expressed maximally in rat colon and renal cortex (Fig. 1). NHE-1 is localized on the basolateral domain, whereas NHE-3 on the apical domain in rat epithelial cells [Khan et al, 2001; Khan, 2002].

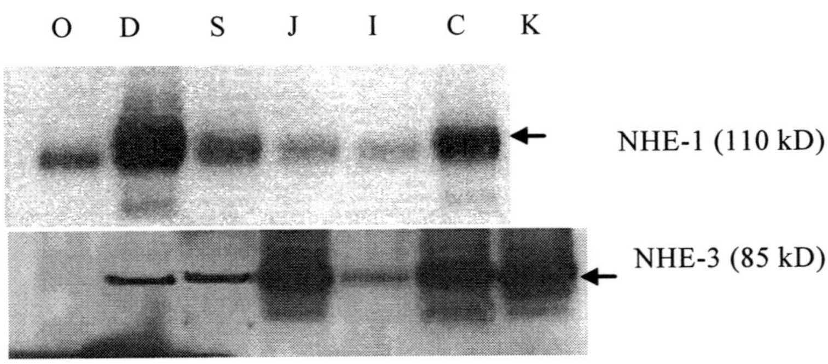

Fig.1. Differential expression of the rat NHE-1 and NHE-3 proteins in oesophagus [O], duodenum [D], stomach [S], jejunum [J], ileum [I], colon [C] and kidney [K] microsomes

Recently, using RT-PCR with human specific primers [Table 5], we have shown the expression of NHE-1, 2, 3, 5, 6 and 7 mRNA isoforms in the large and small bowel [Fig. 2]. To extend these studies further, we are now focusing on the regulation of expression of these isoforms under the normal condition, and their modulation in inflammatory bowel disease. To address this issue, we have made recombinant fusion protein vector using a bacterial GST-expression system [Pharmacia]. The fusion protein constructs were made using a segment of

C-terminal sequence of each human isoform fused in-frame with GST, a bacterial protein in pGEX expression system [Pharmacia]. A representative strategy of making fusion protein is shown in Fig. 3. The fusion proteins will be injected into rabbits to obtain polyclonal antisera. Earlier, we have raised and characterized anti-human NHE-1 antibodies in rabbits [Khan et al, 2003]. A representative picture showing reaction of antiserum raised against human NHE-3 with the fusion protein antigen by ECL western blot analysis is shown [Fig. 4].

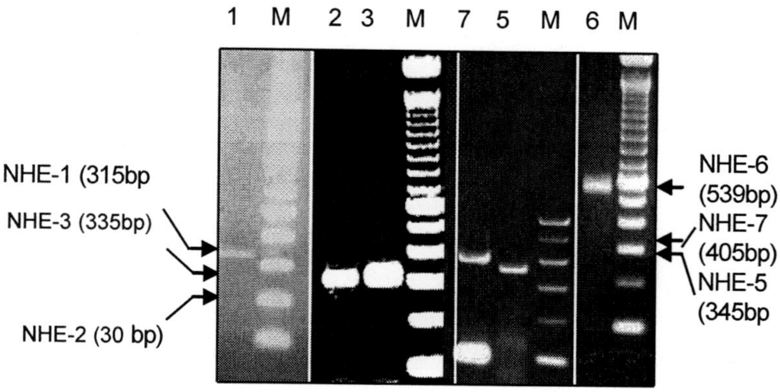

Fig.2. Expression of NHE mRNA isoforms in human large bowel total RNA using RT-PCR amplification for 35 cycle with specific primers [Table 5]. M= 100 bp ladder [Pharmacia].

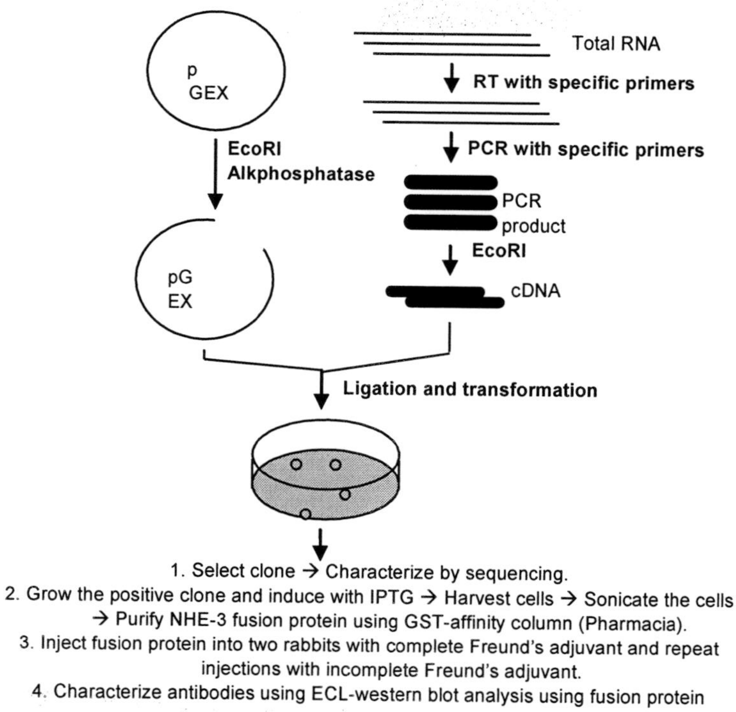

Fig.3. Strategy of cloning and raising fusion protein in bacteria.

Table 5. Primer structure, PCR fragment in base pair (bp) and reference (ref) from which primers were designed are shown. 1=sense primer, 2=antisense primer. Positions of each primer are indicated in parentheses.

Primers	PCR Fragment	Ref
NHE-1 1. 5'- TCACCATCGACCCGGCTT-3' (2203 -2220 bp) 2. 5'- CCTGCTGGCCCTGGCGTT- 3' (2500–2517 bp)	315 bp	Sardet et al, 1989
NHE-2 1. 5'-CAGGCG CTTCTTGCCAGA A-3' (2223-2241 bp) 5'- AAGGCTTCTCACTCCCAAATC –3' (2507-2527 bp)	305 bp	Malakooti et al, 1999
NHE-3 1. 5'- GGATCGAGTTCCTGGCTAGT-3' (2208-2225 bp) 2. 5'- CCGGTTAGCGGCGTGTCG- 3' (2527-2542 bp)	335 bp	Brant et al, 1995
NHE-5 1. 5'-GAATTCTGGAACCAGAGCATCTCA-3' (2436–2453 bp) 2. 5'-GAATTCCCTGCTCCTCCCCGA-3' (2766-2780 bp)	345 bp	Attaphitaya et al, 1999 (GI:14776209)
NHE-6 1. 5'- TGTGGGTATTTGGTGGT GGC -3' (1504-1524 bp) 2. 5'-ACGAGACATGGTCCAGCC-3' (2025-2042 bp)	539 bp	XM_010374
NHE-7 1. 5'-AGAGCGCATGGATATTCA-3' (1786-1802 bp) 2. 5'-AAAGTCAAGCATTATCTT-3' (2173-2190 bp)	405 bp	XM_062645

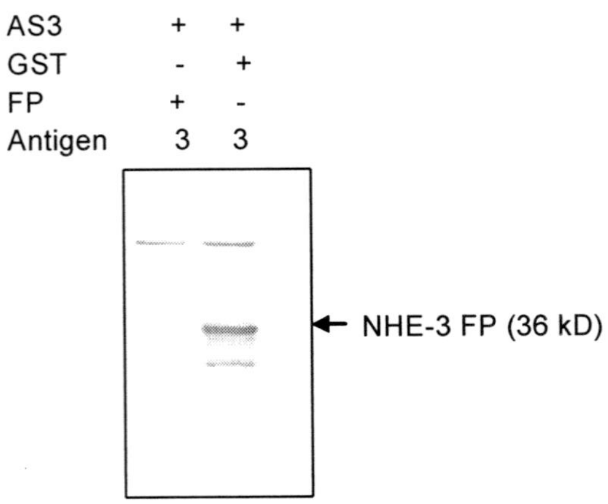

Fig.4. ECL-western blot analysis showing reaction of rabbit antiserum against human NHE-3 (AS3) with the NHE-3 fusion protein antigen (3). Reaction is blocked by NHE-3 fusion protein antigen [FP] in lane 1, but not by the fusion partner [GST] in lane 2.

```
Bovine   -MLLWPGASGLPPPRIFPSLLVVVALVGLLPVLRSYGLQISPTASTIRGSEPPRERSIGD 59
Pig      -MLLWSGICGLSPPRIFPSLLVVVALVGLLPVLRSHGLQPSPTASTIRGSEPPRERSIGD 59
Rabbit   -MLLWSAVRGLSPPRIVPSLLVVVALAGLLPGLRSHGLQLSPTDSTTPDSQPSRERSIGD 59
Human    -MVLRSGICGLSPHRIFPSLLVVVALVGLLPVLRSHGLQLSPTASTIRSSEPPRERSIGD 59
Rat      MMLRWSGIWGLYPPRIFPSLLVVVALVGLLPVLRSHGLQLNPTASTIRGSEPPRERSIGD 60
Mouse    MMLRWSGVWGFHPPRIFPSLLVVVALVGLLPVLRSHGLQHSPTASTIRGSEPPRERSIGD 60
Chinese  MMLRWSGIWGLSPPRIFPSLLVVVALVGLLPVLRSHGLQPSPTANTIRGAEPPRERSIGD 60
         *: .. *: * **.*********.**** ***:*** .** .* .:.*.*******

Bovine   VTTAP--PELAPESRPVNHSFAEHSPKA-RKAFPVLGIDYQHVRIPFEIALWILLACLMK 116
Pig      VTTAP--PELAPESRPVNHSVTEHGMKA-RKAFPVLGIDYTHVRTPFEISLWILLACLMK 116
Rabbit   VTTAP--PEVTPESRPVNRSVTEHGMKP-RKAFPVLGIDYTHVRTPFEISLWILLACLMK 116
Human    VTTAP--PEVTPESRPVNHSVTDHGMKP-RKAFPVLGIDYTHVRTPFEISLWILLACLMK 116
Rat      VTTAPSEPLHHPDDRNLTNLYIEHGAKPVRKAFPVLDIDYLHVRTPFEISLWILLACLMK 120
Mouse    VTTAPSEPLHRPDDHNLTNLIIEHGGKPSRKAFPVLDIDYPHVRTPFEISLWILLACLMK 120
Chinese  VTTAPSEPVHHPDDRNLTNLHIEHGAKTLRKAFPVLDIDYLHVRTPFEISLWILLACLMK 120
         ***** * *.:.:.. :*.*.*******.*** *** ****.**********

Bovine   IGFHVIPTISSIVPESCLLIVVGLLVGGLIKGVGETPPILQSEVFFLFLLPPIILDAGYF 176
Pig      IGFHVIPTISSIVPESCLLIVVGLLVGGLIKAVGETPPFLQSEVFFLFLLPPIILDAGYF 176
Rabbit   IGFHVIPTISSIVPESCLLIVVGLLVGGLIKGVGEKPPFLQSEVFFLFLLPPIILDAGYF 176
Human    IGFHVIPTISSIVPESCLLIVVGLLVGGLIKGVGETPPFLQSDVFFLFLLPPIILDAGYF 176
Rat      IGFHVIPTISSIVPESCLLIVVGLLVGGLIKGVGETPPFLQSDVFFLFLLPPIILDAGYF 180
Mouse    IGFHVIPTISSIVPESCLLIVVGLLVGGLIKGVGETPPFLQSDVFFLFLLPPIILDAGYF 180
Chinese  IGFHVIPTISSIVPESCLLIVVGLLVGGLIKGVGETPPFLQSDVFFLFLLPPIILDAGYF 180
         *******************************.***.**.***.*****************

Bovine   LPLRQFTENLGTILIFAVVGTLWNAFFLGGLMYAVCLVGGEQINNIGLLENLLFGSIISA 236
Pig      LPLRQFTENLGTILIFAVVGTLWNAFFLGGLMYAVCLVGGEQINNIGLLDNLLFGSIISA 236
Rabbit   LPLRQFTENLGTILIFAVVGTLWNAFFLGGLMYAVCLVGGEQINNIGLLDNLLFGSIISA 236
Human    LPLRQFTENLGTILIFAVVGTLWNAFFLGGLMYAVCLVGGEQINNIGLLDNLLFGSIISA 236
Rat      LPLRQFTENLGTILIFAVVGTLWNAFFLGGLLYAVCLVGGEQINNIGLLDTLLFGSIISA 240
Mouse    LPLRQFTENLGTILIFAVVGTLWNAFFLGGLLYAVCLVGGEQINNIGLLDTLLFGSIISA 240
Chinese  LPLRQFTENLGTILIFAVVGTLWNAFFLGGLLYAVCLVGGEQINNIGLLDTLLFGSIISA 240
         *******************************.*****************:.*********

Bovine   VDPVAVLAVFEEIHINELLHILVFGESLLNDAVTVVLYHLFEEFANYDRVGIVDIILGFL 296
Pig      VDPVAVLAVFEEIHINELLHILVFGESLLNDAVTVVLYHLFEEFANYDRVGIVDIVLGFL 296
Rabbit   VDPVAVLAVFEEIHINELLHILVFGESLLNDAVTVVLYHLFEEFANYDHVGIVDIVLGFL 296
Human    VDPVAVLAVFEEIHINELLHILVFGESLLNDAVTVVLYHLFEEFANYEHVGIVDIFLGFL 296
Rat      VDPVAVLAVFEEIHINELLHILVFGESLLNDAVTVVLYHLFEEFASYEYVGISDIFLGFL 300
Mouse    VDPVAVLAVFEEIHINELLHILVFGESLLNDAVTVVLYHLFEEFASYDSVGISDIFLGFL 300
Chinese  VDPVAVVAVFEEIHINELLHILVFGESLLNDAVTVVLYHLFEEFANYDSIGISDIFLGFL 300
         ******.************************************* *.:.:** **.****

Bovine   SFFVVSLGGVFVGVVYGVIAAFTSRFTSHIRVIEPLFVFLYSYMAYLSAELFHLSGIMAL 356
Pig      SFFVVSLGGVFVGVVYGVIAAFTSRFTSHIRVIEPLFVFLYSYMAYLSAELFHLSGIMAL 356
Rabbit   SFFVVALGGVFVGVVYGVIAAFTSRFTAHIRVIEPLFVFLYSYMAYLSAELFHLSGIMAL 356
Human    SFFVVALGGVLVGVVYGVIAAFTSRFTSHIRVIEPLFVFLYSYMAYLSAELFHLSGIMAL 356
Rat      SFFVVSLGGVFVGVVYGVIAAFTSRFTSHIRVIEPLFVFLYSYMAYLSAELFHLSGIMAL 360
Mouse    SFFVVALGGVFVGVVYGVIAAFTSRFTSHIRVIEPLFVFLYSYMAYLSAELFHLSGIMAL 360
Chinese  SFFVVALGGVFVGVVYGVIAAFTSRFTSHIRVIEPLFVFLYSYMAYLSAELFHLSGIMAL 360
         *****.****.****************.********************************

Bovine   IASGVVMRPYVEANISHKSHTTIKYFLKMWSSVSETLIFIFLGVSTVAGSHHWNWTFVIS 416
Pig      IASGVVMRPYVEANISHKSHTTIKYFLKMWSSVSETLIFIFLGVSTVAGSHHWNWTFVIS 416
Rabbit   IASGVVMRPYVEANISHKSHTTIKYFLKMWSSVSETLIFIFLGVSTVAGSHHWNWTFVIS 416
Human    IASGVVMRPYVEANISHKSHTTIKYFLKMWSSVSETLIFIFLGVSTVAGSHHWNWTFVIS 416
```

Fig. 5. Continued on next page

```
Bovine   TLLFCLIARVLGVLGLTWFINKFRIVKLTPKDQFIIAYGGLRGAIAFSLGYLLDKKHFPM 476
Pig      TLLFCLIARVLGVLGLTWFINKFRIVKLTPKDQFIIAYGGLRGAIAFSLGHLLDKNHFPM 476
Rabbit   TLLFCLIARVLGVLGLTWFINKFRIVKLTPKDQFIIAYGGLRGAIAFSLGYLLDKKHFPM 476
Human    TLLFCLIARVLGVLGLTWFINKFRIVKLTPKDQFIIAYGGLRGAIAFSLGYLLDKKHFPM 476
Rat      TLLFCLIARVLGVLVLTWFINKFRIVKLTPKDQFIIAYGGLRGAIAFSLGYLLDKKHFPM 480
Mouse    TLLFCLIARVLGVLVLTWFINKFRIVKLTPKDQFIIAYGGLRGAIAFSLGYLLDKKHFPM 480
Chinese  TLLFCLIARVLGVLVLTWFINKFRIVKLTPKDQFIIAYGGLRGAIAFSLGYLMDKKHFPM 480
         ************** ***********************************:*:**:****

Bovine   CDLFLTAIITVIFFTVFVQGMTIRPLVDLLAVKKKQETKRSINEEIHTQFLDHLLTGIED 536
Pig      CDLFLTAIITVIFFTVFVQGMTIRPLVDLLAVKKKQETKRSINEEIHTQFLDHLLTGIED 536
Rabbit   CDLFLTAIITVIFFTVFVQGMTIRPLVDLLAVKKKQETKRSINEEIHTQFLDHLLTGIED 536
Human    CDLFLTAIITVIFFTVFVQGMTIRPLVDLLAVKKKQETKRSINEEIHTQFLDHLLTGIED 536
Rat      CDLFLTAIITVIFFTVFVQGMTIRPLVDLLAVKKKQETKRSINEEIHTQFLDHLLTGIED 540
Mouse    CDLFLTAIITVIFFTVFVQGMTIRPLVDLLAVKKKQETKRSINEEIHTQFLDHLLTGIED 540
Chinese  CDLFLTAIITVIFFTVFVQGMTIRPLVDLLAVKKKQETKRSINEEIHTQFLDHLLTGIED 540
         ************************************************************
```

```
Bovine   ICGHYGHHHWKDKLNRFNKKYVKKCLIAGERSKEPQLIAFYHKMEMKQAIELVESGGMGK 596
Pig      ICGHYGHHHWKDKLNRFNKKYVKKCLIAGERSKEPQLIAFYHKMEMKQAIELVESGGMGK 596
Rabbit   ICGHYGHHHWKDKLNRFNKKYVKKCLIAGERSKEPQLIAFYHKMEMKQAIELVESGGMGK 596
Human    ICGHYGHHHWKDKLNRFNKKYVKKCLIAGERSKEPQLIAFYHKMEMKQAIELVESGGMGK 596
Rat      ICGHYGHHHWKDKLNRFNKKYVKKCLIAGERSKEPQLIAFYHKMEMKQAIELVESGGMGK 600
Mouse    ICGHYGHHHWKDKLNRFNKKYVKKCLIAGERSKEPQLIAFYHKMEMKQAIELVESGGMGK 600
Chinese  ICGHYGHHHWKDKLNRFNKKYVKKCLIAGERSKEPQLIAFYHKMEMKQAIELVESGGMGK 600
         ************************************************************

Bovine   IPSAVSTVSMQNIHPKSLPAERILPALSKDKEEEIRKILRNNLQKTRQRLRSYNRHTLVA 656
Pig      IPSAVSTVSMQNIHPKTLPTERILPALSKDKEEEIRKILRNNLQKTRQRLRSYNRHTLVA 656
Rabbit   IPSAVSTVSMQNIHPKALPAERILPALSKDKEEEIRKILRNNLQKTRQRLRSYNRHTLVA 656
Human    IPSAVSTVSMQNIHPKSLPSERILPALSKDKEEEIRKILRNNLQKTRQRLRSYNRHTLVA 656
Rat      IPSAVSTVSMQNIHPKSAASERILPALSKDKEEEIRKILRSNLQKTRQRLRSYNRHTLVA 660
Mouse    IPSAVSTVSMQNIHPKAVTSDRILPALSKDKEEEIRKILRSNLQKTRQRLRSYNRHTLVA 660
Chinese  IPSAVSTVSMQNIHPKSMASERILPALSKDKEEEIRKILRSNLQKTRQRLRSYNRHTLVA 660
         ****************: .::********************.*******************

Bovine   DPYEEAWNQMLLRRQKARQLEQKISNYLTVPAHKLDSPTMSRARIGSDPLAYEPKADLPV 716
Pig      DPYEEAWNQMLLRRQKARQLEQKINNHLTVPAHKLDSPTMSRARIGSDPLAYEPKADLPV 716
Rabbit   DPYEEAWNQMLLRRQKARQLEQKINNYLTVPAHKLDSPTMSRARIGSDPLAYEPKADLPV 716
Human    DPYEEAWNQMLLRRQKARQLEQKINNYLTVPAHKLDSPTMSRARIGSDPLAYEPKEDLPV 716
Rat      DPYEEAWNQMLLRRQKARQLEQKITNYLTVPAHKLDSPTMSRARIGSDPLAYEPKADLPV 720
Mouse    DPYEEAWNQMLLRRQKARQLEQKITNYLTVPAHKLDSPTLSRARIGSDPLAYEPKADLPV 720
Chinese  DPYEEAWNQMLLRRQKARQLEQKMSNYLTVPAHKLDSPTMSRARIGSDPLAYEPKADLPV 720
         ***********************:.*:*************.*************** ****
```

```
Bovine   ITIDPASPQSPESVDLVNEELKGKVLGLSRDPGRLAE-EEEDDNGPR-WLTMRTKEPSSP 774
Pig      ITIDPASPQSPESVDLVNEELKGKVLGLSREPGRATE-EEDDDEDHDGGLVMRTKEPSSP 775
Rabbit   ITIDPASPQSPESVDLVNEELKGKVLGLSREPRVAEE-AAEEDEDGG--IVMRPKEPSSP 773
Human    ITIDPASPQSPESVDLVNEELKGKVLGLSRDPAKVAE-EDEDD-DGG--IMMRSKETSSP 772
Rat      ITIDPASPQSPESVDLVNEELKGKVLGLKRGP-RTTP--EEEEEDEDGVIMIRSKEPSSP 777
Mouse    ITIDPASPQSPESVDLVNEELKGKVLGLNRGP-RVTP--EEEEEDEDGIIMIRSKEPSSP 777
Chinese  ITIDPASPQSPESVDLVNEELKAKVLGVNRDPTRLTRGEEDEDEDEDGVIMMRRKEPSSP 780
         **********************.****:.* *      :::.   ::* **.***
```

Note: * = identical bases, - = gap, : = conservative mismatch, . = semi mismatches

Fig. 5. Homology of NHE-1 protein sequence in the indicated species

1.2 Inter Species and Inter Isoforms Homology

NHE-1 protein sequence from different species shows homology ranging from 92%-95% [Fig, 5], whereas it shows least identity [8%] with Drosophila NHE-1 [Table 2]. The aa content of NHE-1 ranges between 815 aa - 822 aa in different species, whereas the aa content of human NHE-1 – NHE-8 isoforms ranges from 595 aa – 896 aa [Table 2], with a predicted molecular weight ranging from 67 kD – 99 kD, and pI from 5.97 – 9.2 [Table 2]. The NHE-1 – NHE-8 proteins are unstable based on their predicted instability index [Table 2]. In addition, the half-life of these proteins ranges from 20-30 hrs predicted using erythrocytes lysate in-vitro [Table 2]. On the other hand, the nucleotide sequences of NHE-1 isoform from different species are 80-90% identical, indicating its conserved nature [Fig. 5, Table 6]. Drosophila NHE-1 contains 1099 aa [Table 2].

Amino acid content of NHE-3 ranges from 831-839 with MW from 92.7-94.8 kD, and pI from 7.32-8.06 units [Table 7]. The inter species homology of NHE-3 aa sequence ranges from 86-88% [Fig. 6, Table 7], while nucleotide sequence identity in different species ranges from 80-88% [Table 6]. Except kangaroo, NHE-3 protein in all other species is characterized as unstable based on its instability index value **[Table 7].**

Table 6. Percent identity of human NHE-1 and NHE-3 nucleotide sequence with indicated species. ND=not done.

Species	Human NHE-1	Human NHE-3
1. Rat NHE-1	80	82
2. Rabbit NHE-1	ND	86
3. Bovine NHE-1	86	ND
4. Mouse NHE-1	82	ND
5. Pig NHE-1	90	ND

Table 7. Physical properties of the Human NHE-3 with indicated species. Analysis was done using expasy.org (Protparam)

Species	Aa	% identity	MW	pI	Instability index	Half-life (hr)
Human NHE-3	834	100	92.9	7.32	40.02	30
Kangaroo NHE-3	839	87	94.8	8.05	38.82*	20
Rabbit NHE-3	832	86	92.7	8.02	40.29	30
Rat NHE-3	831	88	93.2	8.06	40.01	30

Note: aa= amino acid, pI=isoelectric pH, MW=molecular weight, Instability index classifies them unstable proteins except *stable protein.

```
Kangaroo   MP-LGVRGTRREFRFPVWGLLLLALWM-LPRALGVEEIPGPDSHEKQGFQIVTFKWHHVQ 58
Rat        MW-HPALG-------PGWKPLLALAVA-VTSLRGVRGIE-EEPNSGGSFQIVTFKWHHVQ 50
Human      MWGLGARG-------PDRGLLLALALGGLARAGGVEVEPGGAHGESGGFQVVTFEWAHVQ 53
Rabbit     MSGRGGCG-------PCWGLLLALVLA-LGALPWTQGAEQEHHDEIQGFQIVTFKWHHVQ 52
           *      *       *    **      :     ..        .  .**:***:* ***

Kangaroo   DPYIIALWILVASLAKIVFHLSHKVTSVVPESALLIVLGLILGGIVWAADHIASFTLTPT 118
Rat        DPYIIALWILVASLAKIVFHLSHKVTSVVPESALLIVLGLVLGGIVWAADHIASFTLTPT 110
Human      DPYVIALWILVASLAKIGFHLSHKVTSVVPESALLIVLGLVLGGIVWAADHIASFTLTPT 113
Rabbit     DPYIIALWVLVASLAKIVFHLSHKVTSVVPESALLIVLGLVLGGIVLAADHIASFTLTPT 112
           ***:****:******** **********************:***** *************

Kangaroo   VFFFYLLPPIVLDAGYFMPNRLFFGNLGTILLYAVIGTVWNAATTGLSLYGVYLSGIMGD 178
Rat        LFFFYLLPPIVLDAGYFMPNRLFFGNLGTILLYAVIGTIWNAATTGLSLYGVFLSGLMGE 170
Human      VFFFYLLPPIVLDAGYFMPNRLFFGNLGTILLYAVVGTVWNAATTGLSLYGVFLSGLMGD 173
Rabbit     VFFFYLLPPIVLDAGYFMPNRLFFSNLGSILLYAVVGTVWNAATTGLSLYGVFLSGIMGE 172
           :***********************.***:******:**:*************:***:**:

Kangaroo   LSIGLLDFLLFGSLIAAVDPVAVLAVFEEVHVNDVLFIIVFGESLLNDAVTVVLYNVFDS 238
Rat        LKIGLLDFLLFGSLIAAVDPVAVLAVFEEVHVNEVLFIIVFGESLLNDAVTVVLYNVFES 230
Human      LQIGLLDFLLFGSLMAAVDPVAVLAVFEEVHVNEVLFIIVFGESLLNDAVTVVLYNVFES 233
Rabbit     LKIGLLDFLLFGSLIAAVDPVAVLAVFEEVHVNEVLFIIVFGESLLNDAVTVVLYNVFQS 232
           *.************:******************:************************:*

Kangaroo   FVSLGADKVTGVDCVKGIVSFFVVSLGGTLIGIIFAFLLSLVTRFTKHVRIIEPGFVFII 298
Rat        FVTLGGDAVTGVDCVKGIVSFFVVSLGGTLVGVIFAFLLSLVTRFTKHVRIIEPGFVFVI 290
Human      FVALGGDNVTGVDCVKGIVSFFVVSLGGTLVGVVFAFLLSLVTRFTKHVRIIEPGFVFII 293
Rabbit     FVTLGGDKVTGVDCVKGIVSFFVVSLGGTLVGVVFAFLLSLVTRFTKHVRVIEPGFVFII 292
           **:**.* **********************:*::****************:*******:*

Kangaroo   SYLSYLTSEMLSLSAILAITFCGICCQKYVKANISEQSATTVRYTMKMLASGAETIIFMF 358
Rat        SYLSYLTSEMLSLSAILAITFCGICCQKYVKANISEQSATTVRYTMKMLASGAETIIFMF 350
Human      SYLSYLTSEMLSLSAILAITFCGICCQKYVKANISEQSATTVRYTMKMLASSAETIIFMF 353
Rabbit     SYLSYLTSEMLSLSSILAITFCGICCQKYVKANISEQSATTVRYTMKMLASGAETIIFMF 352
           **************:************************************.********

Kangaroo   LGISAVDPAIWTWNTAFILLTLVFISVYRAIGVVLQTWLLNKYRMVQLEIIDQVVMSYGG 418
Rat        LGISAVDPVIWTWNTAFVLLTLVFISVYRAIGVVLQTWILNRYRMVQLETIDQVVMSYGG 410
Human      LGISAVNPFIWTWNTAFVLLTLVFISVYRAIGVVLQTWLLNRYRMVQLEPIDQVVLSYGG 413
Rabbit     LGISAVDPLIWTWNTAFVLLTLLFVSVFRAIGVVLQTWLLNRYRMVQLELIDQVVMSYGG 412
           ******:* ********:****:*:**:**********:**:******* *****:****

Kangaroo   LRGAVAYALVVLLDEKKVKEKNLFVSTTIIVVFFTVIFQGLTIKPLVQWLKVKKSEHREP 478
Rat        LRGAVAYALVVLLDEKKVKEKNLFVSTTLIVVFFTVIFQGLTIKPLVQWLKVKRSEQREP 470
Human      LRGAVAFALVVLLDGDKVKEKNLFVSTTIIVVFFTVIFQGLTIKPLVQWLKVKRSEHREP 473
Rabbit     LRGAVAFALVALLDGNKVKEKNLFVSTTIIVVFFTVIFQGLTIKPLVQWLKVKRSEHREP 472
           ******:***.*** .************:************************:**:***

Kangaroo   KLNEKLHGRAFDHILSAIEDISGQIGHNYLRDKWSNFDRKVLSKLLMRRSAQKSRDRILN 538
Rat        KLNEKLHGRAFDHILSAIEDISGQIGHNYLRDKWSNFDRKFLSKVLMRRSAQKSRDRILN 530
Human      RLNEKLHGRAFDHILSAIEDISGQIGHNYLRDKWSHFDRKFLSRVLMRRSAQKSRDRILN 533
Rabbit     KLNEKLHGRAFDHILSAIEDISGQIGHNYLRDKWANFDRRFLSKLLMRQSAQKSRDRILN 532
           :*********************************::***:.**::***:***********

Kangaroo   VFHELNLKDAISYVAEGERRGSLAFIRSPSTDNIVNVDFSTPRPSTVEASVSYLLRENVS 598
Rat        VFHELNLKDAISYVAEGERRGSLAFIRSPSTDNMVNVDFSTPRPSTVEASVSYFLRENVS 590
Human      VFHELNLKDAISYVAEGERRGSLAFIRSPSTDNVVNVDF-TPRSSTVEASVSYLLRENVS 592
Rabbit     VFHELNLKDAISYVTEGERRGSLAFIRSPSTDNMVNVDFSTPRPSTVEASVSYLLRESAS 592
           **************:******************:***** ***.*********:***..*
```

Figure 6 continued on next page

```
Kangaroo   FHRTMRKRLESFKSTKLGINQTKKTAKLYK-RERGQKRRNSSIPNGKIPMESPTRDFTFK 717
Rat        FHRTMRKRLESFKSAKLGINQNKKAAKLYK-RERAQKRRNSSIPNGKLPMENLAHNFTIK 709
Human      FHRTMRKRLESFKSTKLGLNQNKKAAKLYK-RERAQKRRNSSIPNGKLPMESPAQNFTIK 711
Rabbit     FHRTMRKRLESFKSAKLGLGQSKKATKHKRERERAQKRRNSSVPNGKLPLDSPAYGLTLK 712
           **************:***:.*.**::*  : ***.*******:****:*::. : .:*:*

Kangaroo   EKELEFSDPEET-NEYEAEEMSGGIEFLANVTQDTATDSTTGIDNPVFSPEE--DQSIFT 774
Rat        EKDLELSEPEEA-TNY--EEISGGIEFLASVTKDVASDSGAGIDNPVFSPDEDLDPSILS 766
Human      EKDLELSDTEEP-PNYD-EEMSGGIEFLASVTKDTASDSPAGIDNPVFSPDEALDRSLLA 769
Rabbit     ERELELSDPEEAPDYYEAEKMSGGIEFLASVTKDTTSDSPAGIDNPVFSPDEDLAPSLLA 772
           *::**:*:.**.    *  *::********.**:*.::** :*********:*    *:::

Kangaroo   KVPPWLSPEETVVPSQRARVQIPYSPSNFRRLTPIRLSTKSVDSFLLADSPEERPRSFLP 834
Rat        RVPPWLSPGETVVPSQRARVQIPNSPSNFRRLTPFRLSNKSVDSFLQADGPEEQLQPASP 826
Human      RLPPWLSPGETVVPSQRARTQIPYSPGTFRRLMPFRLSSKSVDSFLQADGPEERPPAALP 829
Rabbit     RVPPWLSPGEAVVPSQRARVQIPYSPGNFRRLAPFRLSNKSVDSFLLAEDGAEH-----P 827
           ::****** *:********.*** **..**** *:***.******* *:.  *:     *

Kangaroo   ESTHM 839
Rat        ESTHM 831
Human      ESTHM 834
Rabbit     ESTHM 832
           *****
```

Note: * = identical bases, - = gap, : = conservative mismatch, . = semi mismatches

Fig.6. Homology of NHE-3 sequence in the indicated species.

1.3 Topology

Various models of topography have been proposed showing 10-13 transmembrane domains depending of the type of algorithm used. For example, Sardet et al, 1989 reported 10 transmembrane domains in NHE-1, whereas Wakabayashi et al, 2000 reported 12 transmembrane domains. There are two distinct domains in these isoforms; the N-terminus and C-terminus. N-terminus consisting of 450-500 aa is hydrophobic and contains 10-13 trans membrane domains [Table 2]. A typical hydropathy plot for rat NHE-1 predicted using expasy.org is shown in Fig. 7. The N-terminus of Rat NHE-1 is predicted to have 13 transmembrane domains. The C-terminus is a regulatory domain and is located in the cytoplasm. However, there is a consensus that the N-terminus [about 500 aa] is hydrophobic, and harbors multiple trans membrane domains [Counillon and Pousseygur, 2000; Orlowski et al, 1992, Siczkowski, et al, 1995].

Transmembrane domain 4 has binding site for amiloride which is a competitive inhibitor of the ion-transport site in this molecule. Responsible residues were identified to be Phe^{165} and Phe^{167} [Counillon et al, 1993]. It is presumed that epithelial NHE-1 resistance to amiloride has probably occurred due to substitutions of these amino acids [Counillon et al, 1993]. Trans membrane domain 4 is rich in proline which creates flexibility allowing free carbonyl groups interaction with the inhibitors and transported cations. Trans membrane domains 8 and 10 were reported to be a major site of interaction with these antagonists, although other regions modestly influence sensitivity to certain drugs [Orlowski and Kandasamy, 1996]. This region occurs between intracellular loop 4 and transmembrane domain 9 in the model proposed by Wakabayashi et al, 2000.

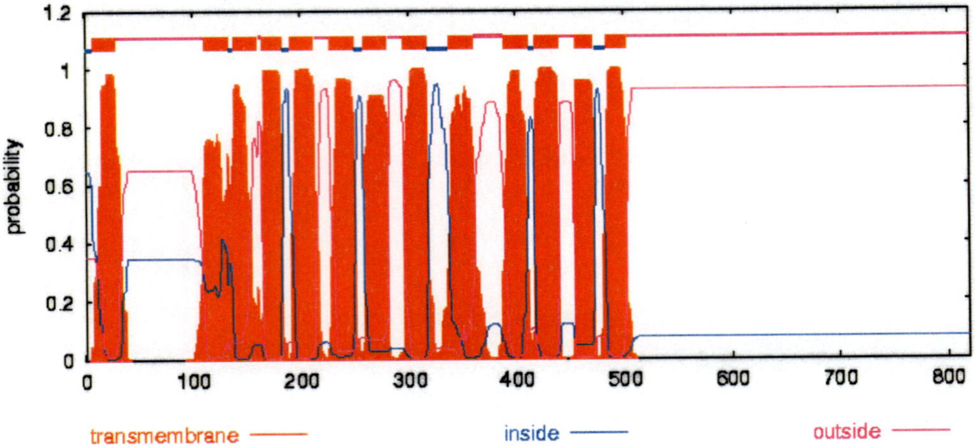

Fig.7. Hydropathy plot showing trans-membrane domains in rat NHE-1 protein [Expacy.org].

The C-terminus is hydrophilic, and does not contain a stretch of hydrophobic aa sequence that can constitute a conventional trans membrane domain [Orlowski et al, 1992]. Although based on the interactions of various cytoplasmic regulatory molecules [Counillon and Pousseygur, 2000; Orlowski et al, 1992] C-terminus is considered to be located entirely in the cytoplasm, prediction of O-glycation sites in the rat NHE-1 C-terminus raised possibility of an extra cellular epitope in this region [Khan, 2001]. To address this we have utilized two different methods, the ELISA and the flow cytometer analysis using the RBC vesicles and intact RBCs [Khan, 2001]. For ELISA, the right side out vesicles were selected by concanavalin-A which binds to the glycosylated moieties present out side the plasma membrane [Steck et al, 1970; Steck 1974; Sharon and Lee, 1972]. Our findings demonstrated presence of both the extra cellular and cytoplasmic epitopes in the C-terminus [Khan, 2001]. Normally a conventional trans membrane domain consist of α helix which is not predicted in this sequence, but it contains β sheets [26.5%] and loop structures [73.5%]. The β pleated sheets have been shown to constitute trans membrane domains in some membrane proteins both in the prokaryotes and eukaryotes [Noda et al, 1983; Schirmer et al, 1995]. Circular dichorism analysis has shown distinct domains consisting of the α, β and random coil in the NHE-1 C-terminus [Gebreselassie et al, 1998]. Thus a new topography of NHE-1 is predicted with 13 trans membrane domains as against 12 trans membrane domains as suggested by Wakabayshi et al, 2000. Similarly, C-terminus of NHE-3 has been shown to contain a trans membrane domain consisting of β-sheets [Biemesderfer et al, 1998].

1.4. Regulation

NHE genes are constitutive, but not inducible. However, they respond to various stimuli by increasing or decreasing the amount of mRNA, protein and activity. A large number of studies have reported a parallel change in the activity, protein and mRNA levels in various conditions. The regulation involves pre- and post- transcriptional and translational events, covalent modification of NHE through various signal transduction pathways, and intracellular internalization of certain isoform e.g. NHE-3. Under normal condition extra-cellular concentration of Na^+ exceeds its saturation, therefore these isoforms are quiescent under normal conditions. However, under hyperosmolarity and acidotic conditions, NHE activity is increased to rescue cells from dehydration and acidic damage. In the following section we will focus on the regulation of NHE-1 and NHE-3 isoforms.

NHE-1 responds to various chemical, physical and hormonal stimuli, growth factor, acidosis, ischemia and cytokines. Activation of NHE-1 is coupled through various signal transduction mechanisms involving G-protein, tyrosine kinase, RhoA or MAP kinases. These pathways produce cAMP and Ca^{2+}, intracellular second messengers leading to activation of PKA, PKC, PLC, IP_3, diacylglycerol (DAG) and Ca-calmodulin [Emmer et al, 1989]. These molecules then directly phosphorylate NHE-1 or several other target molecules that specifically activate NHE- protein, mRNA and activity. For example, increase in intracellular Ca^{2+} activates calmodulin (Ca^{2+}/CaM) which binds to a high affinity calmodulin (CaM) binding site in NHE-1 corresponding to amino acids 636-656 and activates NHE-1 [Wakabayashi et al, 1994]. Site-directed mutagenesis of this region renders NHE-1 constitutively active, indicating that the site is autoinhibitory. Thus any condition or factor that increases intracellular Ca^{2+} should activate NHE-1. Calmodulin appears to be a direct substrate of Jak2 phosphorylation [Mukhin et al, 2001]. Thus bradykinin activation of NHE-1 occurs via Jak-2 mediated phosphorylation of calmodulin at tyrosine residue. Similarly, Ca^{2+}/Calmodulin also regulates ileal Na^+ absorption by NHE-3 via its phosphorylation [Rood et al 1988], as evidenced by an inhibition of NHE-3 by specific peptide inhibitor of Ca^{2+}/CaM kinase II [Cohen et al 1995]. Tyr kinase[s] acting over short time periods stimulate[s] absorption of NaCl and brush-border Na^+-H^+ exchange, and also Ca^{2+}-induced inhibition of NaCl absorption. These studies suggest involvement of brush-border Tyr kinase in short-term signal transduction in intestinal epithelial cells [Donowitz et al, 1994]. NHE-3 activation by metabolic acidosis is mediated by endothein-1 receptor in renal epithelial cells [Laghmani et al, 2001]. NHE-3 activity is also acutely inhibited by parathyroid hormone in the apical membrane of proximal tubules [Girardi et al, 2000]. Thyroid hormones stimulate NHE activity and may play an active role in the recovery from muscular acidosis [Incerpi et al, 1999]. Growth factors such as insulin, PDGF, thrombin, and okadaic acid also activate NHE-1 through its phosphorylation [Counillon and Pousseygur, 1993; Voyno-Yasenetskaya, 1998]. β-adrenergic receptor stimulation, adenosine and phorbol esters, an analogue of DAG also activate NHE-1. Growth factors stimulate NHE-1 by changing the affinity constant for intracellular H^+, K'[Hi^+], whereas NHE-2 and NHE-3 by changing Vmax [Yun et al, 1995]. A specific interaction between the N- and C-termini has also been suggested for protein kinase-induced regulation of Na^+/H^+ exchangers [Yun et al, 1995]. Inhibition of NHE-1 by amiloride suppresses cell growth, supporting its role in cell growth and differentiation [Delvaux et al, 1990]. cAMP and osmotic shrinkage induce NHE-1 activity and regulates cell volume through mechanistically independent pathways [Pedersen et al, 2003]. There are indications that serine/threonine phosphatase inhibits NHE-1 activity [Pedersen et al, 2003].

Abrogation of MAPK activation without any effect on hyperosmolarity-induced NHE-1 activation, suggests that the stimulation of NHE and the activation of Erk, p38 MAPK and SAPK are parallel but independent events [Gillis et al, 2001]. This is important to emphasize that in inflamed colonic muscle inhibition of NHE-1 by amiloride does not reverse the inflammation-induced expression of MAP kinases [Unpublished findings, IK]. AngII stimulates macula densa basolateral Na/H exchange via AT1 receptors [Bell and Peti-Peterdi, 1999].

C terminus of NHE-3 has several discrete subdomains involved with stimulation or inhibition of its activity [Levine et al, 1995]. Regulation of the renal Na/H exchanger NHE-3 by protein kinase A is a key intermediate step in the hormonal regulation of acid-base and salt balance. Although changes in NHE-3 whole protein phosphorylation is difficult to detect in response to cAMP, the tryptic phosphopeptide map of in-vivo phosphorylated NHE-3 showed a complex pattern of constitutive and cAMP-induced phosphopeptides involving phosphorylation of Ser^{552} and Ser^{605} [Zhao et al, 1997]. Glucocorticoids and acid activate NHE-3 expression synergistically by increasing translation and trafficking respectively [Ambuhl et al, 1999].

2. Roles in Diseases

Various diseases are associated with deregulation of electrolyte and intracellular pH homeostasis. For example, colitis is manifested commonly by diarrhea and constipation, hyperplasia, hypertrophy, cellular necrosis and apoptosis. NHE may participate both in constipation and secretary diarrhea by over activation or suppression to rectify volume defect. Except for one report showing an association of genetic susceptibility of end stage renal disease to NHE-5 [Yu et al, 2000], almost in all diseases molecular and biochemical modulation of NHE-1 and NHE-3 has been studied. Therefore in this section we will discuss recent developments related to possible roles of these isoforms in human diseases and experimental models of IBD.

2.1 Inflammatory Bowel Disease

Inflammatory bowel diseases such as Crohn's disease and ulcerative colitis are a common occurrence in every society, which afflict individuals in the prime of life [Chutkan, 2001]. This is an entity of unknown etiology. Identification of a causative gene has been elusive so far, and therefore these are considered as a multigenic disease. Damage in this condition is mainly localized to mucosal and submucosal layers, but smooth muscle apparatus and nerves are also affected [Cook et al, 2000; Schneider et al, 2001; Zhao et al, 2001]. IBD is associated with diarrhea, constipation, pain, ulcers. Pathogenesis of microvillus inclusion disease is due to defects in the apical, but not the basolateral membrane transport systems [Michail et al, 1998]. Inflammation is associated with diarrhea which is caused due to a selective down regulation of NHE-2 and NHE-3 expression and activity by IFN-gamma in intestinal epithelial cells [Rocha et al, 2001]. However, congenital secretory diarrhea, an autosomal recessive disorder is not caused due to mutations in the NHE-1, NHE-2, NHE-3, or NHE-5 genes [Muller et al, 2000]. Thus to date no NHE mutation has been identified as a cause of diarrhea or IBD. The changes in mucosal expression of these isofroms account for secretory and absorptive defects in epithelial cells. Muscle hypertrophy, hyperplasia and contractility attenuation are also commonly observed both in the human IBD and experimental colitis. Gene knockout studies have shown that NHE-3 is the major absorptive

Na^+/H^+ exchanger in kidney and intestine. Loss of the exchanger gene impairs acid-base balance and Na^+-fluid volume homeostasis [Schultheis et al, 1998]. NHE-3 absorbs about 40 % of the filtered load of sodium [Wang *et al.* 1999]. NHE-3 null mice had a severe absorptive defect in the intestine, as indicated by diarrhea and an increase in the volume and alkalinity of the contents in all segments of the intestine. This is consistent with mild renal proximal tubular acidosis resulting from a deficit in HCO_3 re-absorption, although HCO_3 losses in the intestine would also contribute to the acidosis. The occurrence of a severe intestinal absorptive defect in NHE-3 deficient, but not in NHE-2-deficient mice, demonstrated that the functions of these two isoforms are different in the GI tract [Brooks et al, 2001]. These studies demonstrate that NHE-3 is responsible for most of the apical NHE activity in the intestinal epithelium, which mediates NaCl absorption via coupling with the apical Cl/HCO $_3$ exchanger [Shull et al, 2000]. Double-knockout mice with loss of NHE-2 and NHE-3 genes caused no reduction in viability, changes in systemic acid-base status or aldosterone levels, and diarrheal state, supporting that the NaCl, HCO_3, and fluid absorption is dependent on apical NHE-3, with little contribution from NHE-2 [Ledoussal et al, 2001]. In-vivo data suggest that intestinal absorptive defect is part of a homeostatic mechanism mediated by elevated interferon-gamma, activated in response to regulate the fluidity of the GI tract [Woo et al, 2002]. Our studies on experimental colitis have shown induction of NHE-1 and NHE-3 isoforms in rat colon which is not reverted by anti-inflammatory substances [Khan et al, 1998; Khan and Ali, 1999; Khan, 2002]. Since NSAIDs also suppresses INF-γ and other cytokines, we do not believe that this molecule regulates induction of NHE-1 in this model of IBD. We have also reported that renal expression of NHE is also altered in colitis [Khan, 2002]. Renal NHE-3 is selectively regulated by cyclooxygenase-2 via action of TNF-α in experimental colitis [Khan, 2002]. In contrast human IBD condition is associated with the suppression of NHE-1, which might be responsible for diarrhea in this condition [Khan et al, 2003]. This discrepancy in humans and animal model of colitis manifests a differential regulation of this protein in acute and chronic inflammation. Species differences are also likely to cause this discrepancy. Intracellular pHi of colonocytes in Crohn's disease and the luminal pH in active ulcerative colitis is decreased [Fallingborg et al, 1993; Yiangou et al, 2001]. Therefore the suppression of NHE-1 may also be an adaptative response to neutralize the colonocytes and colonic lumen pH in IBD. A recent report has demonstrated that NHE inhibition ameliorates intestinal inflammation in dextran sulfate-treated mice [Nemeth et al, 2002]. In our unpublished findings we have observed that inhibition of NHE-1 reverses contractility, but not the hypertrophy or MAPK induction [Unpublished findings, IK].

NHE over-expressing transgenic animals have been shown to accumulate intracellular Ca_i^{2+} in arterial smooth muscle [Kiraku et al, 2000], therefore it can also contribute to altered muscle contractility by inflammation. Over expression of NHE-1 in the colonic muscle with decreased contractility might lead to accumulation of Ca_i^{2+}, indicating that the contractile apparatus has become insensitive to Ca_i^{2+} in inflamed muscle in colitis. Thus a Ca^{2+}-independent mechanism in reduced contraction in IBD is predicted.

These findings suggest an independent pathway for MAPK and inflammation-induced changes in muscle structure and function in experimental colitis. NHE-1 null mice exhibited ataxia and retarded growth. Loss of NHE-1 gene caused a mild glandular atrophy and thickening of lamina propria without any sign of defect in absorptive function. On the other hand, gastritis induced due to deletion of NHE-2 gene in mouse was characterized by a substantially reduction in the number of parietal cells and infiltration of primarily neutrophils which were replaced by lymphocytes and plasma cells [Boivin et al, 2000; Schultheis et al, 1998]. Since there were no signs of diarrhea in NHE-2 null mice [Boivin et al, 2000], the NHE-1 and NHE-2 gene do not contribute to absorptive and secretory defects in IBD. It is also likely that the loss of a gene is counteracted by induction of other isoforms. Nevertheless, a prominent role of these isoforms in other phenotypes of IBD is predicted.

2.2. Human Diabetic Placenta

Pregnancy-induced hypertension, pre-eclimpsia and diabetes are quite common problems which have serious consequences on the health of mother and fetus. Placental NHE-1 plays an important role in maintaining adequate supply of nutrients and electrolytes to fetus and mother, and is uniquely regulated in this organ. NHE regulates syncytiotrophoblast cell volume and intracellular pH, as well as H^+ loss from and Na^+ transfer to the fetus [Sibley et al, 2002]. The activity and expression of NHE in the microvillous plasma membrane of the syncytiotrophoblast is reduced in placentas of mothers with preterm, growth restricted babies as compared to their normal counterparts [Sibley et al, 2002]. There are differential effects of gestation in normal pregnancy on NHE mRNA, protein and activity, indicating a complex control involving transcriptional, post transcriptional and post translational events [Sibley et al, 2002]. In placentas, the microvilli facing the maternal side facilitate uptake of substances from the maternal circulation [Jones and Fox, 1976; Kulanthaivel et al, 1992; Mayhew and Jairam, 2000]. Diabetes causes an enlargement of the villus surface and vascularisation, and focal thickening of the basement membrane [Mayhew and Jairam, 2000; Jones and Fox, 1976; Kulanthaivel et al, 1992; Younes et al, 1996], which is associated with alterations in the uptake of amino acids and glucose in diabetic pregnancies [Osmond et al, 2000; Vejtrop et al., 1977]. The levels of Na^+ and Mg^{++} are decreased in the diabetic maternal plasma [Husain et al, 2000]. NHE-1 and NHE regulatory factor are present primarily in the microvillus membranes, whereas NHE-3 and NHE-3 kinase, a regulatory protein are localized on the membranes contiguous with the basal membranes in syncytiotrophoblasts. This polarity of NHE expression is necessary to maintain Na^+ and H^+ exchange between the placenta and the maternal and fetal circulations [Pepe et al, 2001]. Interestingly, the Na^+/H^+ exchange mechanism in the placental brush border membrane is not regulated by parathyroid hormone, angiotensin or insulin [Brunette and Leclerc, 1996]. Therefore, NHE-1 in this organ is considered to be different from the isoform present in the renal brush border [Brunette and leclerc, 1996]. We have recently provided cloning evidences that this unique regulation is not, at least, due to differences in the C-terminus regulatory domain of NHE-1 [Khan et al, 2003]. Genetic and biochemical studies have implicated NHE-1 in the development of diabetes [Davies et al, 1995; Dubouix et al, 2000; Mayhem and Jairam, 2000; Mayhew and Morahan et al, 1994; Ng et al, 1994; Siczkowski et al, 1995]. Diabetes is an underlying cause of hypertension and nephropathy, which could be due to an increased expression of NHE protein. Induction of NHE-1 expression causes Na^+ retention by the tissues leading to volume expansion and hypertension. Diabetes is also associated with placental hypervascularization due to fetal hypoxia [Ikagaya et al, 1999; Scano et al, 2000]. Control of hyperglycemia ameliorates the adverse diabetic consequences which supports the idea that hyperglycemia is one of the major factors causing diabetes related malformations in the mothers and babies [Nazer and Ramirez, 2000 Review]. We have reported a selective suppression of NHE-1 protein which seems to be independent of the changes seen in the mothers' and babies' health or changes in the diabetic placentas [Khan et al, 2003; Nazer and Ramirez, 2000]. These findings support the notion that controlling the blood glucose level is beneficial, and the molecular changes that occur in the placental NHE-1 isoform may be necessary to protect the fetus and the mother [Khan et al, 2003]. Although factors such as stability and translation rate remain a distinct possibility for down regulation, the lack of change in mRNA levels supports a post-transcriptional regulation. Elevated NHE-1 activity, but not the mRNA or protein levels in leucocytes have been reported in human preeclimptic women [Lee et al, 2001, 2002]. We have shown PGE-2 mediated suppression of NHE-1 in placentas from preeclimptic pregnancies, which might render the placenta with impaired uptake of water and electrolytes, and therefore may be involved in the pathogenesis of preeclimpsia [Khan et al, 1999].

2.3. Gastroesophageal Reflux Disease

Gastroesophageal reflux disease [GERD] is associated with stomach acid reflux into esophagus which causes epithelial cell damage, and triggers the events that lead to inflammation. Acid reflux in human GERD condition, damages intercellular connections and exposes the basolateral domains leading to cellular acidosis [Carlsson et al, 1999; DeMeester et al, 1980; Dobashi, 2001; Hamilton and Orlando, 1989; Khalbuss et al, 1995; Orlando, 1984, 1997, 2001; Tobey et al, 1992, 1993, 1995, 1998]. We have shown recently an induction of NHE-1 in GERD biopsy samples [Siddique and Khan, 2003]. NHE-1 protects the esophageal mucosa by maintaining ionic and pHi homeostasis [Carlsson et al, 1999; Khalbuss et al, 1995]. Human esophageal NHE-1 responds to decrease in pHi by extruding H^+ efficiently, thus is considered to be an efficient epithelial defense against acid injury [Khalbuss et al, 1995]. Induction of NHE-1, we believe, is an important initiating factor in the GERD pathogenesis. The induction of NHE-1 causes a pH-paradox leading to a decreased luminal pH, and alkalinization of the cell cytoplasm. Similar phenomenon underlying cellular damage following the NHE-1 activation has been suggested in other pathologic condition [Gazmuri et al, 2002]. A study has reported an induction of NHE-1 and suppression of sodium pump, suggesting an accumulation of Na^+, thus leading to cell volume expansion and necrosis in experimental GERD [Orlando et al, 1984]. NHE-1 and sodium pump activity in human GERD appears to be a histamine-dependent mechanism which is important in the disease pathogenesis [Siddique and Khan, 2003].

2.4. Hypertension

Hypertension is associated with volume expansion leading to cardiovascular complications and tissue-damage, indicating that the mechanisms of electrolyte and water homeostasis are impaired. NHE isoforms 1 and 3 are consistently induced in a variety of cell types from hypertensive humans and animal models [Al-Qattan et al, 2003; Khong et al, 2001; LaPointe et al, 2002; Lucchesi et al, 1994; Quinn and Ng, 2001]. There are several experimental models of hypertension including SHR, Goldblatt-clipped kidney(s) and angiotensin II induced-hypertension that mimic human condition. We and others have shown an induction of renal expression of NHE-1 and NHE-3 in a rat 2k-1c Goldblatt model [Aldred et al 2000; Alqattan et al, 2003; Kelley, 1997; Khan et al 2002; LaPointe et al, 2002; Takahashi et al, 1997; Quinn and Ng, 2001]. In angiotensin-II induced hypertension and in SHR rats, we observed an induction of NHE-1 and NHE-3 in rat heart, colon, kidney and brain [Unpublished findings, IK]. In addition, in human preeclimptic placentas, we have also shown an induction of NHE-1 isoform [Khan et al, 1999]. These findings consistently support an induction of NHE-1 and 3 isoforms in different tissues in hypertensive rats. Induction of NHE-1 and NHE-3 causes accumulation of Na^+, and thus volume expansion which might explain cardiomyocyte's hypertrophy, hyperplasia and cell damage. The tissue damage has been predicted to be mediated by induction of certain proteases following NHE-1 induction [Karmazyn, 2001]. In addition, the induction leads to an increase in intracellular Na^+ causing activation of Ca^{+2}/Na^+ exchanger, which leads to an increase in intracellular Ca^{+2} and hence arrhythmia and muscle contractility [Karmazyn, 2001]. A selective inhibition of NHE-1 has been shown to be cardioprotective [Karmazyn, 2001].

2.5. Ischemia-Reperfusion

Ischemia reperfusion is associated with many diseases, and is considered as a main cause of tissue damage. Ischemia is associated with a shift from aerobic to anaerobic metabolism leading to acidotic condition which induces NHE-1. Therefore, ischemia/reperfusion injuries are considered due to induction of NHE-1, and have been reversed following its inhibition [Avkiran, 1999; Chen et al, 2001; Cingolani et al, 2003; Gazmuri et al, 2001; Haist et al, 2003; Karmazyn, 2001; Linz and Busch, 2003; Morris, 2002]. NHE-1 is being considered as a target in human hypertension and tissue damage. Since chemical inhibitors are not so selective, antisense oligo, siRNA and miRNA molecular inhibitors might allow higher level of specificity and effectiveness.

3. Methods Used

Studies on NHE have utilized a wide variety of biochemical, immunological, pharmacologic and molecular biology techniques. Perhaps the most common and the first used techniques include pharmacological tools. In fact pharmacologic techniques have been important instruments in revealing different isozymes/isoforms of various proteins, receptors and pumps etc. Functional assays of NHE are based on the intracellular pH measurement using pH sensitive dyes, and transport studies using ^{23}Na tracer element. These techniques have their own merits and demerits and will not be discussed in this chapter. The molecular biology tools however have been important in elucidating the structure and function of NHE isoforms. In this section, we will briefly describe some common techniques used for studies on NHE. Cloning era for eukaryotic cDNA cloning started with the discovery of reverse transcriptase in early 1970s. Success of cloning and measurement of mRNA levels requires methods that produce an intact cellular RNA of high quality. The commonly used technique for RNA extraction is cesium chloride centrifugation [Chirgwin et al 1979]. This method gives a very high quality RNA free from RNase contamination [Fig. 8]. There are several kits available now a days from different sources, which are based on guanidinium isothiocyanate mediated RNase inhibition followed by phenol-chloroform extraction [Chomczynski and Sachi, 1987]. The kits are quite easy to use, but their high cost and lack of freedom to standardize the conditions limit their significance. Thus we have always relied on the cesium chloride gradient method. Briefly, tissues (up to 100 mg) are homogenized/polytroned in 8M guanidinium isothiocyanate solution. The lysate so obtained is centrifuged at 10,000g for 10 min at 4°C. The pellet is discarded and the supernatant is layered on top of 3 ml of 4M cesium chloride solution. The gradient is run at 110,000g for 18-24 hr at 15°C using a swinging bucket rotor in a Beckman ultracentrifuge. The pellet containing RNA is recovered, and suspended in RNase-free water. RNA so obtained can be used directly or stored at -80°C until use. This RNA is suitable for any enzymatic or non enzymatic procedure.

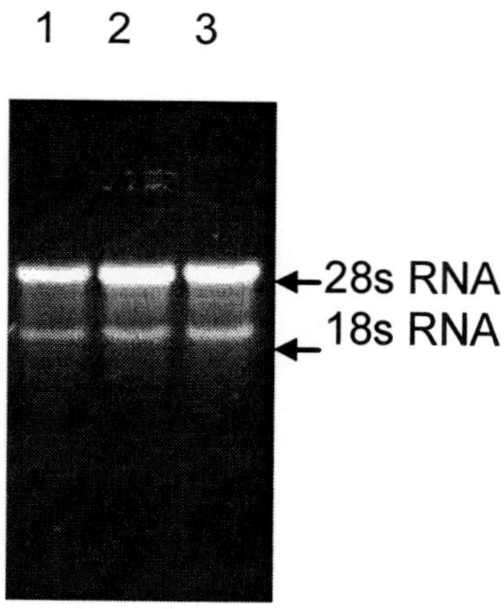

Fig. 8. Ethidium bromide stained agarose gel picture showing total RNA from rat colon extracted using cesium chloride centrifugation method.

However, some times further purification by acid-phenol and chloroform extraction followed by ethanol precipitation is required. Estimation of mRNA is required to investigate regulation of gene expression. This is achieved by several methods including northern blot analysis, S1-protection assay or RT-PCR. While northern blot is reliable, due to limited sensitivity it cannot be used for detection of low abundant messages. Therefore more sensitive method such as RT-PCR is used commonly. RT-PCR requires much more rigorous optimization and precautions than those taken in any other method. We have used RT-PCR routinely for quantitation of mRNA using an internal control or a competitive control [Khan and Al-Awadi, 1997]. Competitive PCR is much more reliable, and should be used for quantitation of mRNA [Khan and Al-Awadi, 1997]. In addition, characterization of PCR fragment is also important in RT-PCR quantitation which can be achieved by using restriction analysis or nucleotide sequencing. Reverse transcription is performed using AMV or MMLV reverse transcriptase and total or polyA^{+} RNA. We have relied on AMV based reverse transcription, and have used it routinely in our studies [Khan et al, 1992, Khan and al-Awadi, 1997]. However, now a days one can use RT-PCR together in one step using single step-RT-PCR kit [Invitrogen].

4. Conclusions

Identification of new genes encoding NHE isoforms and alternative splicing are increasing its structural and functional diversity. These isoforms are reported to be integral membrane protein, but recent computer analysis shows a strong probability of mitochondrial target sequence, at least, in NHE-1 and NHE-3 in addition to NHE-6 isoform. These isoforms are important in maintaining normal absorptive and secretory functions of GI-tract. An alteration in NHE expression leads to pHi and ionic concentration imbalances, hence contributing to pathogenesis of many human diseases. NHE-1 is being considered as a very important therapeutic target in cardiovascular diseases and ischemia/reperfusion injuries. Studies utilizing molecular inhibitors such antipeptides, antisense oligos, siRNA or miRNA should provide added specificity and less side effects, and therefore prove more useful as compared to the presently known chemical inhibitors.

Acknowledgments

Mrs Thomas, N, Haridas S and Al-Shamali A for technical support. Financial support from the Kuwait University Research Administration through grants MB 027, MB 01/99 and MB 01/00.

References

Aldred KL, Harris PJ, Eitle E. Increased proximal tubule NHE-3 and H+-ATPase activities in spontaneously hypertensive rats. *J Hypertens* 2000;18[5]:623-8.

Al-Qattan KK, Khan I, Alnaqeeb MA, Ali M. Mechanism of garlic [Allium sativum] induced reduction of hypertension in 2K-1C rats: a possible mediation of Na/H exchanger isoform-1. *Prostaglandins Leukot Essent Fatty Acids* 2003; 69[4]:217-22.

Ambuhl PM, Yang X, Peng Y, Preisig PA, Moe OW, Alpern RJ. Glucocorticoids enhance acid activation of the Na+/H+ exchanger 3 [NHE3]. *J Clin Invest* 1999;103[3]:429-35.

Attaphitaya S, Park K, Melvin JE. Molecular cloning and functional expression of a rat Na+/H+ exchanger [NHE5] highly expressed in brain. *J Biol Chem* 1999; 274 [7], 4383-4388.

Attaphitaya S, Nehrke K, Melvin JE. Acute inhibition of brain-specific Na[+]/H[+] exchanger isoform 5 by protein kinases A and C and cell shrinkage. *Am J Physiol Cell Physiol* 2001;281[4]:C1146-57.

Avkiran M. Rational basis for use of sodium-hydrogen exchange inhibitors in myocardial ischemia. *Am J Cardiol* 1999;83[10A]:10G-17G

Baird NR, Orlowski J, Szabo EZ, Zaun HC, Schultheis PJ, Menon AG, Shull GE. Molecular cloning, genomic organization, and functional expression of Na+/H+ exchanger isoform 5 [NHE5] from human brain. *J Biol Chem* 1999;274[7]:4377-82

Bell PD, Peti-Peterdi J. Angiotensin II stimulates macula densa basolateral sodium/hydrogen exchange via type 1 angiotensin II receptors. *J Am Soc Nephrol* 1999;10 Suppl 11:S225-9.

Biemesderfer D, DeGray B and Aronson P S. Membrane topology of NHE3. Epitopes within the carboxyl-terminal hydrophilic domain are exoplasmic. *J Biol Chem* 1998; 273[20]:12391-12396.

Boivin GP, Schultheis PJ, Shull GE, Stemmermann GN. Variant form of diffuse corporal gastritis in NHE2 knockout mice. *Comp Med* 2000;50[5]:511-5.

Bookstein C, DePaoli AM, Xie Y, Niu P, Musch MW, Rao MC, Chang EB. Na+/H+ exchangers, NHE-1 and NHE-3, of rat intestine. Expression and localization. *J Clin Invest* 1994;93[1]:106-13.

Bookstein C, Musch MW, DePaoli A, Xie Y, Rabenau K, Villereal M, Rao MC, Chang EB. Characterization of the rat Na+/H+ exchanger isoform NHE4 and localization in rat hippocampus. *Am J Physiol* 1996;271[5 Pt 1]:C1629-38.

Brant SR, Yun CH, Donowitz M, Tse CM. Cloning, tissue distribution, and functional analysis of the human Na+/N+ exchanger isoform, NHE3. *Am J Physiol* 1995;269[1 Pt 1]:C198-206.

Brett CL, Wei Y, Donowitz M, Rao R. Human Na[+]/H[+] exchanger isoform 6 is found in recycling endosomes of cells, not in mitochondria. *Am J Physiol Cell Physiol* 2002;282 [5]:C1031-41.

Brooks HL, Sorensen AM, Terris J, Schultheis PJ, Lorenz JN, Shull GE, Knepper MA. Profiling of renal tubule Na+ transporter abundances in NHE3 and NCC null mice using targeted proteomics. *J Physiol* 2001;530[Pt 3]:359-66

Brunette MG and Leclerc, CD. Na^+ transport by human placental brush border membranes: are there several mechanisms? *J Cell Physiol* 1996;167, 72-80.

Carlsson R, Fandriks L, Jonsson C, Lundell L, Orlando RC: Is the esophageal squamous epithelial barrier function impaired in patients with gastroesophageal reflux disease? *Scand J Gastroenterol* 1999;34: 454-458.

Cavet ME, Akhter S, de Medina FS, Donowitz M, Tse CM. Na[+]/H[+] exchangers [NHE1-3] have similar turnover numbers but different percentages on the cell surface. *Am J Physiol* 1999;277[6 Pt 1]:C1111-21

Chambrey R, Achard JM, Warnock DG. Heterologous expression of rat NHE4: a highly amiloride-resistant Na+/H+ exchanger isoform. *Am J Physiol* 1997;272[1 Pt 1]:C90-8.

Chen L, Gan XT, Haist JV, Feng Q, Lu X, Chakrabarti S, Karmazyn M. Attenuation of compensatory right ventricular hypertrophy and heart failure following monocrotaline-induced pulmonary vascular injury by the Na+-H+ exchange inhibitor cariporide. *J Pharmacol Exp Ther* 2001;298[2]:469-76.

Chirgwin JM, Przybyla AE, MacDonald R, Rutter WJ. Isolation of biologically active ribonucleic acid from sources enriched in ribonuclease. Biochemistry 1979; 18: 5294-5299.

Chomczynski P and Sachi N. Single step method of RNA isolation by acid guanidinium thiocyanate–phenol–chloroform extraction. *Anaytical Biochemistry* 1987;162:156–159

Chow CW. Regulation and intracellular localization of the epithelial isoforms of the Na+/H+ exchangers NHE2 and NHE3. *Clin Invest Med* 1999;22(5):195-206.

Chow CW, Woodside M, Demaurex N, Yu FH, Plant P, Rotin D, Grinstein S, Orlowski J. Proline-rich motifs of the Na+/H+ exchanger 2 isoform. Binding of Src homology domain 3 and role in apical targeting in epithelia. *J Biol Chem* 1999; 274(15):10481-8.

Chutkan RK. Inflammatory bowel disease. Prim Care 2001; 28: 539-56.

Cingolani HE, Rebolledo OR, Portiansky EL, Perez NG, Camilion de Hurtado MC. Regression of hypertensive myocardial fibrosis by Na[+]/H[+] exchange inhibition. *Hypertension.* 2003; 41[2]:373-7.

Claros MG, Vincens P. Computational method to predict mitochondrially imported proteins and their targeting sequences. *Eur J Biochem* 1996; 241, 770-786.

Cohen ME, Reinlib L, Watson AJ, Gorelick F, Rys-Sikora K, Tse M, Rood RP, Czernik AJ, Sharp GW, Donowitz M. Rabbit ileal villus cell brush border Na+/H+ exchange is regulated by Ca2+/calmodulin-dependent protein kinase II, a brush border membrane protein. *Proc Natl Acad Sci (USA)* 1990; 87[22]:8990-4.

Cook TA, Brading AF, Mortensen NJ. Abnormal contractile properties of rectal smooth muscle in chronic ulcerative colitis. *Aliment Pharmacol Ther* 2000; 14: 1287-94.

Counillon L and Pousseygur J. Molecular biology and hormonal regulation of vertebrate Na/H exchanger isoforms. Molecular Biology and function of carrier proteins, edited by L Reuss, JM Russell, Jr and ML Jennings. New York: Rockefeller Univ. Press, p170-185, 1993.

Counillon L, Pouyssegur J. Nucleotide sequence of the Chinese hamster Na+/H+ exchanger NHE1. *Biochim Biophys Acta* 1993; 1172[3]:343-5.

Counillon L, Noel J, Reithmeier RA, Pouyssegur J. Random mutagenesis reveals a novel site involved in inhibitor interaction within the fourth transmembrane segment of the Na+/H+ exchanger-1. *Biochemistry* 1997; 36(10):2951-9.

Counillon, L., Poussegur, J. The expanding family of tilizatio Na[+]/H[+] exchangers. *J Biol Chem* 2000; 275[1]:1-4.

Counillon L, Franchi A, Pouyssegur J.A point mutation of the Na+/H+ exchanger gene (NHE1) and amplification of the mutated allele confer amiloride resistance upon chronic acidosis. *Proc Natl Acad Sci USA* 1993; 90(10):4508-12.

Davies JE, Siczkowski M, Sweeney FP, Quinn PA, Krolewski B, Krolewski AS, Ng LL Glucose-induced changes in turnover of Na^+/H^+ exchanger of immortalized lymphoblasts from type I diabetic patients with nephropathy. *Diabetes* 1995; 44, 382-388.

Delvaux M, Bastie MJ, Chentoufi J, Cragoe EJ Jr, Vaysse N, Ribet A.Amiloride and analogues inhibit Na[+]-H+ exchange and cell proliferation in AR42J pancreatic cell line. *Am J Physiol* 1990; 259[5 Pt 1]:G842-9.

DeMeester TR, Wang CI, Wernly JA, Pellegrini CA, Little AG, Klementschitsch P, Bermudez G, Johnson LF, Skinner DB. Technique, indications, and clinical use of 24 hour esophageal pH monitoring. *J Thorac Cardiovasc Surg* 1980;79: 656-701.

DeSilva, MG, Elliott K, Dahl, HH, Fitzpatrick E, Wilcox S, Delatycki M,Williamson R, Efron D, Lynch M, Forrest S. Disruption of a novel member of a sodium/hydrogen exchanger family and DOCK3 is associated with an attention deficit hyperactivity disorder-like phenotype. *J Med Genet* 2003; 40[10], 733-740.

Dibrov P, Fliegel L. Comparative molecular analysis of Na+/H+ exchangers: a unified model for Na+/H+ antiport. *FEBS LETT.* 1998; 424:1-5.

Dobashi K. GERD [gastroesophageal reflux disease]. Nippon Rinsho 2001; 59: 2045-2050.

Donowitz M, Montgomery JL, Walker MS, Cohen ME. Brush-border tyrosine phosphorylation stimulates ileal neutral NaCl absorption and brush-border Na[+]-H+ exchange. *Am J Physiol* 1994; 266[4 Pt 1]:G647-56.

Dubouix A, Gennero I, Nieto M, Ser N, Hannaire-Broutin H, Tauber JP, Pourrat J, Fauvel J, Barthe P, Chap H, Salles JP. Polymorphism of the 5' untranslated region of NHE1 gene associated with type-I diabetes. *Mol Cell Biol. Res Commun,* [2000]; 3: 141-144.

Dudeja PK, Rao DD, Syed I, Joshi V, Dahdal RY, Gardner C, Risk MC, Schmidt L, Bavishi D, Kim KE, Harig JM, Goldstein JL, Layden TJ, Ramaswamy K. Intestinal distribution of human Na+/H+ exchanger isoforms NHE-1, NHE-2, and NHE-3 mRNA. *Am J Physiol* 1996; 271[3 Pt 1]:G483-93.

Emmer E, Rood RP, Wesolek JH, Cohen ME, Braithwaite RS, Sharp GW, Murer H, Donowitz M. Role of calcium and calmodulin in the regulation of the rabbit ileal brush-border membrane Na+/H+ antiporter. *J Membr Biol* 1989;108[3]:207-15.

Fallingborg J, Christensen LA, Jacobsen BA, Rasmussen SN. Very low intraluminal colonic pH in patients with active ulcerative colitis. *Dig Dis Sci* 1993; 38: 1989-1993.

Fliegel L, Sardet C, Pousseygur J, Barr A. Identification of the protein and cDNA of the cardiac Na+/H+ exchanger. *FEBS Lett* 1991; 279:25-29.

Gazmuri RJ, Ayoub IM, Kolarova JD, Karmazyn M. Myocardial protection during ventricular fibrillation by inhibition of the sodium-hydrogen exchanger isoform-1. *Crit Care Med.* 30[4 Suppl]: S166-71, 2002 Review.

Gazmuri RJ, Ayoub IM, Hoffner E, Kolarova JD. Successful ventricular defibrillation by the selective sodium-hydrogen exchanger isoform-1 inhibitor cariporide. *Circulation.* 2001; 104[2]:234-9.

Gebreselassie D, Rajarathnam K, Fliegel L. Expression, purification, and characterization of the carboxyl-terminal region of the Na+/H+ exchanger. *Biochem Cell Biol* 1998; 76[5]:837-42.

Gillis D, Shrode LD, Krump E, Howard CM, Rubie EA, Tibbles LA, Woodgett J, Grinstein S. Osmotic stimulation of the Na+/H+ exchanger NHE1: relationship to the activation of three MAPK pathways. *J Membr Biol* 2001; 181[3]:205-14.

Girardi AC, Titan SM, Malnic G, Reboucas NA. Chronic effect of parathyroid hormone on NHE3 expression in rat renal proximal tubules. *Kidney Int* 2000; 58[4]:1623-31.

Goyal S, Vanden Heuvel G, Aronson PS. Renal expression of novel Na+/H+ exchanger isoform NHE-8. *Am J Physiol Renal Physiol* 2003; 284 [3]:F467-73.

Haist JV, Hirst CN, Karmazyn M. Effective protection by NHE-1 inhibition in ischemic and reperfused heart under preconditioning blockade. *Am J Physiol Heart Circ Physiol* 2003; 284[3]:H798-803.

Hamilton BH, Orlando RC. In vivo alkaline secretion by mammalian esophagus. *Gastroenterology* 1989; 97:640-648

Hildebrandt F, Pizzonia JH, Reilly RF, Reboucas NA, Sardet C, Pouyssegur J, Slayman CW, Aronson PS, Igarashi P. Cloning, sequence, and tissue distribution of a rabbit renal Na+/H+ exchanger transcript. *Biochim Biophys Acta* 1991;1129[1]:105-8

Hoogerwerf WA, Tsao SC, Devuyst O, Levine SA, Yun CH, Yip JW, Cohen ME, Wilson PD, Lazenby AJ, Tse CM, Donowitz M. NHE2 and NHE3 are human and rabbit intestinal brush-border proteins. *Am J Physiol* 1996; 270[1 Pt 1]:G29-41.

Husain SM, Mughal MZ, Sibley CP. Effects of acute maternal hyperglycaemia and hyperosmolality on maternofetal transfer of calcium and magnesium across the in situ perfused rat placenta. *Magnes Res* 2000; 13:239-247.

Ikegaya N, Yonemura K, Suzuki T, Kato-Ohishi H, Taminato T, Hishida A. Impairment of ventilatory response to metabolic acidosis in insulin-dependent diabetic patients with advanced nephropathy. *Ren Fail* 1999; 21[5]:495-8.

Incerpi S, Luly P, De Vito P, Farias RN. Short-term effects of thyroid hormones on the Na/H antiport in L-6 myoblasts: high molecular specificity for 3,3',5-triiodo-L-thyronine. *Endocrinology* 1999; 140[2]:683-9.

Jones CJ and Fox H. An ultrastructural and ultrahistochemical study of the placenta of the diabetic woman. *J Pathol* 1976; 119, 91-99.

Karmazyn M. Therapeutic potential of Na-H exchange inhibitors for the treatment of heart failure. *Expert Opin Investig Drugs*. 2001;10[5]: 835-43. Review

Kelly MP, Quinn PA, Davies JE, Ng LL. Activity and expression of Na[+]-H+ exchanger isoforms 1 and 3 in kidney proximal tubules of hypertensive rats. *Circ Res.* 1997; 80[6]:853-60.

Khalbuss WE, Marousis CG, Subramanyam M, Orlando RC. Effect of HCl on transmembrane potentials and intracellular pH in rabbit esophageal epithelium. *Gastroenterology* 1995;108:662-672

Khan I, Al-Awadi FM. Colonic muscle enhances the production of interleukin-1 beta messenger RNA in experimental colitis. *Gut* 1997; 40, 307-312.

Khan I, Tabb T, Garfield RE, Grover AK. Polymerase chain reaction assay of mRNA using 28S rRNA as internal standard. Neuroscience Letters 1992; 147[1]:114-117.

Khan I, al-Awadi FM, Abul H. Colitis-induced changes in the expression of the Na+/H+ exchanger isoform NHE-1. *J Pharmacol Exp Ther* 1998; 285[2]:869-75

Khan I, Ali M. Altered expression of the Na+/H+ exchanger isoform-3 in experimental colitis: effect of garlic. *Mol Cell Biochem* 1999; 200[1-2]:77-84.

Khan I, Al-Yatama M, Nandakumaran M. Expression of the Na[+]-H+ exchanger isoform-1 and cyclooxygenases in human placentas: their implications in preeclampsia. *Biochem Mol Biol Int* 1999; 47[4]:715-22.

Khan I, Cheng B. Selective suppression of renal Na+/H+ exchanger isoform-3 by prolonged stimulation of rats with adrenocorticotropic hormone. Endocrine 2001; 16(3):189-94

Khan I. Topology of the C-terminus of Sodium Hydrogen Exchanger Isoform-1: Presence of an Extracellular Epitope. *Arch Biochem Biophys* 2001; 391: 25-29.

Khan I, Thomas N, Haridas S. Expression and sub cellular localization of the sodium hydrogen exchanger isoform-1 in rat tissues: A possible functional relevance. *Molecular Cellular Biochemistry* 2001; 219:153-161.

Khan I. Antisense inhibition of cyclooxygenase-2 causes a selective suppression of the Na+-H+ exchanger isoform 3 in rat kidney in experimental colitis. *Nephron* 2002; 91[1]:120-128.

Khan I, Al-Qattan KK, Alnaqeeb MA, Ali M. Altered expression of Na[+]/H[+] exchanger isoforms 1 and 3 in clipped and unclipped kidneys of a 2-kidney-1-clip Goldblatt model of hypertension. *Nephron* 2002 Oct;92[2]:346-55

Khan I, Siddique I, Al-Awadi FM, Mohan K. Role of Na+/H+ exchanger isoform-1 in human inflammatory bowel disease. *Can J Gastroenterol* 2003; 17(1):31-36.

Khan I, Al-Awadi FM, Nandakumaran M, Abul H, Al-Azemi M. Suppression of Na+-H+ exchanger-1 in placentas of type 2 diabetic pregnant women: possible functional implication. *Acta Diabetol* 2003; 40[1]:28-36.

Siddique I and Khan I. Regulation of Na/H exchanger-1 in gastroesophageal reflux disease: possible interaction of histamine receptor. *Dig Dis Sci* 2003; 48(9):1832-8.

Khong TK, McIntyre DJ, Sagnella GA, Markandu ND, Miller MA, Baker EH, Griffiths JR, MacGregor GA. In-vivo intracellular pH at rest and during exercise in patients with essential hypertension. *J Hypertens* 2001; 19(9):1595-600.

Kiraku J, Sugiyama T, Ashida T, Takahashi N, Fujii J, Kuro-o M, Nagai R. Increases in intracellular calcium of arterial smooth muscle cells in transgenic mice overexpressing Na+/H+ exchanger. *J Cardiovasc Pharmacol* 2000; 35[3]:511-513.

Klanke CA, Su YR, Callen DF, Wang Z, Meneton P, Baird N, Kandasamy RA, Orlowski J, Otterud BE, Leppert M, et al. Molecular cloning and physical and genetic mapping of a novel human Na+/H+ exchanger (NHE5/SLC9A5) to chromosome 16q22.1. *Genomics* 1995; 25(3):615-22.

Kulanthaivel P, Furesz TC, Moe AJ, Smith CH, Mahesh VB, Leibach FH, Ganapathy V. Human placental syncytiotrophoblast expresses two pharmacologically distinguishable types of Na^+/H^+ exchangers, NHE-1 in the maternal-facing [brush border] membrane and NHE-2 in the fetal-facing [basal] membrane. *Biochem J* 1992; 284: 33-38.

Laghmani K, Preisig PA, Moe OW, Yanagisawa M, Alpern RJ. Endothelin-1/endothelin-B receptor-mediated increases in NHE3 activity in chronic metabolic acidosis. *J Clin Invest* 2001; 107[12]:1563-9.

LaPointe MS, Sodhi C, Sahai A, Batlle D. Na+/H+ exchange activity and NHE-3 expression in renal tubules from the spontaneously hypertensive rat. *Kidney Int* 2002; 62[1]:157-65.

Ledoussal C, Woo AL, Miller ML, Shull GE. Loss of the NHE2 Na[+]/H[+] exchanger has no apparent effect on diarrheal state of NHE3-deficient mice. *Am J Physiol Gastrointest Liver Physiol* 2001; 281[6]:G1385-96.

Lee VM, Quinn PA, Jennings SC, Halligan AW, Ng LL. Altered Na+/H+ exchanger isoform 1 activity in immortalized lymphoblasts from women with pre-eclampsia: evidence for an intermediate phenotype. *Clin Sci (Lond)* 2002; 103(5):503-9.

Lee VM, Halligan AW, Ng LL. Leucocyte intracellular pH and Na+/H+ exchanger isoform-1 activity in postpartum women with pre-eclampsia. *BJOG.* 2001; 108(6):615-22.

Levine SA, Nath SK, Yun CH, Yip JW, Montrose M, Donowitz M, Tse CM. Separate C-terminal domains of the epithelial specific brush border Na+/H+ exchanger isoform NHE3 are involved in stimulation and inhibition by protein kinases/growth factors. *J Biol Chem* 1995; 270[23]:13716-25.

Lifton RP, Sardet C, Pouyssegur J, Lalouel JM. Cloning of the human genomic amiloride-sensitive Na+/H+ antiporter gene, identification of genetic polymorphisms, and localization on the genetic map of chromosome 1p. *Genomics.* 1990; 7[1]:131-135.

Linz WJ, Busch AE, NHE-1 inhibition: from protection during acute ischaemia/reperfusion to prevention/reversal of myocardial utilizatio. *Naunyn Schmiedebergs Arch Pharmacol* 2003; 368[4]:239-246.

Lucchesi PA, DeRoux N, Berk BC. Na(+)-H+ exchanger expression in vascular smooth muscle of spontaneously hypertensive and Wistar-Kyoto rats. *Hypertension* 1994; 24(6):734-8.

Maher MM, Gontarek JD, Bess RS, Donowitz M, Yeo CJ. The Na+/H+ exchange isoform NHE3 regulates basal canine ileal Na+ absorption in vivo. *Gastroenterology* 1997; 112[1]:174-183.

Malakooti J, Dahdal RY, Schmidt L, Layden TJ, Dudeja PK, Ramaswamy K. Molecular cloning, tissue distribution, and functional expression of the human Na[+]/H[+] exchanger NHE2. *Am J Physiol* 1999; 277[2 Pt 1]:G383-390

Mayhew TM, Jairam IC. Stereological comparison of 3D spatial relationships involving villi and intervillous pores in human placentas from control and diabetic pregnancies. *J Anat* 2000; 197: 263-274.

Michail S, Collins JF, Xu H, Kaufman S, Vanderhoof J, Ghishan FK. Abnormal expression of brush-border membrane transporters in the duodenal mucosa of two patients with microvillus inclusion disease. *J Pediatr Gastroenterol Nutr* 1998; 27 [5]:536-42.

Miyazaki E, Sakaguchi M, Wakabayashi S, Shigekawa M, Mihara K. NHE6 protein possesses a signal peptide destined for endoplasmic reticulum membrane and localizes in secretory organelles of the cell. *J Biol Chem* 2001; 276[52]:49221-7

Morahan G, McClive P, Huang D, Little P, Baxter A. Genetic and physiological association of diabetes susceptibility with raised Na^+/H^+ exchange activity. *Proc Natl Acad Sci [USA]* 1994; 91: 5898-5902.

Morris K. Targeting the myocardial sodium-hydrogen exchange for treatment of heart failure. *Expert Opin Ther Targets* 2002; 6[3]:291-8.

Mukhin YV, Vlasova T, Jaffa AA, Collinsworth G, Bell JL, Tholanikunnel BG, Pettus T, Fitzgibbon W, Ploth DW, Raymond JR, Garnovskaya MN. Bradykinin B2 receptors activate Na+/H+ exchange in mIMCD-3 cells via Janus kinase 2 and Ca2+/calmodulin. *J Biol Chem* 2001; 276[20]:17339-17346.

Muller T, Wijmenga C, Phillips AD, Janecke A, Houwen RH, Fischer H, Ellemunter H, Fruhwirth M, Offner F, Hofer S, Muller W, Booth IW, Heinz-Erian P. Congenital sodium diarrhea is an autosomal recessive disorder of sodium/proton exchange but unrelated to known candidate genes. *Gastroenterology* 2000; 119[6]:1506-13.

Nagase T, Ishikawa K, Suyama M, Kikuno R, Hirosawa M, Miyajima N. Prediction of the coding sequences of unidentified human genes. The complete sequences of 100 new cDNA clones from brain which code for large proteins in vitro. *DNA Res* 1999; 6 (1): 63-70

Nazer J, Ramirez R. Congenital malformations in the offspring of diabetic mothers. *Rev Med Chil* 2000; 128:1045-52. Review.

Nemeth ZH, Deitch EA, Szabo C, Mabley JG, Pacher P, Fekete Z, Hauser CJ, Hasko G. Na+/H+ exchanger blockade inhibits enterocyte inflammatory response and protects against colitis. *Am J Physiol Gastrointest Liver Physiol* 2002; 283[1]:G122-32.

Ng LL, Davies JE, Siczkowski M, Sweeney FP, Quinn PA, Krolewski B, Krolewski AS. Abnormal Na^+/H^+ antiporter phenotype and turnover of immortalized lymphoblasts from type 1 diabetic patients with nephropathy. *J Clin Invest* 1994; 93: 2750-2757.

Nielsen H, Engelbrecht J, Brunak S, vonHeijne G. Identification of prokaryotic and eukaryotic signal peptides and prediction of their cleavage sites. *Protein Engineering* 1997; 10: 1-6.

Noda M, Takahashi H, Tanabe T, Toyosato M, Kikyotani S, Hirose T, Asai M, Takashima H, Inayama S, Miyata T, Numa S. Primary structures of beta- and delta-subunit precursors of Torpedo californica acetylcholine receptor deduced from cDNA sequences. *Nature* 1983; 301:251-255.

Numata M, Petrecca K, Lake N, Orlowski J. Identification of a mitochondrial Na+/H+ exchanger. *J Biol Chem* 1998; 273[12]:6951-9.

Numata M. and Orlowski J. Molecular Cloning and Characterization of a Novel [Na+,K+]/H+Exchanger Localized to the trans-Golgi Network. *J Biol Chem* 2001; 276: 17387-17394

Orlando RC, Bryson JC, Powell DW. Mechanisms of H^+ injury in rabbit esophageal epithelium. Am J Physiol 246: G718-724, 1984

Orlando RC. The pathogenesis of gastroesophageal reflux disease: the relationship between epithelial defense, dysmotility, and acid exposure. *Am J Gastroenterol* 1997; 92[4 Suppl]: 3S-5S.

Orlando RC. Overview of the mechanisms of gastroesophageal reflux. *Am J Med* 2001; *111* Suppl 8A: 174S-177S, Review.

Orlowski, J., Ramani, A., Kandasamy, R. A., and Shull, G. E. Molecular cloning of putative members of the Na/H exchanger gene family. cDNA cloning, deduced amino acid sequence, and mRNA tissue expression of the rat Na/H exchanger NHE-1 and two structurally related proteins. *J Biol Chem* 1992; 267[13]:9331-9339.

Orlowski J, Kandasamy RA.J Delineation of transmembrane domains of the Na+/H+ exchanger that confer sensitivity to pharmacological antagonists. *J Biol Chem* 1996; 271(33):19922-7.

Orlowski J, Grinstein S. Diversity of the mammalian sodium/proton exchanger SLC9 gene family. *Pflugers Arch* 2004; 447[5]:549-65.

Osmond DT, Nolan CJ, King RG, Brennecke SP, Gude NM. Effects of gestational diabetes on human placental glucose uptake, transfer, and utilization. *Diabetologia* 2000; 43: 576-582.

Pedersen SF, King SA, Rigor RR, Zhuang Z, Warren JM, Cala PM. Molecular cloning of NHE1 from winter flounder RBCs: activation by osmotic shrinkage, cAMP, and calyculin A. *Am J Physiol Cell Physiol* 2003; 284[6]:C1561-1576.

Pepe GJ, Burch MG, Sibley CP, Davies WA, Albrecht ED. Expression of the mRNAs and Proteins for the Na(+)/H(+) exchangers and their regulatory factors in baboon and human placental syncytiotrophoblast. *Endocrinology* 2001; 142(8):3685-92.

Quinn PA, Ng LL. Potassium and activity of the sodium hydrogen exchanger isoform 1 in vascular smooth muscle of hypertensive rats. *Metabolism* 2001; 50[7]:778-82.

Reilly RF, Hildebrandt F, Biemesderfer D, Sardet C, Pouyssegur J, Aronson PS, Slayman CW, Igarashi P. cDNA cloning and immunolocalization of a Na[+]-H+ exchanger in LLC-PK1 renal epithelial cells. *Am J Physiol* 1991; 261[6 Pt 2]:F1088-94.

Rocha F, Musch MW, Lishanskiy L, Bookstein C, Sugi K, Xie Y, Chang EB. IFN-gamma downregulates expression of Na[+]/H[+] exchangers NHE-2 and NHE-3 in rat intestine and human Caco-2/bbe cells. *Am J Physiol Cell Physiol* 2001; 280 [5]:C1224-32.

Rood RP, Emmer E, Wesolek J, McCullen J, Husain Z, Cohen ME, Braithwaite RS, Murer H, Sharp GW, Donowitz M Regulation of the rabbit ileal brush-border Na+/H+ exchanger by an ATP-requiring Ca2+/calmodulin-mediated process. *J Clin Invest* 1988; 82[3]:1091-7.

Rossmann H, Sonnentag T, Heinzmann A, Seidler B, Bachmann O, Vieillard-Baron D, Gregor M, Seidler U. Differential expression and regulation of Na(+)/H(+) exchanger isoforms in rabbit parietal and mucous cells. *Am J Physiol Gastrointest Liver Physiol* 2001; 281(2):G447-58.

Sardet C, Franchi A, Pouyssegur J. Molecular cloning of the growth-factor-activatable human Na+/H+ antiporter. *Cold Spring Harb Symp Quant Biol* 1988; 53 Pt 2:1011-8.

Sardet C, Franchi A, Pouyssegur J. Molecular cloning, primary structure, and expression of the human growth factor-activatable Na+/H+ antiporter. *Cell* 1989; 56[2]:271-80

Scano G, Filippelli M, Romagnoli I, Mancini M, Misuri G, Duranti R, Rosi E. Hypoxic and hypercapnic breathlessness in patients with type I diabetes mellitus. *Chest* 2000; 117[4]:960-7.

Schirmer, T., Keller, T. A., Wang, Y. F., and Rosenbusch, J. P. Structural basis for sugar translocation through maltoporin channels at 3.1 A resolution. *Science* 1995; 267:512-514.

Schneider J, Jehle EC, Starlinger MJ, Neunlist M, Michel K, Hoppe S, Schemann M. Neurotransmitter coding of enteric neurones in the submucous plexus is changed in non-inflamed rectum of patients with Crohn's disease. *Neurogastroenterol Motil* 2001; 13: 255-264

Schultheis PJ, Clarke LL, Meneton P, Miller ML, Soleimani M, Gawenis LR, Riddle TM, Duffy JJ, Doetschman T, Wang T, Giebisch G, Aronson PS, Lorenz JN, Shull GE. Renal and intestinal absorptive defects in mice lacking the NHE3 Na+/H+ exchanger. *Nat Genet* 1998; 19[3]:282-5.

Schultheis PJ, Clarke LL, Meneton P, Harline M, Boivin GP, Stemmermann G, Duffy JJ, Doetschman T, Miller ML, and Shull GE. Targeted disruption of the murine Na^+/H^+ exchanger isoform 2 gene causes reduced viability of gastric parietal cells and loss of net acid secretion. *J Clin Invest* 1998; 101:1243-1253,

Sharon, N, Lis H. Lectins: Cell-agglutinating and sugar-specific proteins. *Science.* 1972; 177[53]:949-959.

Shugrue CA, Obermuller N, Bachmann S, Slayman CW, Reilly RF. Molecular cloning of NHE3 from LLC-PK1 cells and localization in pig kidney. *J Am Soc Nephrol.* 1999; 10[8]:1649-57.

Sibley CP, Glazier JD, Greenwood SL, Lacey H, Mynett K, Speake P, Jansson T, Johansson M, Powell TL. Regulation of placental transfer: the Na(+)/H(+) exchanger--a review. *Placenta.* 2002; 23 Suppl A:S39-46.

Siczkowski M, Davies JE, Sweeney FP, Kofoed-Enevoldsen A, Ng LL. Na^+/H^+ exchanger isoform-1 abundance in skin fibroblasts of type I diabetic patients with ephropathy. *Metabolism* 1995; 44: 791-795.

Siczkowski M, Davies JE, Ng LL. Na[+]-H+ exchanger isoform 1 phosphorylation in normal Wistar-Kyoto and spontaneously hypertensive rats. *Circ Res* 1995; 76[5]:825-31.

Siddique I, Khan I. Regulation of Na/H exchanger-1 in gastroesophageal reflux disease: possible interaction of histamine receptor. *Dig Dis Sci* 2003; 48:1832-8.

Steck TL, Weinstein RS, Strauss JH, Wallach DFH. Inside-out red cell membrane vesicles: preparation and purification. *Science* 1970; 168[928]:255-7.

Steck TL. The organization of proteins in the human red blood cell membrane. A review. *J Cell Biol* 1974; 62[1]:1-19.

Szaszi K, Paulsen A, Szabo EZ, Numata M, Grinstein S, Orlowski J. Clathrin-mediated endocytosis and recycling of the neuron-specific Na+/H+ exchanger NHE5 isoform. Regulation by phosphatidylinositol 3'-kinase and the actin cytoskeleton. *J Biol Chem* 2002; 277(45): 42623-32.

Takaichi K, Wang D, Balkovetz DF, Warnock DG. Cloning, sequencing, and expression of Na[+]-H+ antiporter cDNAs from human tissues. *Am J Physiol* 1992; 262[4 Pt 1]:C1069-76.

Takahashi E, Abe J, Berk BC. Angiotensin II Stimulates $p90^{rsk}$ in Vascular Smooth Muscle Cells A Potential Na^+-H^+ Exchanger Kinase. *Circulation Research.* 1997; 81:268-273

Tobey NA, Koves G, Orlando RC. Human esophageal epithelial cells possess an Na^+/H^+ exchanger for H^+ extrusion. *Am J Gastroenterol* 1998; 93: 2075-2081

Tobey NA, Reddy SP, Keku TO, Cragoe EJ Jr, Orlando RC. Studies of pHi in rabbit esophageal basal and squamous epithelial cells in culture. *Gastroenterology* 1992; 103: 830-839.

Tobey NA, Reddy SP, Khalbuss WE, Silvers SM, Cragoe EJ Jr, Orlando RC. Na[$^+$]-dependent and –independent Cl-/$HCO3^+$ exchangers in cultured rabbit esophageal epithelial cells. *Gastroenterology* 1993; 104:185-195.

Tobey NA, Reddy SP, Keku TO, Cragoe EJ Jr, Orlando RC. Mechanisms of HCl-induced lowering of intracellular pH in rabbit esophageal epithelial cells. *Gastroenterology* 1993; 105:1035-1044..

Tobey NA. Systemic factors in esophageal mucosal protection. *Digestion* 1995; 56: 38-44,

Tse CM, Ma AI, Yang VW, Watson AJ, Levine S, Montrose MH, Potter J, Sardet C, Pouyssegur J, Donowitz M. Molecular cloning and expression of a cDNA encoding the rabbit ileal villus cell basolateral membrane Na+/H+ exchanger. *EMBO J* 1991; 10[8]:1957-67.

Tse CM, Levine SA, Yun CH, Montrose MH, Little PJ, Pouyssegur J, Donowitz M. Cloning and expression of a rabbit cDNA encoding a serum-activated ethylisopropylamiloride-resistant epithelial Na+/H+ exchanger isoform [NHE-2]. *J Biol Chem* 1993; 268[16]:11917-24.

Tse M, Levine S, Yun C, Brant S, Counillon LT, Pouyssegur J, Donowitz M. Structure/function studies of the epithelial isoforms of the mammalian Na+/H+ exchanger gene family. *J Membr Biol* 1993; 135[2]:93-108. Review.

Vejtorp M, Pedersen J, Klebbe JG, Lund E. Low concentration of plasma amino acids in newborn babies of diabetic mothers. *Acta Paediatr Scand* 1977; 66: 53-58.

Voyno-Yasenetskaya TA. G proteins and Na+/H+ exchange. *Biol Signals Recept* 1998; 7[2]:118-24.

Wakabayashi, S., Bertrand, B., Ikeda, T., Pouyssegur, J., and Shigekawa, M. Mutation of calmodulin-binding site renders the Na+/H+ exchanger [NHE1] highly H[+]-sensitive and Ca2+ regulation-defective. *J Biol Chem* 1994; 269[18]:13710-13715.

Wakabayashi S, Pang T, Su X, Shigekawa M. A novel topology model of the human Na(+)/H(+) exchanger isoform 1. *J Biol Chem* 2000; 275(11):7942-9

Wang T, Yang CL, Abbiati T, Schultheis PJ, Shull GE, Giebisch G, Aronson, PS. Mechanism of proximal tubule bicarbonate absorption in NHE3 null mice. *American Journal of Physiology* 1999; 277: F298-302

Woo AL, Gildea LA, Tack LM, Miller ML, Spicer Z, Millhorn DE, Finkelman FD, Hassett DJ, Shull GE. In vivo evidence for interferon-gamma-mediated homeostatic mechanisms in small intestine of the NHE3 Na+/H+ exchanger knockout model of congenital diarrhea. *J Biol Chem*. 2002; 277[50]:49036-46.

Yao W, Qian G, Yang X. Roles of NHE-1 in the proliferation and apoptosis of pulmonary artery smooth muscle cells in rats. *Chin Med J [Engl]*. 2002; 115[1]:107-9.

Yiangou Y, Facer P, Smith JA, Sangameswaran L, Eglen R, Birch R, Knowles C, Williams N, Anand P. Increased acid-sensing ion channel ASIC-3 in inflamed human intestine. *Eur J Gastroenterol Hepatol* 2001; 13: 891-896.

Younes B, Baez-Giangreco A, al-Nuaim L, Al-Hakeem A, Talib ZA. Basement membrane thickening in the placentas from diabetic women. *Pathol Int* 1996; 46: 100-104.

Yu H, Freedman BI, Rich SS, Bowden DW. Human Na+/H+ exchanger genes: identification of polymorphisms by radiation hybrid mapping and analysis of linkage in end-stage renal disease. *Hypertension* 2000; 35(1 Pt 1):135-43.

Yun CH, Tse CM, Nath SK, Levine SA, Brant SR, Donowitz M. Mammalian Na+/H+ exchanger gene family: structure and function studies. *Am J Physiol* 1995; 269[1 Pt 1]:G1-11.

Zhao H, Wiederkehr MR, Fan L, Collazo RL, Crowder LA, Moe OW. Acute inhibition of Na/H exchanger NHE-3 by cAMP. Role of protein kinase A and NHE-3 phosphoserines 552 and 605. *J Biol Chem* 1999; 274[7]:3978-87

Zhao A, Bossone C, Pineiro-Carrero V, Shea-Donohue T. Colitis-induced alterations in adrenergic control of circular smooth muscle in vitro in rats. *J Pharmacol Exp Ther* 2001; 299:768-774.

Zerbini G, Maestroni A, Breviario D, Mangili R, Casari G. Alternative splicing of NHE-1 mediates Na-Li counter transport and associates with activity rate. *Diabetes.* 2003; 52[6]:1511-1518.

In: Trends in RNA Research
Editor: Peter A. McNamara, pp. 225-251

ISBN: 1-59454-506-5

Chapter X

16S rRNA Gene Sequencing: Interest and Limits for Identification and Characterization of Novel Taxa within the Family *Acidaminococcaceae*

Hélène Marchandin[1,2] and Estelle Jumas-Bilak[2]

[1] Laboratoire de Bactériologie, Hôpital Arnaud de Villeneuve, 371, avenue du doyen Gaston Giraud, 34295 Montpellier Cedex 5, France

[2] Laboratoire de Bactériologie, EA3755, Faculté de Pharmacie, 15, avenue Charles Flahault, BP 14491, 34093 Montpellier Cedex 5, France

Abstract

The most useful molecular chronometers used today for the determination of phylogenetic relationships are the 16S ribosomal RNA genes (16S rDNA genes) and their gene products, the 16S rRNA. Genotypic identification procedures have increasingly received attention as an alternative or complement to conventional phenotypic methods in a polyphasic approach. Broad-range primers that recognize 16S rDNA sequences conserved among a wide variety of bacteria are used to amplify species-specific variable regions of interest. Sequence determination and comparative database searches allow the unknown isolates to be assigned to a bacterial taxon.

In this chapter, application of 16S rRNA gene sequencing to the members of the family *Acidaminococcaceae*, formerly *Sporomusa* sub-branch, will be described. First of all, the family *Acidaminococcaceae* and its anaerobic, Gram-negative members will be presented as well as the particular taxonomic position of this family in the bacterial phylogenetic tree within the Gram-positive bacteria phylum. This chapter will focus on the genera recovered in humans in a medical practice, i.e., *Acidaminococcus*, *Anaeroglobus*, *Megasphaera*, *Veillonella*, *Dialister* and *Selenomonas*. Based on recent studies, two main interesting aspects of sequencing 16S rDNA genes will be developed. First, we underlined the interest in identification of the members of the family *Acidaminococcaceae* when phenotypic characteristics are poor or not discriminating enough to affiliate a bacterial isolate to a definite species. This part is supported by a study conducted over a 3-year period in a teaching hospital demonstrating the powerful capacity of identification of this method. Secondly, we highlighted the interest in detection and characterization of hitherto unknown taxa. The underestimated biodiversity of this family will be underlined here based on culture-dependent and culture-independent studies. The latter part will be illustrated by examples of polyphasic characterization of novel taxa within the family *Acidaminococcaceae* like the genus *Anaeroglobus* and the species *Veillonella montpellierensis* and *Megasphaera micronuciformis*. Despite the efficiency of the 16S rDNA-based methods, it is important to know their limits, particularly when its discriminating power is not high enough in some genera like *Selenomonas* or impaired by a high level of heterogeneity between *rrn* copies in a genome like in the genus *Veillonella*.

To avoid pitfalls of 16S rRNA gene sequencing, developing alternative molecular markers, like *dnaK* gene sequencing in the genus *Veillonella*, is still needed as well as constituting larger computerized databases for sequences of these markers. In a more global approach, the considerations presented herein could be extended to all bacterial species.

Introduction

Molecular rRNA-Based Strategies to Identify Bacterial Species and 16S rRNA-Based Phylogeny

The 16S ribosomal RNA genes (rDNA) and their gene products, the 16S rRNA, are the most useful molecular chronometers used today for the determination of phylogenetic relationships between bacteria [Woese]. The 16S rRNA genes exist in all bacteria and accumulate mutations at a slow, constant rate over time, hence they have been called the 'ultimate molecular clocks' by Woese [Woese]. Highly variable portions of the gene sequence provide unique signatures to any bacterium and useful information about relationships between them. Therefore, phylogenetically informative target, such as 16S rDNA, allow the construction of a phylogenetic taxonomy, and then the genotypic identification of bacteria.

In the last twenty years, genotypic identification procedures have increasingly received attention as an alternative or complement to conventional phenotypic methods. Broad-range primers that recognize 16S rDNA sequences conserved among a wide variety of bacteria are used to amplify species-specific variable regions of interest. Sequencing ribosomal rRNA genes generally provides the most accurate means for identifying microorganisms, particularly when phenotypic identification cannot affiliate a strain to a definite taxon. Sequence determination and comparative database searches allow the unknown isolate to be assigned to a group of bacteria. Broad-range PCR primers are also used to detect microorganisms in virtually any source, for instance pathogens in clinical specimens, even when these pathogens are unculturable microorganisms. Our knowledge of microbial diversity has been greatly extended by comparative 16S rDNA sequence analysis performed directly on DNA extracted from different environments. This approach named metagenomics has proven to be a powerful tool for detecting the presence of a high diversity of microorganisms, independently of their ability to grow in culture media [Paster et al.].

In the present chapter, we will consider different aspects of the use of 16S rDNA in bacterial identification in the light of recent works we performed on the bacterial family *Acidaminococcaceae*.

The Family Acidaminococcaceae

The family *Acidaminococcaceae*, formerly named *Sporomusa* sub-branch, belongs to the phylum Firmicutes or low GC Gram-positive bacteria phylum and is classified in the class Clostridia and in the order Clostridiales [Garrity and Holt, Stackebrandt et al., 1985; Schleifer et al., Carlier et al., 2002]. This order comprises the eight following families: *Clostridiaceae, Lachnospiraceae, Peptostreptococcaceae, Eubacteriaceae, Peptococcaceae, Heliobacteriaceae, Acidaminococcacee,* and *Syntrophomonadaceae.* According to the Bergey's Manual of Systematic Bacteriology [Garrity and Holt] and to the NCBI taxonomy database available from www.ncbi.nlm.nih.gov/Taxonomy/Browser, members belonging to the family *Acidaminococcaceae* are Gram-negative anaerobes classified into 27 genera including Gram-negative cocci: *Acidaminococcus*, *Anaeroglobus*, *Megasphaera*, and *Veillonella* and Gram-negative bacilli: *Acetonema*, *Allisonella*, *Anaeroarcus*, *Anaeromusa*, *Anaerosinus*, *Anaerovibrio*, *Centipeda*, *Dendrosporobacter*, *Dialister*, *Mitsuokella*, *Papillibacter*, *Pectinatus*, *Phascolarctobacterium*, *Propionispira*, *Propionispora*, *Quinella*, *Schwartzia*, *Selenomonas*, *Sporomusa*, *Succiniclasticum*, *Succinispira*, *Thermosinus* and *Zymophilus*.

This family groups members with various morphological characteristics, various types of metabolism and various habitats [Holt *et al.*]. Indeed, it includes cocci and motile or nonmotile rods, which can be regular, ovoid, curved, crescent-shaped, spiral-shaped or serpentine. Members of this family have also various habitats but most of them have been isolated from the mouth or intestinal tract of humans and animal, from rumen of cattle and sheep, from the gut of termites, from anoxic muds and sewage sludge or from breweries. If some genera like *Dialister* comprises non-fermentative bacteria, most of the members of the family *Acidaminococcaceae* have a fermentative type of metabolism but the substrate differ according to the genus, carbohydrates (wide or very limited range) or amino-acids; some others are lipolytic or can attack glycerol. Taxonomic and epidemiological data are summarized in Table 1 for the eight genera including human isolates. Members of these genera are part of the normal human flora of the mouth, upper respiratory tract, gastrointestinal tract or vagina, and when isolated from clinical specimens in aerobic-anaerobic polymicrobial cultures, *Veillonella* spp. and *Dialister* spp. are usually regarded as commensal or contaminant organisms. However, some reports have identified these bacteria as pathogenic, causing serious human infections such as bacteremia, endocarditis, osteomyelitis and prosthetic joint infections for *Veillonella* spp. [Liu *et al.*, 1998; Houston *et al.*, Singh and Yu, Brook and Frazier, Pouchot *et al.*, Beumont *et al.*, Brook, Marchandin *et al.*, 2001] and periodontitis, refractory periodontitis, acute necrotizing ulcerative gingivitis and brain abscesses for *D. pneumosintes* [Doan *et al.*, Paster *et al.*, Rousée *et al.*].

Table 1. Taxonomic and epidemiological data for species belonging to the eight genera of the family *Acidaminococcaceae* encountered in human clinical samples [Holt *et al.*, Krieg andHolt]

Genus [reference]	Species (Type [T])	Principal habitat	Pathogenic potential
Acidaminococcus [Rogosa, 1969]	*A. fermentans* [T]	Intestinal tract of pigs and humans	Ovine foot rot; human bacteremia, abdominal and pulmonary abscesses
Anaeroglobus [Carlier *et al.*, 2002]	*A. geminatus* [T]	Gastrointestinal tract and mouth of humans	Unknown
Centipeda [Lai *et al.*]	*C. periodontii* [T]	Unknown	Human periodontal lesions
Megasphaera [Rogosa, 1971; Sugihara *et al.*; Brancaccio & Legendre; Marchandin *et al.*, 2003a]			
	M. elsdenii [T]	Intestinal contents and feces of cattle, sheep and pig Normal human feces Conjunctiva	Lung abscess Endocarditis Gastrointestinal disorders Foot rot lesions in goats
	M. cerevisiae	Spoiled bottle beer	Unknown
	M. micronuciformis	Unknown (found in human samples)	Unknown
Putative species:	*M. hominis*	Human colonic biota	Unknown
Mitsuokella [Lan *et al.*]			
	M. multacida [T]	Feces of humans and pigs	Dental root canal infections in humans
	M. jalaludinii	Rumen of cattle	Unknown

Table 1 continued

Genus / Species	Habitat	Pathogenicity
Dialister [Moore & Moore, Rousée *et al.*, Downes *et al.*, Contreras *et al.*]		
D. pneumosintes [T]	Gingival crevices, upper respiratory tracts	Brain abscesses, bacteremia, infections of respiratory tract, head and neck.
D. invisus	Isolated from the human oral cavity in patients with endodontic and periodontal infections.	
Selenomonas [Dighe *et al.*, Moore *et al.*]		
S. sputigena [T]	Human gingival crevice	Periodontal pathogens, Bacteremia, Septicemia
S. dianae	Human gingival crevice	
S. flueggei	Human gingival crevice	
S. infelix	Human gingival crevice	
S. noxia	Human gingival crevice	
S. ruminantium	Rumen and pig cecum	Unknown
S. lacticifex	Brewery yeast	Unknown
S. lipolytica	Tropical lagoon receiving water from an oil mill	Unknown
Veillonella [Jumas-Bilak *et al.*, 2004]		
V. parvula [T]	Buccal or intestinal human, rat, rabbit	Bacteremia, prosthetic joint infection, myositis, endocarditis, osteomyelitis
V. atypica	Human, rarely rodent, buccal	
V. dispar	Human: mouth and respiratory tract	
V. ratti	Rat: mouth and intestine, Human: semen	Unknown
V. caviae	Mouth, guinea pig	
V. rodentium	Buccal or intestinal: hamster,rat, rabbit	
V. criceti	Hamster: mouth	
V. montpellierensis	Human: amniotic fluid, gastric fluid of a newborn	

Phylogenetic analyses have revealed the relationship between the species of genera belonging to the family *Acidaminococcaceae* [Both *et al.*, Willems and Collins, Marchandin *et al.*, 2003a] (Fig. 1). Although the bacteria of this family belong to the Gram-positive phylum, their negative Gram staining is not an artifact as previously suggested [Gupta]. Electron microscopy observations of the outer membrane showed that the bacterial cell wall was indeed of the Gram-negative type, as previously described for *Veillonella* spp. [Bladen and Mergenhagen] and *Selenomonas ruminantium* [Kamio and Takahashi, Kalmokoff *et al.*], and more recently for *Megasphaera micronuciformis* [Marchandin *et al.*, 2003a] (Fig. 2), *Dialister micraerophilus* and *Dialister propionicifaciens* [Jumas-Bilak *et al.*, 2005]. The type of cell wall is considered to be a reliable criterion for bacterial classification, as the dichotomy of Gram-positive and Gram-negative species was confirmed by molecular phylogenies based on 16S rRNA and heat-shock protein 70 (HSP70) gene (*dnaK*) sequences [Gupta]. The family *Acidaminococcaceae* is an exception to this general rule that correlates the cell-wall structure with the phylogenetic placement of a bacterium.

Further aspects developed in this chapter will focus on the main genera recovered in humans in a medical practice, i.e., *Acidaminococcus*, *Anaeroglobus*, *Megasphaera*, *Veillonella*, *Dialister* and *Selenomonas*. Following examples will underline the difficulty of phenotypic identification in clinically relevant genera of the family *Acidaminococcaceae* and the usefulness of 16S rRNA gene sequencing for bacterial identification, detection and characterization of unknown taxa within this family.

Methods for Bacterial Identification by 16S rDNA Sequencing

Cell Lysis, Extraction of DNA and PCR Amplification

In most cases, crude DNA extracts were obtained by boiling for 10 min and freezing for 5 min a highly concentrated bacterial suspension. Alternatively, DNA was extracted as follows: bacteria were lysed using a 3-h incubation at 37°C in 100 µl of Tris-EDTA buffer (10 mM Tris, 1 mM EDTA, 0.1 M NaCl, pH 8.0) and lysosyme followed by a 1-h incubation at 55°C in a solution of 25 mg of proteinase K per ml and 10% sodium dodecyl sulfate. Next, 200 µl of protein precipitation solution (Promega) was added to each tube and left for half an hour on ice. Extraction of nucleic acids was carried out by isopropyl precipitation followed by washing with ethanol 70°. Three microliters of extracts were used as the template in the PCR experiments. 16S rDNA was selectively amplified by PCR using 5'-GTGCTGCAGAGAGTTTGATCCTGGCTCAG-3' (27f), position 8-36 (*E. coli* numbering), as the forward primer and 5'-CACGGATCCTACGGGTACCTTGTTACGACTT-3' (1492r), position 1478-1508 (*E. coli* numbering) as the reverse primer. Alternatively, primers 63f: 5'-CAGGCCTAACACATGCAAGTC-3', position 43-63 (*E. coli* numbering) and 1387r: 5'-GGGCGGWGTGTACAAGGC-3', position 1387-1403 (*E. coli* numbering) were used. The PCR was carried out in 50 µl containing approximately 0.5 µg DNA template, 200 mM each primer, 200 mM each dNTP and 1 U *Taq* polymerase (Promega) in the appropriate reaction buffer. Temperature cycling was done by using 30 cycles of 1 min at 94 °C, 1 min at 65 °C and 2 min at 72 °C.

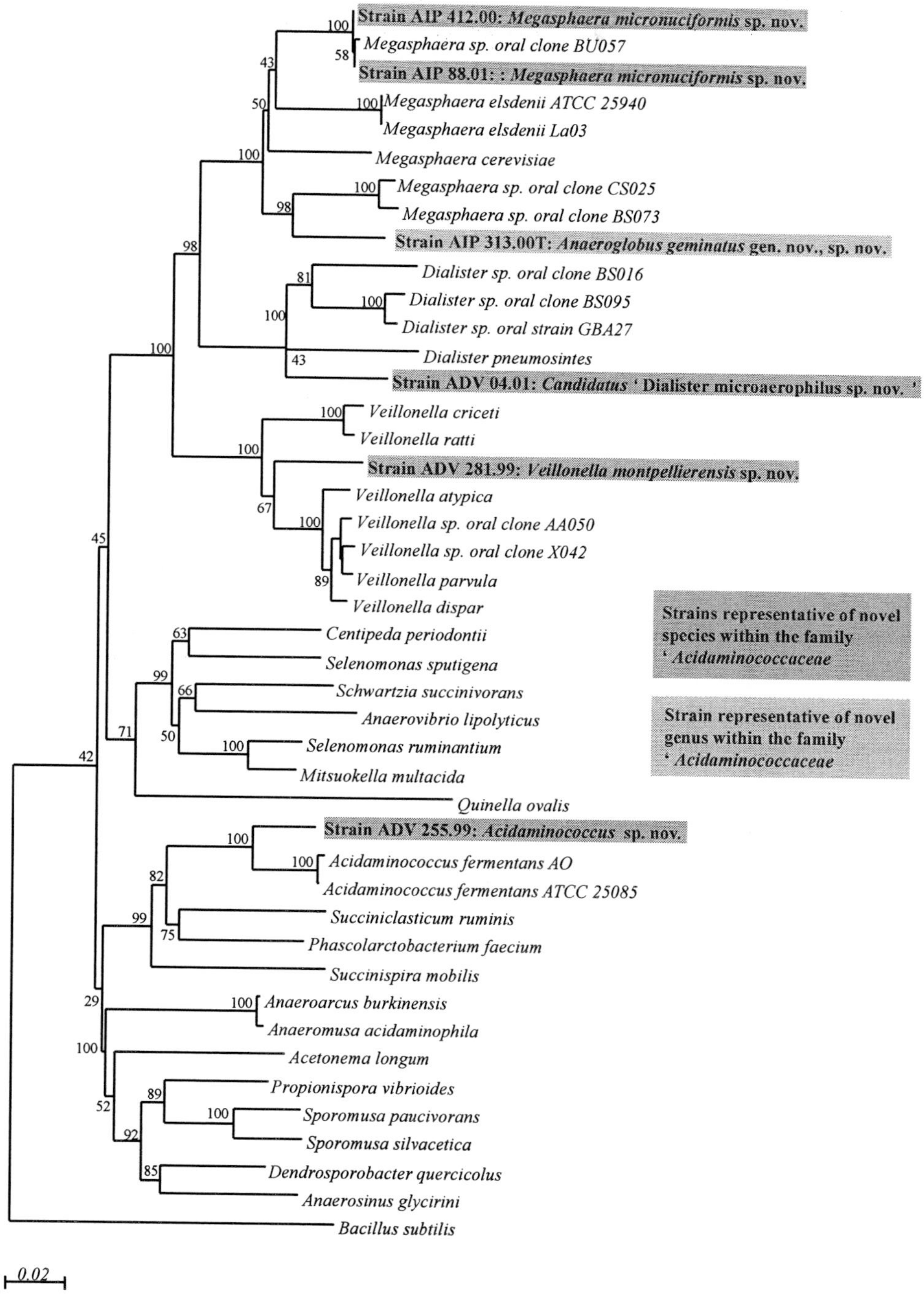

Figure 1. Consensus phylogenetic tree based on 16S rRNA gene sequences, computed using the Jukes-Cantor equation and neighbour joining, indicating the relationships between potential novel taxa within the family '*Acidaminococcaceae*' and other members of this family. Numbers given at the nodes represent bootstrap percentage values (1000 analyses). Bar, 2 nucleotide substitutions per 100 nucleotides.

The selective amplification of one 16S rDNA copy was performed by cutting *Sma*I macrorestriction fragments obtained after PFGE migration in a low-melting agarose gel. Intact genomic DNA was extracted in agarose plugs as described previously and submitted to PFGE after restriction by 40 U of the endonuclease *Sma*I [Marchandin *et al.*, 2001; Marchandin *et al.* 2003b]. The macrorestriction fragments were separated with a contour-clamped homogeneous field electrophoresis apparatus, CHEF DRII (Bio-Rad) in 0.5X TBE at 10°C. The PFGE parameters for resolution of *Sma*I fragments were 42 h at 4.5 V cm^{-1}, with switch times ramped from 1 to 15 s in a 1% agarose gel. In order to determine the 16S rDNA copy number and to find the *Sma*I fragments carrying 16S rDNA, the PFGE gels were transferred onto Nytran N (Scheilcher and Schull) nylon membrane by vacuum blotting in 20X SSC and hybridized with a 16S rDNA probe. The probe was obtained by PCR as described above with a dNTP mixture containing 0.1 mM digoxigenin-labeled dUTP (Roche). Hybridization of the probe was detected by using chromogenic substrate NBT/BCIP (Roche). Then, 2 µl melted agarose corresponding to each fragment hybridized by 16S rDNA probe were used as the PCR template.

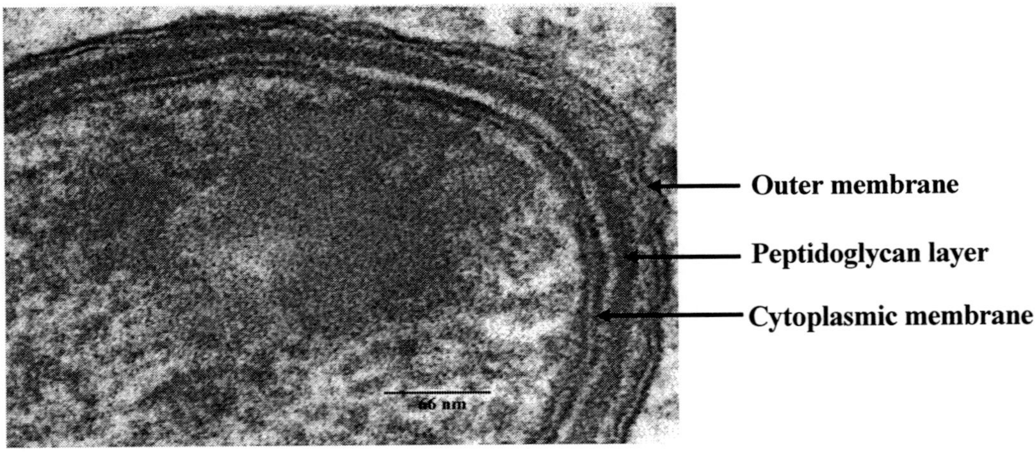

Figure 2. Ultrastructure of *Megasphaera micronuciformis*: cell wall and membranes after electron microscopy of ultrathin sections.

Sequencing and Analysis of 16S rDNA Sequence Data

The presence of a single 1.4 kb PCR product was checked by electrophoresis on a 1.5% agarose gel in TBE 0.5X. PCR products were then directly sequenced on an Applied Biosystems Automatic Sequencer (Genome Express) by using the forward primer 63f.

For bacterial identification, partial 16S rDNA sequences ranging from 500 to 700 bp containing <1% undetermined positions were compared with sequences deposited in GenBank, EMBL, and DJB databases using the program standard nucleotide-nucleotide BLAST (http://www.ncbi.nlm.nih.gov/BLAST/) [Altschul]. The sequences were also matched against the RDPII database (Ribosomal Database Project II at http://rdp.cme.msu.edu) [Maidak] and the European Ribosomal database (http://oberon.rug.ac.be:8080/rRNA/). Finally, sequences were matched against our own sequence database constituted by non-deposited 16S rDNA sequences compiled during a routine activity of identification of clinical isolates. The occurrence of chimeric sequences was checked on-line using the program Chimera Check of the RDPII web site.

For the description of novel taxa, an almost complete 16S rDNA sequence (more than 1300 bp with less than 0.5% ambiguity) was obtained as recommended [Stackebrandt *et al.*, 2002] by sequencing each strand using primers 27f and 1492r.

16S rDNA-Based Phylogenetic Analysis of Unidentified Isolates

For isolates that were not identified after 16S rDNA sequence analysis, taxonomic relationships were inferred from 16S rDNA sequence comparison. Sequences were obtained from the GenBank database and aligned by using the multisequence alignment program ClustalW [Thompson *et al.*]. Phylogenetic relationships were inferred from this alignment by using programs of the PHYLIP software package [Felsenstein]. A distance matrix was generated using DNADIST under the assumptions of either Jukes & Cantor or Kimura 2-parameters. Phylogenetic trees were derived from this matrix using neighbor joining. Isolates were assigned to the taxonomic group of the two bacterial strains forming the taxonomic frame of the unidentified isolate. The significance of the nodes was tested by bootstrap analysis to produce 1000 trees and then a consensus tree was compiled using the CONSENSE program.

Interest of 16S rDNA Sequencing in Identification of the Members of the Family *Acidaminococcaceae*

Efficiency of 16S rDNA sequencing for the routine identification of members of the family *Acidaminococcaceae* was investigated over a 3-year period based on a study of 16S rDNA gene sequencing in identification of clinical anaerobes in a French hospital, the University Hospital of Montpellier (This work has been presented at the 7th Biennial Congress of the Anaerobe Society of the Americas, Annapolis, July 19-21, 2004).

Criteria for Identification

A 97% similarity in 16S rDNA sequence has been proposed for the bacterial species delineation based on correlation study between DNA-DNA hybridization results and similarity values for 16S rDNA sequences [Stackebrandt and Goebel] but this recommendation has been questioned [Drancourt *et al.*, 2000; Woo *et al.*]. In our routine exercise of bacterial identification by 16S rDNA sequencing, we observed that most bacteria belonging to the same species shared at least 99% of their 16S rDNA nucleotides. However, a similarity level between 98 and 99% should not exclude the link of two strains in the same species. Similarity levels between 97 and 98% were rarely observed suggesting that this range of values might correspond to a cutoff zone. Thus, in the absence of a generally accepted value, we retained a 98% similarity as cutoff for identification to the species level. No cutoff value has been defined for identification to the genus level. The criteria we retained for novel genus detection are similarity values below 92% and equivalent similarities when the strain was compared with members of different genera of a same family.

The use of these criteria was necessary to rapidly analyze a large collection of unidentified isolates but ideally it may be necessary to use different cutoff values depending on the bacterial group under investigation. Indeed, all bacteria do not evolve at the same speed resulting in variable level of intra-taxa polymorphism according to the bacterial group analyzed.

Results

Over a 3-year period, we performed 16S rDNA molecular identification of 107 anaerobic clinical isolates, which could not be identified by conventional methods in a routine practice. The 16S rDNA sequences obtained were compared to sequences available in GenBank database according to the criteria of interpretation presented above. Among them, 60 (56.1%) were shown to belong to the family *Acidaminococcaceae*. They were more precisely affiliated to 6 known genera, i.e. *Acidaminococcus*, *Dialister*, *Megasphaera*, *Veillonella*, *Selenomonas* and *Succiniclasticum*. Percents of identity > 98%, together with a good discrimination between species, allowed good species identification for 25 of these 60 isolates (41.7%): *A. fermentans* (n=2), *D. invisus* (n=1), *D. pneumosintes* (n=21), and *M. elsdenii* (n=1). For 29 isolates (48.3%), identification to the species level could not be achieved since percents of identity in sequences were below 97%. However, 28 of these isolates could be affiliated to a known genus within the family *Acidaminococcaceae*, suggesting that they may represent novel species within the following genera: *Acidaminococcus* (n=2), *Dialister* (n=20), *Megasphaera* (n=2), *Succiniclasticum* (n=1), and *Veillonella* (n=3). The remaining isolate could be affiliated to the family *Acidaminococcaceae* but not to any known genus, suggesting that it could represent a novel genus within this family. For 6 of the 60 isolates (10%), discriminating power of 16S rDNA gene sequencing was insufficient to identify species in a known genus (*Veillonella* and *Selenomonas*). These results are summarized in Table 2. Finally, 16S rDNA sequencing proves its efficiency to identify human isolates, particularly those displaying only few positive phenotypic characteristics, like *D. pneumosintes* and to detect unknown taxa.

Moreover, some of the isolates identified by 16S rDNA gene sequencing represented species rarely or never reported from human samples like a clinical isolate belonging to the *Veillonella ratti*-*Veillonella criceti* group.

Table 2. Results of 16S rDNA-based identification for 60 clinical isolates belonging to the family *Acidaminococcaceae*

Identification (number of strains)	Origin of the isolates	% of 16S rDNA sequence with that of the closest species/isolate
Strains identified to the species level (n=25)		
A. fermentans (2)	Drain	99.1-99.2% with *A. fermentans*
D. invisus (1)	Face fistula	99.8 % with *D. invisus*
D. pneumosintes (21)	Various samples*	> 99% with *D. pneumosintes*
M. elsdenii (1)	Gastric fluid	100% with *M. elsdenii*
Strains affiliated to a known genus (n=28)		
Acidaminococcus sp. (2)		
Strain ADV 255.99	Peritoneal fluid	95.6% with *A. fermentans*
Strain ADV 5206.02	Rectum	99.9% with strain ADV 255.99
Dialister sp. (16) [a]	Various samples **	
including Strain ADV 04.01	Amniotic fluid	93.2% with *D. pneumosintes*
Dialister sp. (4) [b]	Semen, wound,	93.1% with D. pneumosintes and
	abscess, scab	94.6% with strain ADV 04.01
Megasphaera sp. (2) [c]		
Strain AIP 49.01	Whitlow	94.5% with *M. cerevisiae*
StrainAIP 412.00	Abscess	94.4% with *M. cerevisiae*
Succiniclasticum sp. (1)	Abscess	90% with *S. ruminis*
Veillonella sp. (3) [d]		
Strain ADV 281.99	Gastric fluid	94.7% with *V. atypica*
Strain ADV 2216.03	Amniotic fluid	94.9% with *V. atypica*
Strain ADV 3198.03	Amniotic fluid	94.8% with *V. atypica*
Strain affiliated to an unknown genus (n=1)		
Strain AIP 313.00 [e]	Fluid collection	91% with *M. cerevisiae*
Strains not identified due to poor discrimination between two species (n=6)		
Veillonella sp. (5)		
Strain ADV 4313.2	Semen	100% with *V. ratti* and
		98.9% with *V. criceti*
Strain ADV 360.1	Bone	99.5% with *V. dispar* and
		98.9% with *V. parvula*
Strain LBVCM 101	Whitlow	98.5 with *V. dispar* and
V. parvula		
Strains LBVCM 119 (2)	Throat, tonsil	98.5 with *V. dispar* and
		98.8 with *V. parvula*
Selenomonas sp. (1)		
Strain ADV 2288.02	Abscess	97% with both *S. infelix* and *S. dianae*

* including peritoneal fluid, blood, wounds, bronchoalveolar lavage fluid, sinus, abscesses.

** including amniotic fluid samples, pylonidal cysts, abscesses, wounds, Bartholin gland.

[a] Further characterized as *Dialister micraerophilus* [Jumas-Bilak *et al.*, 2005]

[b] Further characterized as *Dialister propionicifaciens* [Jumas-Bilak *et al.*, 2005]

[c] Further characterized as *Megasphaera micronuciformis* [Marchandin *et al.*, 2003a]

[d] Further characterized as *Veillonella montpellierensis* [Jumas-Bilak *et al.*, 2004]

[e] Further characterized as *Anaeroglobus geminatus* [Carlier *et al.*, 2002]

Example of 16S rDNA-Based Identification of the First Human Isolate belonging to the *Veillonella ratti-Veillonella criceti* phylogenetic group [Marchandin *et al.*, 2005]

A human isolate belonging to the *V. ratti-V. criceti* phylogenetic group was characterized after 16S rDNA sequencing (accession number AY211542) since it showed 100% of identity with the 16S rDNA sequence of the type strain *V. ratti* ATCC 17746^{T} (accession number AF186071) and 98.9% with that of *V. criceti* ATCC 17747^{T} (accession number AF186072). The isolate was recovered in mixed aerobic-anaerobic cultures from semen in a 24-year-old man attending the urology unit of our hospital for infertility.

The genus *Veillonella* is subdivided into eight species (Table 1). Among them, *V. parvula*, *V. atypica* and *V. dispar* were frequently isolated from human clinical specimens in aero-anaerobic polymicrobial cultures or less frequently in pure culture during infectious processes [Houston *et al.*; Liu *et al.*, 1998; Marchandin *et al.*, 2001; Singh and Yu]. The recently described species *V. montpellierensis* has been isolated from polymicrobial clinical samples and then as the causative agent of endocarditis [Jumas-Bilak *et al.*, 2005; Rovery *et al.*]. The four other species have never been isolated from man but only from buccal or intestinal flora of rodents [Rogosa, 1984]. *V. ratti* and *V. criceti* were characterized and named *V. alcalescens* subspecies *ratti* and *V. alcalescens* subspecies *criceti* by Rogosa in 1965 [Rogosa, 1965]. His study included two oral strains isolated five years apart from different rats corresponding to *V. ratti* and 21 strains recovered from the mouth of the hamster corresponding to *V. criceti*. Since then, no additional isolation of *V. ratti* or *V. criceti* has been reported, so these two species have never been isolated from any human sample. The eight *Veillonella* species were created on the basis of serological reactions [Rogosa, 1965], DNA/DNA homology experiments [Mays *et al.*] or 16S rDNA and *dnaK* sequences analysis [Jumas-Bilak et al.] and can not be differentiated by conventional phenotypic and biochemical tests [Kolenbrander and Moore]. Owing to the lack of discriminating characteristic that can be tested in routine practice to differentiate the *Veillonella* species, identification of the clinical isolate as belonging to the *V. ratti-V. criceti* group could only be achieved by using molecular methods.

This strain is the first clinical strain belonging to the *V. ratti-V. criceti* group isolated from a human sample and has been deposited in the Collection de l'Institut Pasteur (=CIP 107810). This strain is also of interest because of its infrequency of isolation since only twenty-three strains belonging to this group have been reported so far. Moreover, despite the increasing number of 16S rDNA sequences available in databases including those deposited for uncultured *Veillonella* sp. clones, no sequence other than that of *V. ratti* and *V. criceti* type strains matched with the sequence of the clinical isolate studied. Our observation suggests also that host restriction limited to rodents may not exist for *V. ratti* and *V. criceti*.

Inclusion of 16S rDNA Sequencing in a Polyphasic Approach for the Description of Unknown Taxa Belonging to the Family *Acidaminococcaceae*

All of the examples detailed hereafter will follow the concept of polyphasic taxonomy, which is presently recommended by the majority of the authors and could be defined as follows according to Vandamme *et al.*: "This polyphasic taxonomy takes into account all available phenotypic and genotypic data and integrates them in a consensus type of classification, framed in a general phylogeny derived from 16S rRNA sequence analysis".

16S rDNA and Analysis of Metabolic End Products for the Description of the Novel Genus *Anaeroglobus* [Carlier *et al.*, 2002]

The hitherto unknown anaerobic coccus named strain AIP 313.00 was isolated from a post-operative fluid collection in a 70-year-old woman hospitalized in a surgical unit of the Montpellier University Hospital. The strain was unreactive in most of the conventional biochemical tests. Further determination of the major metabolic end products by quantitative gas chromatography [Carlier, 1985] was not helpful for identification. Indeed, the strain produced acetic acid (A), propionic acid (P), isobutyric acid (iB), butyric acid (B) and isovaleric acid (iV). This metabolic profile was not compatible with the affiliation of the isolate to any of the previously known genera grouping Gram-negative cocci: *Veillonella*, *Acidaminococcus* and *Megasphaera*, which produced A, P; A, B; and A, P, B, valeric acid and caproic acid, respectively. Finally, based on phenotypic criteria, this strictly anaerobic coccus could not be classified in any of the above genera.

The 16S rDNA sequence analysis revealed a high level of sequence similarity with some oral clone sequences of *Megasphaera* spp. described by Paster *et al.* [Paster *et al.*] and contained in the GenBank database under designations BB166, CS025 and BS073 (approximately 99, 93 and 93%, respectively). These three sequences originated from cloning and sequencing of 16S rDNA from oral flora without isolation of bacterial strains. Sequence similarity calculations revealed that the closest known phylogenetic relatives of strain AIP 313.00 were the Gram-negative cocci *Megasphaera* and *Veillonella*, two genera belonging to the family *Acidaminococcaceae* (approximately 87-91% in sequence similarity). 16S rDNA-based phylogenetic analysis showed that these clones and the unknown bacterium formed a well-separated phylogenetic branch that may represent a novel lineage within the family *Acidaminococcaceae* (Fig. 1). The topology of the phylogenetic tree showed that the closest relative of strain AIP 313.00 is the oral clone BB166 (data not shown). These taxa formed a cluster with a high bootstrap value that was related to another group consisting of oral clones CS025 and BS073. Together, they formed a well-separated branch, which suggests that these sequences represent organisms that belong to the same genus but represent at least two different species.

On the basis of both phenotypic and phylogenetic evidence, a novel genus, *Anaeroglobus* gen. nov., has been created for the unknown bacterium, with one species, *Anaeroglobus geminatus* gen. nov., sp. nov. It was also suggested that the oral clones BB166, CS025 and

BS073 belong to the genus *Anaeroglobus* and that the clone BB166 probably represents another strain of *Anaeroglobus geminatus*.

16S rDNA, Phenotype and Large-Scale Genome Structure for the Description of the Novel Species *Megasphaera micronuciformis* [Marchandin *et al.*, 2003a]

Strains AIP 49.01 and AIP 412.00 were two isolates of Gram-negative, catalase-negative and strictly anaerobic cocci, which remained unidentified by phenotypic testing. They were isolated from two independent human samples, a pus sample from a 71-year-old woman admitted to the emergency unit for a whitlow and a liver abscess in a 32-year-old woman, respectively.

Comparative 16S rRNA gene sequencing demonstrated that these two organisms displayed 99.8% sequence identity and that they were members of the family *Acidaminococcaceae.* The most closely related 16S rDNA sequences were from *Megasphaera* sp. oral clone BU057 (99.8%) and from isolates of *Megasphaera cerevisiae* and *Megasphaera elsdenii* (94.5 and 93.8%, respectively). Phylogenetic analysis based on 16S rDNA sequences showed that these two strains were most closely related to *M. elsdenii* (Fig. 1). However, clear differences from previously described *Megasphaera* species ifn terms of size, biochemical tests, particularly the analysis of metabolic end products, gas production and DNA G+C content questioned the inclusion of these two isolates in a novel species of the genus *Megasphaera.* For example, the two isolates displayed a small cell size below 1 μm (0.4-0.6 μm) never observed for a member of the genus *Megasphaera* [Rogosa, 1971; Rogosa, 1984; Engelmann and Weiss]. Moreover, some constant characteristics of the genus were not observed in the two strains studied like acid production from fructose, lactate fermentation and gas production. However, creating a novel genus within the family *Acidaminococcaceae* was turned down because of two main arguments: (i) creating a new genus would have led to split *M. elsdenii* and *M. cerevisiae* into a paraphyletic genus; (ii) phylogenetic data indicated uncertainty in the branching order between strains AIP 49.01, AIP 412.00, *A. geminatus*, *M. elsdenii* and *M. cerevisiae* (bootstrap values of 43 and 50 at the corresponding nodes) (Fig. 1). Thus, it was more cautious to include the two strains in a novel species in the existing genus *Megasphaera.* On the other hand, inclusion of these two isolates, sharing 99.8% in 16S rDNA gene sequences, in a same species was supported by a genomic organization study conducted on undigested and I-*Ceu*I-restricted DNA by PFGE. Indeed, large-scale chromosome structure has been described as a sensitive indicator of phylogenetic relationships between bacteria [Liu *et al.*, 1999]. The chromosomes of strains AIP 49.01 and AIP 412.00 sized 1780 and 1837 kb in size, respectively. A difference of 3.1% in chromosome size was commonly observed between strains of the same species [Bergthorsson and Ochman]. Mapping experiments with the restriction enzyme I-*Ceu*I showed that the two clinical isolates possessed four *rrn* operons in the same locations on the chromosome. In contrast, the chromosome of *M. elsdenii* was 2578 kb in size, and carried seven *rrn* copies. The equivalent genome size and identical *rrn* skeleton, together with the high percentage of identity in 16S rDNA sequence and phenotypic data, suggested that strains AIP 49.01 and AIP 412.00 belonged to the same species. The name *Megasphaera micronuciformis* sp. nov. has been assigned to this novel species referring to the small size and the walnut-shaped morphology of bacterial cells (Fig. 3).

Finally, this novel species can be differentiated from other *Megasphaera* species by its small size, its DNA G+C content of 46.4 mol%, its metabolic end products and its 16S rDNA gene sequence. As a consequence, an emendation of the description of the genus *Megasphaera* was required [Marchandin *et al.* 2003a].

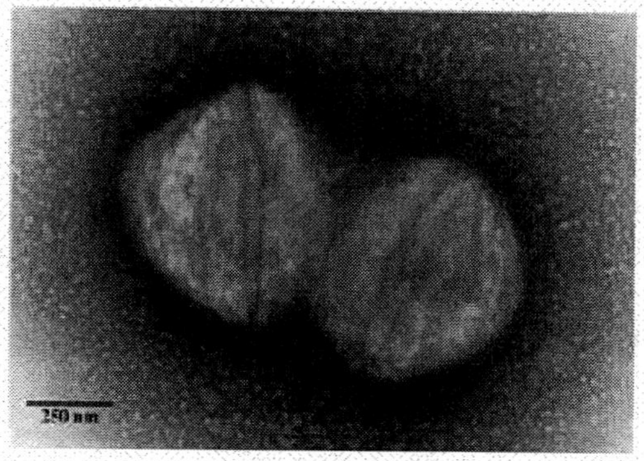

Figure 3. General morphology of *Megasphaera micronuciformis* after negative staining.

16S rDNA and *dnak* Gene Sequences for the Description of the Novel Species *Veillonella montpellierensis* (Jumas-Bilak *et al.*, 2004)

Three strains (strain ADV 3198.03, strain ADV 281.99 and strain ADV 2216.03) of a hitherto unknown Gram-negative and anaerobic coccus were isolated in mixed aerobic-anaerobic cultures from clinical samples of three different patients, a gastric fluid sample in a newborn and two amniotic fluid samples from two women. At a phenotypic level, the isolates displayed all the characteristics of bacteria belonging to the genus *Veillonella.* This identification to the genus level is relatively straightforward since *Veillonella* spp. are the only anaerobic, Gram-negative cocci for which nitrate reduction is observed. In contrast, the identification of *Veillonella* strains to the species level remains uncertain and inconvenient owing to the lack of conventional phenotypic and biochemical discriminating tests previously underlined in this chapter [Kolenbrander and Moore]. The species described so far in the genus *Veillonella* [Mays *et al.*, Rogosa, 1984] can not be differentiated by phenotypic testing and therefore, molecular methods based on 16S rDNA sequences have been increasingly used to identify *Veillonella* isolates [Sato *et al.*, 1997a, b]. In the present case, 16S rDNA sequence analysis revealed that the three clinical strains shared more than 99.5% identity in 16S rDNA sequence and that they were also closely related to another strain at that time unidentified to the species level, *Veillonella* sp. 2001-112662 [Drancourt *et al.*, 2004]. However, 16S rDNA-based identification within the genus *Veillonella* has been recently criticized due to the low level of sequence variation among some species, particularly between *V. dispar* and *V. parvula.* Moreover, a frequent occurrence of intra-chromosomal heterogeneity between 16S rRNA gene copies has been observed in human isolates of *Veillonella* spp. impairing their

16S rDNA-based identification [Marchandin *et al.*, 2003b]. As a consequence and despite the robustness of the 16S rDNA results obtained for the three strains, another genetic marker, the heat-shock protein gene (*dnaK*) was analyzed. A 700-bp *dnaK* fragment was amplified and sequenced as previously described [Marchandin *et al.*, 2003b]. More than 98.4% identity in *dnaK* sequence was observed between the three isolates and the closest species shared only 86.4% with these isolates. 16S rDNA-based phylogenetic analyses showed that the three unknown strains formed a separated subclade clearly remote from known *Veillonella* spp. of human and animal origins (Fig. 4) and this was in accordance with results obtained from *dnaK*-based phylogeny.

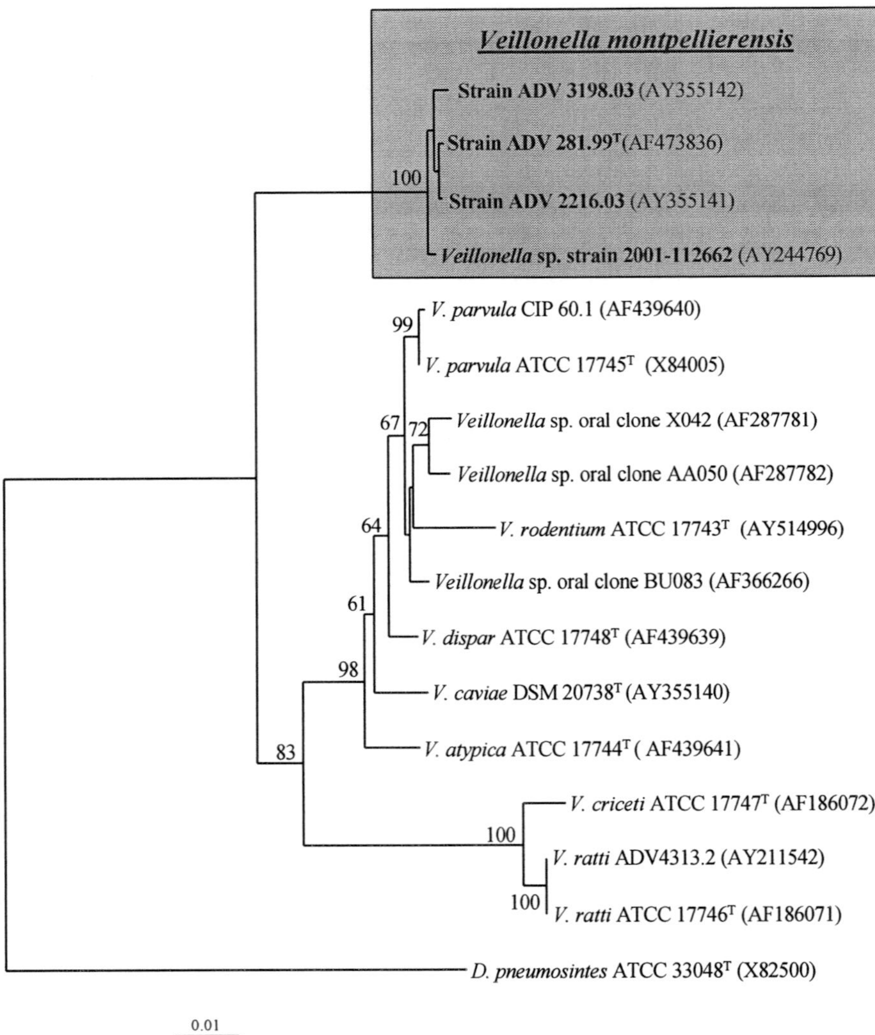

Figure 4. Neighbour-joining tree based on 16S rDNA sequences showing relationships between *V. montpellierensis* strains ADV 281.99^T, ADV 2216.03 and ADV 3198.03 and representatives of the genus *Veillonella*. *Dialister pneumosintes* was the out-group organism. GenBank accession numbers are given in parentheses Bootstrap values higher than 60% are indicated at corresponding nodes. Bar, 0.01 substitution per site.

Both molecular and phylogenetic data obtained for 16S rDNA and *dnaK* sequences were congruent and suggested that the three clinical strains characterized herein, despite the absence of discriminating phenotypic characteristic, represented a new species within the genus *Veillonella*. The name *Veillonella montpellierensis* has been proposed pertaining to the town called Montpellier in South of France, where the type strain and the two other strains supporting the description of this novel species were isolated and characterized. Moreover, we suggested that the strain *Veillonella* sp. 2001-112662 [Drancourt *et al.*, 2004] could be a potential member of this species. *V. montpellierensis* can only be differentiated from the other species of the genus *Veillonella* by 16S rDNA and *dnaK* sequencing. Considering that the three strains of our study were the only anaerobic bacteria recovered in the three corresponding clinical samples and that *Veillonella* sp. strain 2001-112662 was recently reported as the causative organism of endocarditis [Rovery *et al.*, 2005], the new species *V. montpellierensis* is considered as pathogenic for humans.

16s rDNA Sequencing Underlined the Underestimated Biodiversity of the Family *Acidaminococcaceae*

The above study that led to the description of one novel genus and 3 novel species over a 3-year period clearly showed that the biodiversity of the family *Acidaminococcaceae* is actually underestimated. Moreover, the description of additional novel taxa, which is today in process in our laboratory, confirmed the important role of member of *Acidaminococcaceae* in the diversity of the human microbiota.

Particularly, twenty isolates of a tiny coccobacillus were affiliated to the genus *Dialister* on the basis of 16S rDNA sequences analysis (Table 2) but remained unidentified to the species level since all of them showed 16S rDNA gene sequences below 94% comparatively to those of *D. pneumosintes* and *D. invisus*. Two groups of isolates each displaying more than 99% of sequence similarity could be differentiated. The first group includes 16 of the 20 isolates and is represented by the strain ADV 04.01 in Figure 1 and the second one groups the four remaining isolates. These two groups of strains shared no more than 93% of 16S rDNA sequence identity with each other, but there is a lack of phenotypic differences between these isolates, as previously observed for *D. pneumosintes* and *D. invisus* [Downes *et al.*]. Phylogenetic analysis conducted on the isolate ADV 04.01 confirmed that it might represent a novel lineage within the genus *Dialister* (Fig. 1). To further characterize these two groups of strains some additional genetic and genomic studies should be performed in a polyphasic approach. Preliminary results showed that 16S rDNA-based groupings are congruent with results obtained with another molecular marker, *dnaK* and genomic size determined after PFGE analysis of undigested DNAs. Notwithstanding the absence of clearly distinct phenotypic characteristics between the different isolates of the genus *Dialister*, the two groups of strains probably represent two novel species within the genus *Dialister*. It is has been therefore previously proposed that the first group of 16 isolates including *Dialister* sp. strain ADV 04.01 could be named *D. micraerophilus* sp. nov. referring to its capacity to grow in a microaerophilic atmosphere containing 10% of carbon dioxide [Jumas-Bilak *et al.*, 2005].

A potential novel species within the genus *Acidaminococcus* could be represented by the two isolates *Acidaminococcus* sp. strain ADV 255.99 and strain ADV 5206.02 (Tab. 2), which showed 99.9% sequence identity with each other. Indeed, these clinical isolates displayed only 95.6% sequence similarity with *Acidaminococcus fermentans* DSM 20731^T. Proposal of a novel species is also supported by phylogenetic analysis confirming the inclusion of *Acidaminococcus* sp. strain ADV 255.99 into the genus *Acidaminococcus* but revealing that this strain formed a well-separated phylogenetic branch from *A. fermentans*. However, further phenotypic and genetic analyses are needed in order to search for additional discriminating characteristics between this potential novel species and the type species *A. fermentans*.

Strain ADV 1145.03 has been designated as *Succiniclasticum* sp (Tab. 2). However, percent of sequence similarity observed with the closest species, *S. ruminis*, is very low (90%). This suggests that additional isolates have to be studied to better understand whether this isolate should be considered as the first representative of a novel species within the genus *Succiniclasticum*, a genus never reported from humans so far, or a novel genus within the family *Acidaminococcaceae*.

Taken together with the relatively high number of novel species cultured and characterized within the family *Acidaminococcaceae* during the past few years, the results presented above underlined that the diversity among this family is still underestimated.

Culture-independent studies [Paster *et al.*, Rolph *et al.*] led to the same conclusion. Indeed, in recent studies, methods based on 16S rDNA cloning and sequence analysis have been used to determine the bacterial diversity of oral flora. Consequently, an increasing number of sequences corresponding to uncultivated bacteria belonging to the family *Acidaminococcaceae* have been deposited in databases. Number of these 16S rRNA gene sequences corresponded to unidentified oral clones and finally, all these results suggest that many members of this phylogenetic branch remain unknown [Hutter *et al.*, Paster *et al.*, Rolph *et al.*].

Ribosomal Operons Heterogeneity: A Pitfall for Bacterial Identification

If the majority of genera and species created within the family *Acidaminococcaceae* since 1990 have been supported by 16S rRNA gene sequence analysis and therefore can be differentiated using 16S rRNA gene sequencing, some limits of 16S rDNA-based identification have been clearly identified. The 16S rDNA could not be discriminating enough to allow identification of closely related species. Particularly, in the genus *Veillonella*, 16S rDNA-based identification failed to discriminate between *V. dispar* and *V. parvula* for numerous clinical isolates, including four isolates of *Veillonella* sp. unidentified to the species level presented in Table 2. *V. dispar* and *V. parvula* have been described on the basis of serological reactions [Rogosa, 1965; Rogosa, 1984] and DNA-DNA homology studies [Mays *et al.*]. Analysis of their 16S rDNA sequences highlighted the high level of identity (99% identity) between these two species (Fig. 5b), corresponding to only 9 variable positions distributed along the *rrs* gene [Marchandin *et al.*, 2003b].

A

Strain *Veillonella* sp. ADV 360.1

1 2

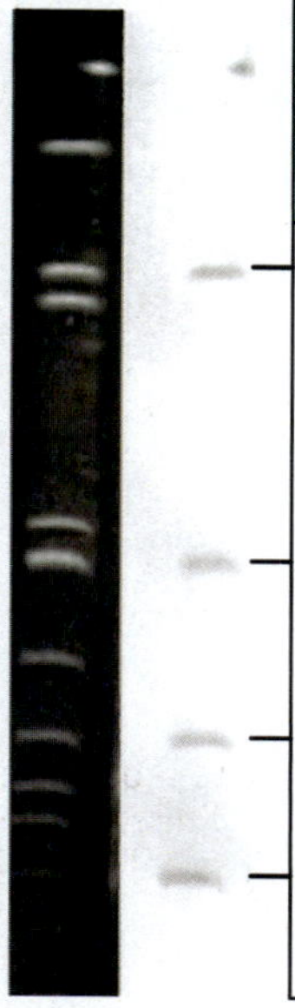

B

	V. dispar	*V. parvula*	*Veillonella* sp. strain ADV 360.1			
			*rrs*A	*rrs*B	*rrs*C	*rrs*D
V. parvula	99.0	100				
Veillonella sp. strain ADV 360.1	99.5	98.9				
rrsA (AF439642)	99.5	98.9	100			
rrsB (AF439643)	98.7	99.7	98.6	100		
rrsC (AF439644)	98.6	99.5	98.5	99.8	100	
rrsD (AF439645)	99.5	98.9	100	98.6	98.5	100

Figure 5. A. Determination of *rrs* copy number for *Veillonella* sp. strain ADV 360.1: 1. PFGE migration of *Sma*I-restricted DNA; 2. Hybridization of the PFGE profile with a 16S rDNA probe showing the presence of 4 *rrs* copies. B. Percents of 16S rDNA sequences similarity between *V. dispar* ATCC 17748^T,*V. parvula* CIP 60.1, *Veillonella* sp. strain ADV 360.1 and each of the four *rrs* copies of *Veillonella* sp. strain ADV 360.1 (*rrs*A to *rrs*D).

On the other hand, the direct sequencing of total 16S rDNA PCR products obtained from several clinical isolates of *Veillonella* sp. gave ambiguous results consisting of undetermined positions. Moreover, visual inspection of the chromatograms showed positions with a double sequencing signal (Fig. 6). These observations suggested that some heterogeneity in sequences between *rrs* copies could exist within some *Veillonella* sp. isolates. We first investigated the *rrs* copy number in the genus *Veillonella* and showed that four 16S rDNA copies were present in the genome of *V. parvula* CIP 60.1, *V. dispar* ATCC 17748^T and 27 clinical isolates unidentified to the species level. The four *rrs* copies were named *rrs*A, *rrs*B, *rrs*C and *rrs*D following the sizes of the *Sma*I hybridizing fragments (Fig. 5A). Each 16S rDNA copy was sequenced for a clinical isolate, *Veillonella* sp. strain ADV 360.1. This isolate could not be affiliated to a definite species on the basis of its 16S rDNA sequence since 99.5% and 98.9% of sequence identity were observed with *V. dispar* ATCC 17748^T and *V. parvula* CIP 60.1, respectively. A relatively high level of heterogeneity in sequences between the four copies was found (1.43%). The sequences of *rrs*A and *rrs*D were identical whereas *rrs*B and *rrs*C varied in three nucleotide positions. *rrs*A and *rrs*D had a maximum identity of 99.5% with the sequence of *V. dispar* compared to 98.9% with that of *V. parvula.* In contrast, *rrs*B and *rrs*C had a maximum identity of 99.5 and 99.7% with *V. parvula* compared to 98.6 and 98.7% with *V. dispar*.

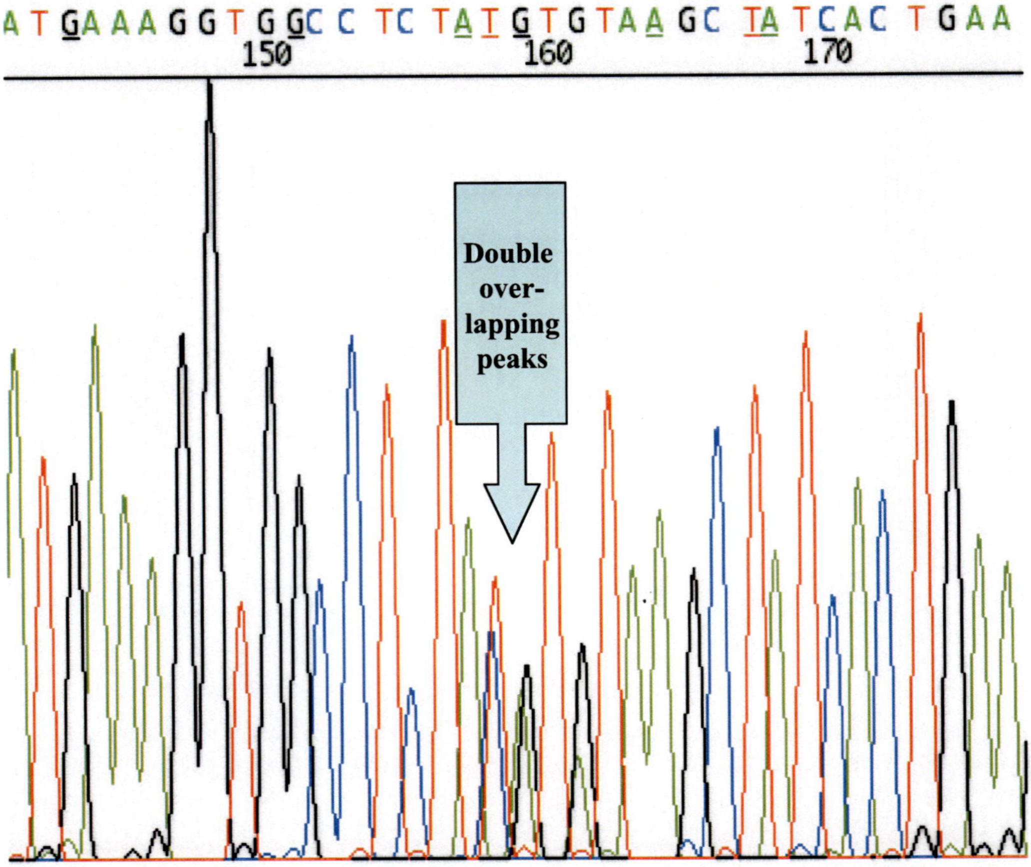

Figure 6. Part of the chromatogram obtained for *Veillonella* sp. strain ADV 360.1 showing multiple overlapping peaks.

These two groups of *rrs* showed 98.5 or 98.6% of sequence similarity with each other. Finally, the *V. dispar* and *V. parvula* 16S rDNA sequences displayed a higher percentage of identity than two *rrs* copies of the *Veillonella* sp. isolate ADV 360.1. Among the 9 variable positions discriminating between *V. dispar* and *V. parvula*, eight were proved to be polymorphic between *rrs* copies and corresponded to the double sequencing signals observed on the chromatogram. The ninth position was shown to differ between two deposited sequences of *V. dispar*. Consequently, none of the variable positions between *V. dispar* and *V. parvula* could be retained as species-specific and therefore, 16S rDNA gene sequencing should no longer considered as suitable for the identification of *V. dispar* and *V. parvula* due to an intra-operon variability higher than inter-species variability. This heterogeneity should be further investigated in the other genera of the family *Acidaminococcaceae* for a better understanding of 16S rDNA gene sequencing limits for identification of bacterial isolates.

While the copy number of 16S rRNA genes may vary from 1 to 15 among eubacterial genomes, it is generally believed that all the copies in an organism are identical or nearly identical in nucleotide sequence since *rrn* operons are subject to homogenization processes. Thus, rRNA sequences generally show low variability between copies in a genome and within species or subspecies [Liao]. However, scattered nucleotide differences between 16S rDNA copies (so-called micro-heterogeneity) have been described in a few cases. Micro-heterogeneity has been identified in other eubacteria, like *Escherichia coli* [Cilia *et al.*, Martinez-Murcia *et al.*]. The general extent of 16S rRNA heterogeneity within a bacterial species has received little attention, except in few works [Cilia *et al.* Clayton *et al.*, Martinez-Murcia *et al.*, Ueda *et al.*]. Clayton *et al.* observed that slightly different 16S rRNA sequences were deposited in the databases for strains belonging to the same species. They thought most of the differences were not caused by sequencing errors but by real intra-specific variations. Part of these variations could be due to inter-operon heterogeneity. Thus, the authors showed that 18% of *rrs* sequence pairs deposited for the same strain exhibited 1-5% heterogeneity [Clayton *et al.*].

To avoid pitfalls of 16S rDNA-based identification, other molecular markers should be studied and determined for a large number of bacteria including members of the family *Acidaminococcaceae*. The 70 kDa heat-shock protein gene (*dnaK*), which is generally present in one copy in bacterial genomes and show higher inter-species variability than 16S rRNA genes, has been already studied for some genera of *Acidaminococcaceae*.

Conclusion

The 16S rRNA gene constitutes a versatile tool for bacterial identification in the clinical laboratory. In a medical practice, 16S rDNA-based identification constitutes a mean for identification of bacterial species that could not be differentiated at a phenotypic level, and this undoubted identification is needed to further study the epidemiology and pathogenic potential of each species. This technology is now applied in an increasing number of clinical laboratories for routine identification of bacterial isolates [Bosshard *et al.*; Drancourt *et al.*, 2004]. Previous studies have shown that 16S rDNA-based sequence identification is particularly useful for slow-growing or fastidious microorganisms, unusual isolates, and bacteria that are poorly differentiated by conventional methods.

In the family *Acidaminococcaceae*, some species yielded poor growth on usual media impairing the correct inoculation of commercialized identification strips like rapid ID32A (bioMérieux) for example. Phenotypic testing could not differentiate between these species in a routine approach. Among these isolates, some could not be further differentiated by gas chromatography. In all of these cases, sequence analysis of the 16S rRNA gene is powerful for identifying known species encountered in human clinical specimens, except for *V. dispar* and *V. parvula*. Identification of strains that are rarely encountered in medical practice or which displayed atypical features when compared to previously reported isolates should be confirmed using a molecular method, like it was the case for the first human clinical isolate belonging to the *V. ratti-V. criceti* group [Marchandin *et al.*, 2005]. Such an accurate identification is needed to report cases involving infrequently isolated bacteria to avoid any misidentification [Janda and Abbott]. The 16S rRNA gene sequencing has been used mainly for aerobic bacteria in clinical laboratories [Ferroni *et al.*, Drancourt *et al.* 2004, Patel]. This technique is likely to be more widely applied in the clinical setting of anaerobes and this will increase our knowledge on their pathogenic power either when they are isolated in pure culture or in mixed aerobic-anaerobic flora [Rousée *et al.*].

The usefulness of 16S rRNA gene sequencing for bacterial identification, detection and characterization of unknown taxa within the family *Acidaminococcaceae* was clearly demonstrated by the studies we present herein. The efficiency of 16S rDNA-based methods is confirmed by countless number of works in other taxa and in various fields such as phylogeny, taxonomy, identification and studies of the microbial diversity whatever its origin, human or environmental. However, 16S rDNA should not be considered as an infallible marker and its failures or pitfalls need to be underlined.

Generally, 16S rDNA is a good and useful identification tool when the taxa to be identified has been described at least partly on the basis of 16S rDNA sequencing. However the current concept of bacterial species is not mainly based on the 16S rDNA sequence but correspond to a "genomically coherent" group of strains sharing many common traits [Rosselló-Mora and Amann]. Consequently, analysis of DNA-DNA hybridization values remains the "gold standard" for defining bacterial species [Stackebrandt *et al.*, 2002]. This method is expensive, time-consuming and difficult to reproduce in routine. Thus, the current taxonomic « gold standard » does not fit to the identification tools actually used.

Despite the recommendations, sequencing of selected bacterial genes and most often 16S rRNA gene has taken an increasingly important role for the description of new taxa in recent years. As a consequence, an ad hoc committee for the re-evaluation of the bacterial species definition in bacteria recommended the analysis of at least five genes of diverse chromosomal loci and wide distribution [Stackebrandt *et al.*, 2002]. The question that remains open is which genes should be analyzed. Moreover, the importance of 16S rDNA comparatively to others markers has not been discussed by this ad hoc committee. A recent study identified individual gene sequences those diverged at a rate reflecting the overall rate of genome divergence. Therefore, these genes could serve as predictors of the whole genome relatedness [Ziegler]. The author found that the individual gene candidate with the poorest ability to predict genome relatedness on a genus or species level was 16S rDNA because it is too highly conserved to differentiate reliably between closely related taxa.

However, in a near future, 16S rDNA will still be the reference marker because it is the sole to be largely represented in the databases. When 16S rRNA is proposed as key word in a bibliographic server such as Medline, 1352 issues are found for the year 2003. Most of these publications concerned the use of 16S rDNA as a molecular tool without critical analysis about its robustness. Issues on the general characteristic of 16S rRNA such as intra genomic heterogeneity, lateral transfer, recombination between species or molecular clock behavior remain very sparse but are desirable for a deeper understanding and a rational use of this molecule. These works should be encouraged in order to propose a rational approach for the development of identification devices such as 16S rDNA-based DNA chips. Beside the necessary improvement of our knowledge on the 16S rDNA, other molecular markers should be urgently used at a large scale in order to complete the databases and then to allow reliable identification of bacteria belonging to related taxa.

Acknowledgements

The authors are very grateful to Dr Hélène Jean-Pierre from the Anaerobe Laboratory of the Montpellier Teaching Hospital and to Jean-Philippe Carlier from the Centre National de Référence des bactéries anaérobies et du botulisme, Institut Pasteur, Paris for their faithful collaboration on all the parts of the work presented in this chapter. We also wish to thank Bernard Gay for electron microscopy, Josiane Campos, Corinne Teyssier and Isabelle Zorgniotti for their excellent technical assistance.

References

Altschul, SF; Madden, TL; Schaffer, AA; Zhang, J; Zhang, Z; Miller,W; Lipman, DJ. Gapped blast and psi-blast: a new generation of protein database search programs. *Nucleic Acids Res*, 1997 25, 3389-3402.

Bergthorsson, U; Ochman, H. Distribution of chromosome length variation in natural isolates of *Escherichia coli*. *Mol Biol Evol*, 1998 15, 6-16.

Beumont, MG; Duncan, J; Mitchell, SD; Esterhai, JL Jr; Edelstein, PH: *Veillonella* myositis in an immunocompromised patient. *Clin Infect Dis*, 1995 21, 678-679.

Bladen, HA; Mergenhagen, SE. Ultrastructures of *Veillonella* and morphological correlation of an outer membrane with particles associated with endotoxic activity. *J Bacteriol*, 1964 88, 1482-1492.

Bosshard, PP; Abels,S; Zbinden, R; Böttger, EC; Altwegg, M. Ribosmal DNA sequencing for identification of aerobic gram-positive rods in the clinical laboratory (an 18-month evaluation). *J Clin Microbiol*, 2003 41, 4134-4140.

Both, B; Buckel, W; Kroppenstedt, R; Stackebrandt, E. Phylogenetic and chemotaxonomic characterization of *Acidaminococcus fermentans*. *FEMS Microbiol Lett*, 1992 76, 7-11.

Brancaccio, M; Legendre, GG. *Megasphaera elsdenii* endocarditis. *J Clin Microbiol*, 1979, 10, 72-74.

Brook I: *Veillonella* infections in children. *J Clin Microbiol*, 1996 34, 1283-1285.

Brook, I; Frazier, EH: Infections caused by *Veillonella* species. *Infect Dis Clin Pract*, 1992 1, 377-381.

Carlier, J-P. Gas Chromatography of fermentation products: its application in diagnosis of anaerobic bacteria. *Bull Inst Pasteur*, 1985 83, 57-69.

Carlier, J-P; Marchandin, H; Jumas-Bilak, E; Lorin, V; Henry, C; Carrière, C, Jean-Pierre, H. *Anaeroglobus geminatus* gen. nov., sp. nov., a novel member of the family *Veillonellaceae*. *Int J Syst Evol Microbiol*, 2002 52, 983-986.

Cilia, B; Lafay, B; Christen, R. Sequence heterogeneities among 16S ribosomal RNA sequences and their effect on phylogenetic analyses at the species level. *Mol Biol Evol*, 1996 13, 451-461.

Clayton, RA; Sutton, G; Hinkle, PS Jr; Bult, C; Fields, C. Intraspecific variation in small-subunit rRNA sequences in GenBank: why single sequences may not adequately represent prokaryotic taxa. *Int J Syst Bacteriol*, 1995 45, 595-599.

Contreras, A; Doan, N; Chen, C; Rusitanonta, T; Flynn, MJ; Slots, J. Importance of *Dialister pneumosintes* in human periodontitis. *Oral Microbiol Immunol*, 2000 15, 269-272.

Dighe, AS; Shouche, YS; Ranade, DR. *Selenomonas lipolytica*, sp. nov., an obligately anaerobic bacterium possessing lipolytic activity. *Int J Syst Evol Microbiol*, 1998 48, 783-791.

Doan, N; Contreras, A; Flynn, J; Slots, J; Chen, C. Molecular identification of *Dialister pneumosintes* in subgingival plaque of humans. *J Clin Microbiol*, 2000 38, 3043-3047.

Downes, J; Munson, M; Wade, WG. *Dialister invisus* sp. nov., isolated from the human oral cavity. *Int J Syst Evol Microbiol*, 2003 53, 1937-1940.

Drancourt, M; Bollet, C; Carlioz, A; Martelin, R; Gayral, JP; Raoult, D. 16S ribosomal sequence analysis of a large collection of environmental and clinical unidentifiable bacterial isolates. *J Clin Microbiol*, 2000 38, 3623-3630.

Drancourt, M; Berger, P; Raoult D. Systematic 16S rRNA gene sequencing of atypical clinical isolates identified 27 new bacterial species associated with humans. *J Clin Microbiol*, 2004 42, 2197–2202.

Engelmann, U; Weiss, N. *Megasphaera cerevisiae* sp. nov.: a new gram-negative obligately anaerobic coccus isolated from spoiled beer. *Syst Appl Microbiol*, 1985 6, 287-290.

Felsenstein, J. Phylip (Phylogeny Inference Package) version 3.5c. 1993. Distributed by the author. Department of Genetics, University of Washington, Seattle.

Ferroni, A; Sermet-Gaudelus, I, Abachin, E; Quesne, G; Lenoir, G; Berche, P; Gaillard, JL. Use of 16S rRNA gene sequencing for identification of nonfermenting gram-negative bacilli recovered from patients attending a single cystic fibrosis center. *J Clin Microbiol*, 2002 40, 3793-3797.

Garrity, GM; Holt, JG. Taxonomic outline of the Archaea and Bacteria. In: Boone DR, Castenholz RW, editors. Bergey's Manual of Systematic Bacteriology, 2nd edn. New York: Springer; 2001; vol. 1; p. 155-166.

Gupta, RS. Protein phylogenies and signature sequences: a reappraisal of evolutionary relationships among Archaebacteria, Eubacteria, and Eukaryotes. *Microbiol Mol Biol Rev*, 1998 62, 1435-1491.

Holt, JG; Krieg, NR; Sneath, PHA; Staley, JT; Williams, ST. Bergey's Manual of Determinative Bacteriology, Ninth Edition. Baltimore: Williams and Wilkins; 1994.

Houston, S; Taylor, D; Rennie, R: Prosthetic valve endocarditis due to *Veillonella dispar*: successful medical treatment following penicillin desensitization. *Clin Infect Dis*, 1997 24, 1013-1014.

Hutter, G; Schlagenhauf, U; Valenza, G; Horn, M; Burgemeister, S; Claus, H; Vogel, U. Molecular analysis of bacteria in periodontitis: evaluation of clone libraries, novel phylotypes and putative pathogens. *Microbiology UK*, 2003 149, 67-75.

Jumas-Bilak, E; Carlier, J-P; Jean-Pierre, H; Teyssier, C; Gay, B; Campos, J; Marchandin, H. *Veillonella montpellierensis* sp. nov, a novel anaerobic Gram-negative coccus isolated from human clinical samples. *Int J Syst Evol Microbiol*, 2004 54, 1311-1316.

Jumas-Bilak, E; Jean-Pierre, H; Carlier, J-P; Teyssier, C; Bernard, K; Gay, B; Campos, J; Morio, F; Marchandin, H. *Dialister micraerophilus* sp. nov. and *Dialister propionicifaciens* sp. nov., isolated from human clinical samples. *Int J Syst Evol Microbiol*, 2005 55, 2471-2478.

Juvonen, RJ; Suihko, M-L. *Megasphaera paucivorans* sp. nov., *Megasphaera sueciensis* sp. nov. and *Pectinatus haikarae* sp. nov., isolated from brewery samples, and emended description of the genus *Pectinatus*. *Int J Syst Evol Microbiol* (in press as of 11 November 2005). Published online 9 December 2005. doi:10.1099/ijs.0.63699-0

Kalmokoff, ML; Austin, JW; Whitford, MF; Teather, RM. Characterization of a major envelope protein from the rumen anaerobe *Selenomonas ruminantium* OB268. *Can J Microbiol*, 2000 46, 295-303.

Kamio, Y; Takahashi, H. Outer membrane proteins and cell surface structure of *Selenomonas ruminantium*. *J Bacteriol*, 1980 141, 899-907.

Kolenbrander, PE; Moore, LVH. The genus *Veillonella*. In: Balows, HG; Trüper, M; Dworkin, W; Harder, W; Schleifer, KH, editors. The *Prokaryotes*, 2nd edn. New York: Springer; 1992; p. 2034-2047.

Krieg NR, Holt JG. Bergey's Manual of Systematic Bacteriology. Baltimore: Williams and Wilkins; 1984.

Lai, CH; Males, BM; Dougherty, PA; Berthold, PA; Listgarten, MA. *Centipeda periodontii* gen. nov., sp. nov., from human periodontal lesions. *Int J Syst Bacteriol*, 1983 33, 628-635.

Lan, GQ; Ho, YW; Abdullah, N. *Mitsuokella jalaludinii* sp. nov., from the rumens of cattle in Malaysia. *Int J Syst Evol Microbiol*, 2002 52, 713-718.

Liao, D. Gene conversion drives within genic sequences: concerted evolution of ribosomal RNA genes in bacteria and archaea. *J Mol Evol*, 2000 51, 305-317.

Liu, JW; Wu, JJ; Wang, LR; Teng, LJ; Huang, TC: Two fatal cases of *Veillonella* bacteremia. *Eur J Clin Microbiol Infect Dis*, 1998 17, 62-64.

Liu, SL; Schryvers, AB; Sanderson, KE; Johnston, RN. Bacterial phylogenetic clusters revealed by genome structure. *J Bacteriol*, 1999 181, 6747-6755.

Maidak, B. L., Cole, J. R., Lilburn, T. G. and 7 other authors. The RDP-II (Ribosomal Database Project). *Nucleic Acids Res*, 2001 29, 173-174.

Marchandin, H; Jean-Pierre, H; Carrière, C; Canovas, F; Darbas, H; Jumas-Bilak, E. Prosthetic joint infection due to *Veillonella dispar*. *Eur J Clin Microbiol Infect Dis*, 2001 20, 340-342.

Marchandin, H; Carlier, JP; Teyssier, C; Gay, B; Jean-Pierre, H; Jumas-Bilak, E. Phylogenetic analysis of some *Sporomusa* sub-branch members isolated from human clinical specimens: description of *Megasphaera micronuciformis* sp. nov. *Int J Syst Evol Microbiol*, 2003 53, 547-553.

Marchandin, H; Teyssier, C; Siméon de Buochberg, M; Jean-Pierre, H; Carrière, C; Jumas-Bilak, E. Intra-chromosomal heterogeneity in the four 16S rRNA gene copies in the genus *Veillonella*: implications in phylogeny and taxonomy. *Microbiology UK*, 2003 149, 1493-1501.

Marchandin, H; Teyssier, C; Jumas-Bilak, E; Robert, M; Artigues, A-C; Jean-Pierre, H. Molecular identification of the first human isolate belonging to the *Veillonella ratti*-

Veillonella criceti group based on 16S rDNA and *dnaK* gene sequencing. *Res Microbiol,* 2005 156, 603-607.

Martinez-Murcia, AJ; Anton, AI; Rodriguez-Valera, F. Patterns of sequence variation in two regions of the 16S rRNA mulltigene family of *Escherichia coli. Int J Syst Bacteriol*, 1999 49, 601-610.

Mays, TD ; Holdeman, LV ; Moore, WEC ; Rogosa, M ; Johnson, JL. Taxonomy of the genus *Veillonella. Int J Syst Bacteriol*, 1982 32, 28-36.

Moore, LVH; Johnson, JL; Moore, WEC. *Selenomonas noxia* sp. nov., *Selenomonas flueggei* sp. nov., *Selenomonas infelix* sp. nov., *Selenomonas dianae* sp. nov., and *Selenomonas artemidis* sp. nov., from the human gingival crevice. *Int J Syst Bacteriol*, 1987 37, 271-280.

Moore, LVH; Moore, WEC. *Oribaculum catoniae* gen. nov., sp. nov.; *Catonella morbi* gen. nov., sp. nov.; and *Dialister pneumosintes* gen. nov., comb. nov., nom. rev., anaerobic gram-negative bacilli from the human gingival crevice. *Int J Syst Bacteriol*, 1994 44, 187-192.

Paster, BJ; Boches, SK; Galvin, JL; Ericson, RE; Lau CN, Levanos, VA; Sahasrabudhe, A; Dewhirst, FE. Bacterial diversity in human subgingival plaque. *J Bacteriol*, 2001 183, 3770-3783.

Patel, JB. 16S rRNA gene sequencing for bacterial pathogen identification in the clinical laboratory. *Mol Diagn*, 2001 6, 313-321.

Pouchot, J; Vinceneux, P; Michon, C; Mathieu, A; Boussougant, Y: Pyogenic sacroiliitis due to *Veillonella parvula. Clin Infect Dis*, 1992 15, 175.

Rogosa, M. The genus Veillonella, IV. Serological groupings, and genus and species emendations. *J Bacteriol*, 1965 90, 704-709.

Rogosa, M. *Acidaminococcus* gen. n., *Acidaminococcus fermentans* sp. n., anaerobic gram-negative diplococci using amino acids as the sole energy source for growth. *J Bacteriol*, 1969 98, 756-766.

Rogosa, M. Transfer of *Peptostreptococcus elsdenii* Gutierrez *et al.* to a new genus, *Megasphaera* [*M. elsdenii* (Gutierrez et al.) comb. nov.]. *Int J Syst Bacteriol*, 1971 21, 187-189.

Rogosa,M. Family I. *Veillonellaceae* Rogosa 1971. In: Krieg NR, Holt JG. *Bergey's Manual of Systematic Bacteriology*, vol. 1. Baltimore: Williams and Wilkins; 1984; p. 680-685.

Rolph, HJ; Lennon, A; Riggio, MP; Saunders, WP; MacKenzie, D; Coldero, L; Bagg, J. Molecular identification of microorganisms from endodontic infections. *J Clin Microbiol*, 2001 39, 3282-3289.

Rossello-Mora, R; Amann, R. The species concept for prokaryotes. *FEMS Microbiol Rev*, 2001 25, 39-67.

Rousée, JM; Bermond, D; Piémont, Y; Tournoud, C; Heller, R; Kehrli, P; Harlay, ML; Monteil, H; Jaulhac, B. *Dialister pneumosintes* associated with human brain abscesses. *J Clin Microbiol*, 2002 40, 3871-3873.

Rovery, C; Etienne, A; Foucault, C; Berger, P; Brouqui, P. *Veillonella montpellierensis* endocarditis. *Emerg Infect Dis*, 2005 11, 1112-1114.

Sato, T; Matsuyama, J; Sato, M; Hoshino, E. Differentiation of *Veillonella atypica*, *Veillonella dispar* and *Veillonell parvula* using restricted fragment-length polymorphism analysis of 16S rDNA amplified by polymerase chain reaction. *Oral Microbiol Immunol*, 1997 12, 350-353.

Sato, T; Sato, M; Matsuyama, J; Hoshino, E. PCR restriction fragment length polymorphism analysis of genes coding for 16S rRNA in *Veillonella* spp. *Int J Syst Bacteriol*, 1997 47, 1268-1270.

Schleifer, KH; Leuteritz, M; Weiss, N; Ludwig, W; Kirchof, G; Seidel-Rüfer, H. Taxonomic study of anaerobic, gram-negative, rod-shaped bacteria from breweries: emended description of *Pectinatus cerevisiiphilus* and description of *Pectinatus frisingensis* sp. nov., *Selenomonas lacticifex* sp. nov., *Zymophilus raffinosivorans* gen. nov., sp. nov., and *Zymophilus paucivorans* sp. nov. *Int. J. Syst. Bacteriol*, 1990 40, 19-27.

Singh, N; Yu, VL: Osteomyelitis due to *Veillonella parvula*: case report and review. *Clin Infect Dis*, 1992 14, 361-363.

Stackebrandt, E; Goebel, BM. Taxonomic note: a place for DNA-DNA reassociation and 16S rRNA sequence analysis in the present species definition in bacteriology. *Int J Syst Bacteriol*, 1994 44, 846-849.

Stackebrandt, E; Pohla, H; Kroppenstedt, R; Hippe, H; Woese, CR. 16S rRNA analysis of *Sporomusa*, *Selenomonas*, and *Megasphaera*: on the phylogenetic origin of Gram-positive Eubacteria. *Arch Microbiol*, 1985 143, 270-276.

Stackebrandt, E; Frederiksen, W; Garrity, GM; Grimont, PAD; Kämpfer, P; Maiden, MCJ; Nesme, X; Rosselló-Mora, R; Swings, J; Trüper, HG; Vauterin, L; Ward, AC; Whitman WB. Report of the ad hoc committee for the re-evaluation of the species definition in bacteriology. *Int J Syst Evol Microbiol*, 2002 52, 1043-1047.

Sugihara, PT; Sutter, VL; Attebery, HR; Bricknell, KS; Finegold, SM. Isolation of *Acidaminococcus fermentans* and *Megasphaera elsdenii* from normal human feces. *Appl Microbiol*, 1974 27, 274-275.

Thompson, JD; Higgins, D; Gibson, TJ. Clustal w: improving the sensitivity of progressive multiple sequence alignment through sequence weighting, position specific gap penalties and weight matrix choice. *Nucleic Acids Res*, 1994 22, 4673-4680.

Ueda, K; Seki, T; Kudo, T; Yoshida, T; Kataoka, M. Two distinct mechanisms cause heterogeneity of 16S rRNA. *J Bacteriol*, 1999 181, 78-82.

Vandamme, P; Pot, B; Gillis, M; de Vos, P; Kersters, K; Swings, J. Polyphasic taxonomy, a consensus approach to bacterial systematics. *Microbiol Rev*, 1996 60, 407-438.

Willems, A; Collins, MD. Phylogenetic placement of *Dialister pneumosintes* (formerly *Bacteroides pneumosintes*) within the *Sporomusa* subbranch of the *Clostridium* subphylum of the gram-positive bacteria. *Int J Syst Bacteriol*, 1995 45, 403-405.

Woese, CR. Bacterial evolution. Microbiol Rev, 1987 51, 221-271.

Woo, PCY; Cheung, EYL, Leung, KW; Yuen, KY. Identification by 16S ribosomal RNA gene sequencing of an *Enterobacteriaceae* species with ambiguous biochemical profile from a renal transplant recipient. *Diagn Microbiol Infect Dis*, 2001 39, 85-93.

Ziegler, D. Gene Sequences Useful for Predicting Relatedness of Whole Genomes in Bacteria. *Int J Syst Evol Microbiol*, 2003 53, 1893-1900.

In: Trends in RNA Research
Editor: Peter A. McNamara, pp. 253-275

ISBN: 1-59454-506-5

Chapter XI

16S rRNA as a Tool for Bacterial Phylogeny and Identification: Some Global Concepts and their Illustration through Applications to the Genus *Bifidobacterium*

Catherine J. Mullié
Faculté de Pharmacie d'Amiens – Université de Picardie Jules Verne
1, rue des Louvels
80037 Amiens Cedex 1
France

Abstract

In 1987, rRNAs were coined by Woese as the ultimate molecular chronometers for determining the phylogenetic relationships between bacteria. Therefore, the primary aim of 16S rRNA gene sequencing was to assess bacterial evolution (with the building of phylogenetic trees). Ever since, other applications have been derived from 16S rDNA sequence knowledge such as identification tools (mainly probes and primers for polymerase chain reactions), description of uncultured bacterial ecosystems through the variety of 16S rDNA they contain or, in taxonomy, the definition of a new species on the sole basis of its 16S rRNA gene sequence being less than 99% similar to any known one. After a brief overview of 16S rRNA structure and its many uses, we will illustrate them by taking the genus *Bifidobacterium* as an example for the determination of a global phylogenetic position, phylogenetic studies of a genus and identification techniques derived from 16S rRNA sequences. Special attention will be devoted to a multiplex PCR approach we developed for the identification of bifidobateria from human origin. This technique has successfully been applied to the identification of isolated bifidobacterial strains as well as to the direct detection of bifidobacterial human species in stool samples. We will also point out some of the limits that arose from 16S rRNA almost universal use, i.e. the existing *continua* within the genus *Bifidobacterium* (limits in the species definition) and the presence of multiple and non similar copies of 16S rRNA genes within a single bacterial cell.

1. Description of 16S rRNA Structure and Functions

Escherichia coli 16S rRNA was the first to be completely sequenced and consists in 1542 nucleotides (bases). This sequence is encoded in the bacterial genome by *rrn*B rRNA operon in *E. coli*. This operon is organised into a polycistronic transcription unit containing rRNA genes in the following order: 5'-16S-23S-5S-3' (Bingen *et al.*, 1994). Depending on the genus and even on the species in a same genus, rRNA operon can be found in a variable number of copies (from 2 up to 11). For example, in the genus *Bifidobacterium*, the number of rRNA operons ranges from 1 to 5 copies. Four copies have been found in the complete genome of *B. longum* NCC2705 (Schell et al., 2002). Multiple copies have also been detected in the species *B. breve* and *B. adolescentis*. *Bifidobacterium adolescentis* ATCC 15703^{T}, type strain of the latter species, was proven to exhibit 5 copies of rRNA operon. However, two different rRNA gene sequences were detected among those 5 copies (Requena *et al.*, 2001). The two rDNA had a 99.4% homology of sequences but showed differences in eight positions (namely positions 219, 467, and several bases between positions 77 to 80 and 93 to 96) (Requena *et al.*, 2001).

Derived from this gene, the 16S rRNA molecule forms part, along with proteins, of the 30S subunit of bacterial ribosomes, these latter being in charge of translation of mRNA in proteins in the bacterial cell (Noller and Woese, 1981). The primary sequence of 16S rRNA soon proved to exhibit major conserved regions through comparative analysis of prokaryotic 16S rRNAs (Woese *et al.*, 1975). Initially, this work was undertaken through the comparison of catalogues of T1 ribonuclease digests of RNAs and, afterwards, by direct matching of entire 16S rRNA sequences, as sequencing of complete sequences became routinely available. Woese *et al.*, (1975) first described nine highly conserved regions (termed universals) that are mostly (six out of nine) found in the 3' end of the molecule. According to Woese (1987), Each one of the primary kingdoms has a particuliar form of 16S rRNA, proceeding from a rRNA sequence signature. In the family *Bifidobacteriaceae*, the 16S rDNA signature consists in conserved nucleotides in positions 122-139 (G-U), 128-233 (C-G), 450-483 (C-G), 602-636 (G-C), 681-709 (C-G), 688-699 (A-U), 823-877 (A-U), 1118-1155 (C-G) and 1311-1326 (A-U) (Stackebrandt *et al.*, 1997).

The secondary structure of 16S rRNA was elucidated in the late seventies, early eighties and is formed of about 50 helical stalks (Noller and Woese, 1981; Woese, 1987). Nuclear Magnetic Resonance (NMR) and Electron Microscopy (EM), along with neutron diffraction and molecular modelling, further helped unravelling the three-dimensional structures of ribosomes.

Implied in the translation process, rRNAs were soon found not only to play a structural role in the ribosome but possibly also a central role in the process reactions themselves. Indeed, mRNA is thought to pass through the 30S subunit, in which a decoding site region, characterised by NMR, has been detected on the 16S rRNA region (Green and Noller, 1997).

Moreover, fixation of tRNA to two binding sites with crucial nucleotide positions on the 16S rRNA moiety has been put forward (for more details on these essential nucleotides, see Green and Noller, 1997). These vital functions help explaining the rather conserved sequences of some 16S rRNA regions whilst other are more "flexible", depending on the bacterial genus, the species or even the strain considered. This duality is of utmost importance in the various uses this molecule has been subjected to in the past three decades, as will be developed in the following paragraphs.

2. Use of 16S rRNA Gene Sequence as a Molecular Chronometer for Bacterial Evolution

Ribosomal RNAs have provided an evolutionary framework for bacterial phylogeny which had so far been lacking in the microbiology world. Indeed, morphological and physiological traits as well as other properties of the bacterial whole cell did not provide sufficiently reliable information to build phylogenetic relationships upon them (Olsen and Woese, 1993). The deciphering of portions of the genome encoding molecules implied in the bacterial cell central functions provided a much greater number of definable characters, enabling microbiologists to take a further step in bacterial phylogeny. In this regard, rRNA gene sequences proved excessively useful, as the comprehensive phylogeny built upon their comparisons led to the discovery of the so far unravelled kingdom Archaeabacteria, which forms part of the former prokaryote taxon.

One of the major reasons for the phylogenetic use of 16S and 23S rRNAs is that they possess a high degree of functional constancy, which implies a good clocklike behaviour, an indispensable feature for a molecular chronometer of evolution, according to Woese (1987). Furthermore, they occur in all organisms and different positions in their sequences change at highly variable rates, which makes them suitable for the measure of even distant phylogenetic relationships. Finally, their size is relatively large and, as seen above, their secondary structure consists in a quite large number of domains, fairly independent of one another (therefore, a mutation on one of them will not systematically alter the sequence of other domain sequences). Moreover, 16S rRNA genes can now be sequenced completely and most of these sequences obtained through one of the major online databases such as Genbank or EMBL which renders them suitable for high throughput phylogenetic (and other) analysis.

Based on 16S rRNA gene sequences, phylogenetic trees can thus be inferred. Various mathematical methods have been designed to draw these trees and test their reliability. Among them, the most widely used are the distance matrix method, the maximum parsimony analysis and the cluster analysis (for more details, see Woese, 1987). At the highest level, the phylogenetic analysis of living organisms through 16S rRNA sequences divides them into three major kingdoms : Archaebacteria, Eubacteria and Eucaryotes. In each of these three kingdoms, 16S rRNAs have slightly different secondary structures due to variations in key nucleotide positions, providing each of them with a so-called signature (a collection of specific oligonucleotides) (Woese, 1987).

If we go back to the genus *Bifidobacterium*, bacteria belonging to this genus are phenotypically defined as being strictly anaerobic, non motile, non-spore forming, Gram positive bacteria that can morphologically display peculiar V or Y shapes under microscopic analysis. They also exhibit the characteristic fructose 6-phosphate phosphoketolase activity (F6PPK). This enzyme is implied in the typical sugar fermentation pattern kown as "bifid-shunt" (Scardovi and Trovatelli, 1965). The 16S rRNA sequence analysis of bifidobacteria has enabled their classification in the kingdom Eubacteria. Within this kingdom, bifidobacteria are found in the group of Gram-positive bacteria. Interestingly, this phylum is homogenous, gathers all Gram-positive bacteria and forms a phylogenetically coherent group. On the contrary, Gram-negative bacteria are split into several (10) groups belonging to the kingdom Eubacteria (Woese, 1987). Eubacterial groups are divided in phyla, themselves split into sub-divisions. Bifidobacteria form the deeper branch included in the subdivision of high G+C species, grouping species whose DNA contains more than 55% guanine plus cytosine (G+C).

To illustrate the kind of results which can be obtained at the genus and species levels, a phylogenetic tree based on 16S rRNA gene sequences of the genus *Bifidobacterium* and related species is presented in Figure 1.

3. Use Of 16S rRNA Gene Sequence for the Design of Identification Tools

As for all other genera, the technique used as a reference for the attribution of a *Bifidobacterium* strain to a certain species remains DNA/DNA hybridisation. Nevertheless, as it is tedious and time consuming, several other identification methods have been described, some of which imply 16S rRNA encoding gene.

3. 1. Sequence Alignments

Sequencing of certain genes of interest and comparison through multiple alignments with sequences stored in online databases (EMBL or Genbank) have been applied to bifidobacteria. One of these genes is the 16S rRNA encoding one. For bifidobacteria, sequences or partial sequences of a combination of 16S rRNA and pyruvate kinase genes have been employed (Vaugien *et al.*, 2002). Nevertheless, this method remains quite expensive, tedious and largely dependent on gene sequences of well-characterised strains being already available in databases. However, multiple alignments provide a basis for the design of other identification tools. Indeed, it enables the comparison of sequences and identification of conserved or variable regions within 16S rRNA genes. This allows for the description of oligonucleotidic probes or primers either universal, genus or species specific, according to the target region chosen for hybridisation of the oligonucleotidic sequences designed.

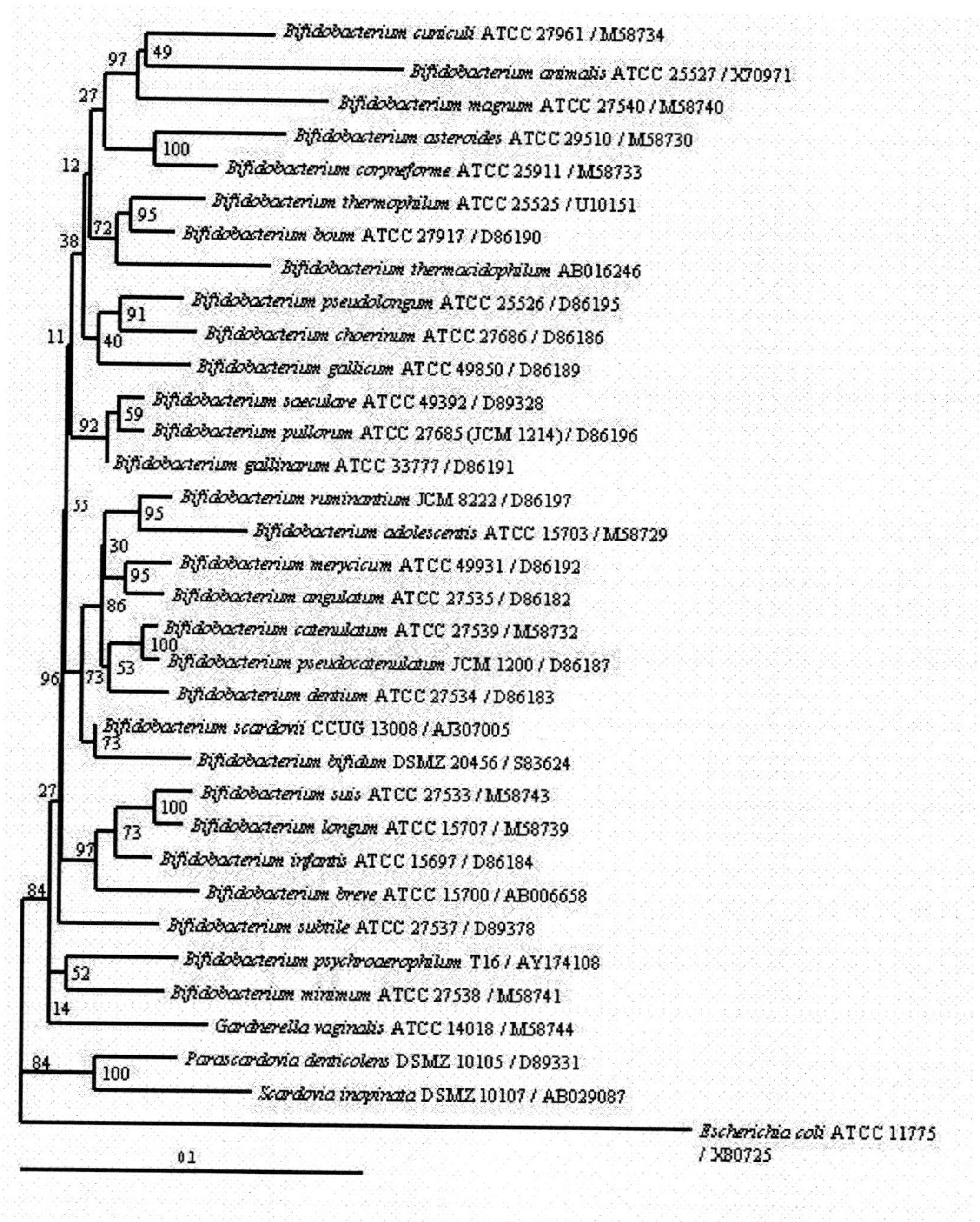

Figure 1. Phylogenetic tree of the genus Bifidobacterium and related species based on 16S rDNA sequences retrieved from Genbank. The tree was rooted with Escherichia coli and constructed using the neighbour-joining method. Bootstrap values at each branch point represent the percentage of bootstrap support (calculated from 1000 trees). The bar represents 0.1% of sequence divergence.

Table 1. Examples of probes used for eubacteria, the genus *Bifidobacterium* and some of its species (5'→ 3').

Bacterial target	Probe	Target site[a]	Use	Reference
Eubacteria	GTA TTA CCG CGG CTG CTG	519-536	FISH[b]	Langendijk *et al*., 1995
	ACT CCT ACG GGA GGC AGC	321-338	FISH	Kaufmann *et al*., 1997
Genus *Bifidobacterium*	GAT AGG ACG CGA CCC CAT	ND[c]		Marteau *et al*., 2001
	GAT TCT GGC TCA GGA TGA ACG	15-35	FISH	Kaufmann *et al*., 1997
	CGG GTG CTT CCC ACT TTC ATG	1412-1432	FISH	Kaufmann *et al*., 1997
	GAG GAC CTT TGC CCA CCA	ND	FISH	Matsuki *et al*., 2004
	CAT CCG GCA TTA CCA CCC	164-181	FISH	Langendijk *et al*., 1995
	CCA CCG TTA CAC CGG GAA	662-679	FISH	Langendijk *et al*., 1995
	CCG GTT TTC AGG GAT CC	1278-1294	FISH	Langendijk *et al*., 1995
B. adolescentis	GCT CCC AGT CAA AAG CG	ND	FISH	Yamamoto *et al*, 1992
	GCG AAA ACT GAC CCT CG	ND	FISH	Matsuki *et al*., 2004
	GGA TCG GCT GGA GCT TGC TCC GGC CGT GAG AGT GGC GAA	42	Microarrays	Wang *et al*., 2002
	CTC CAG TTG GAT GCA TGT CCT TCT GGG AAA GAT TCT ATC GGT	156	Microarrays	Wang *et al*., 2002
	CAA CGG GAT GCG ACC TCG TGA GGG GGA GCG GAT CCC TT	1207	Microarrays	Wang *et al*., 2002
B. bifidum	GCA GGC TCC GAT CCG A	ND	FISH	Yamamoto *et al*, 1992
	CCA CAA TCA CAT GCG ATC ATG	ND	FISH	Matsuki *et al*., 2004
B. breve	AAG GTA CAC TCA ACA CA	ND	FISH	Yamamoto *et al*, 1992
***B. catenulatum* group**	ACA CCC CAT GCG AGG AGT	ND	FISH	Matsuki *et al*., 2004
B. dentium	ACG TCA CGG TGG GAA CTC A	1132-1151	FISH	Nebra *et al*., 2003

Table 1. Continued

Bacterial target	Probe	Target site[a]	Use	Reference
B. infantis	TCA CGC TTG CTC CCC GAT A	ND	FISH	Yamamoto *et al*, 1992
	ACC GGA TGC TCC GCT CCA TCG CAT GGT GGG GTG GGA AAT	154	Microarrays	Wang *et al*., 2002
	GGC TTG ACA TGT GCC GGA TCG CCG TGG AGA CAC GGT TTC C	967	Microarrays	Wang *et al*., 2002
	GGT AGA CAC CCG AAG CCG GTT GGC CCG ACC CTT GTT GGG	1404	Microarrays	Wang *et al*., 2002
B. longum	TCT CGC TTG CTC CCC GAT A	ND	FISH	Yamamoto *et al*, 1992
	AGC CGT ATC TCT ACG ACC GT	ND	FISH	Matsuki *et al*., 2004
	CTC CAT CAC ACC GCA TGG TGT GTT GGG AAA GCC TTT GCG GC	165	Microarrays	Wang *et al*., 2002
	GGC TTG ACA TGT TCC CGA CGA TCC CAG AGA TGG GGT TTC C	959	Microarrays	Wang *et al*., 2002
	AGC CGG TGG CCT AAC CCC TTG CGG GAG GGA GCC GTC TAA TG	1407	Microarrays	Wang *et al*., 2002

[a]: Target region for hybridisation in the 16S rRNA molecule, numbered according to the homologous *E. coli* sequence (Noller and Woese, 1981).
[b]: Fluorescent *in situ* hybridisation
[c]: Not Determined.

3. 2. Oligonucleotidic Probes and Hybridisation

Briefly, identification of bacteria through oligonucleotidic probes is based on the hybridisation (recognition) of a given short sequence of the genome by short oligonucleotides. Usually, their size is set around 20 bases but can be increased to 40 bases for some applications, like the microarray technique. These probes are marked with various fluorescent or radioactive compounds, allowing for their detection. Their use is quicker and also much cheaper than DNA/DNA hybridisation.

The interest of 16S rRNA molecule here lies in its conserved regions as well as in the more unstable ones, depending on the target one is willing to identify. Oligonucleotidic probes can be targeted to recognise almost all of the bacterial species, as it is done in fluorescent *in situ* hybridisation (FISH) techniques using the so-called EUB probes (theoretically able to detect all eubacteria), for example (Table 1)

On the other hand, probe specificity can be refined by selecting 16S rRNA regions specific for a whole genus or for a given species. If we take the genus *Bifidobacterium* as an example, genus- and species-specific oligonucleotidic probes have been designed to identify bifidobacteria and even quantify those isolated from food (See Table 1 and references therein).

The latest development with these labelled hybridisation probes is their use in microarrays, as shown for intestinal bacteria (Wang *et al.*, 2002 ; Cerniglia *et al.*, 2002) but which could also be suitable for food products.

Up to this day, the major disadvantage of these probes remains that, even though they are now mostly labelled with non-radioactive tracers, not all of the species of the genus *Bifidobacterium* can be traced.

3. 3. Ribotyping

Ribotyping can be defined as the analysis of DNA polymorphism (i.e. variations of the nucleotide sequences between individuals of the same species) in the chromosomal regions containing the highly conserved rRNA genes (Bingen *et al.*, 1994). Ribotyping is not only aimed at 16S rRNA gene but at the whole *rrn* operon. After chromosomal DNA extraction, the first step of ribotyping is the digestion of DNA by endonucleases recognizing specific nucleotidic sequences of 4 to 6 bases, known as restriction sites. Those endonucleases are therefore also called restriction enzymes. In a second step, DNA fragments generated this way are separated according to their size by electrophoresis on agarose gels. The profiles thus obtained can be visualised by ethidium bromide staining and are referred to as Restriction Fragment Length Polymorphism (RFLPs). Most of the time, a single base pair variation, accounting for the presence or not of a restriction site, is responsible for RFLPs. However, RFLP can also be due to insertions, deletions or rearrangements of DNA in restriction sites. The choice of the restriction enzyme used to generate RFLPS is of utmost importance. Indeed, an endonuclease recognising at frequently occurring restriction sites will produce a high number of DNA fragments. The electrophoretic profiles thus obtained will be extremely difficult to compare and interpret without the help of a computerised system. To overcome this impediment, two options can be taken. The first consists in choosing a rare cutting endonuclease, which will yield fewer fragments, therefore fewer bands in the electrophoretic profile one wishes to study, hence easing comparisons and interpretations. The second requires, once the fragments are electrophoretically separated, to transfer DNA fragments to a nylon membrane (classic Southern procedure) and selectively highlight some of the fragments produced by hybridisation with tagged (radioactively or not) rDNA probes (universal or specific for a genus, a species...), which will also finally result in fewer detectable bands in the electrophoretic profile. A great diversity of probes have been designed for this type of technique, including 16S rRNA specific probes and some of the RFLP thus highlighted are considered to be species specific while others can be described as strain specific. For this technique, the number of *rrn* operons present in a given species is important as the potential discriminative power increases with the number of copies (Bingen *et al.*, 1994).

This renders ribotyping a suitable tool for both taxonomic and epidemiological studies. Indeed, McCartney *et al.* (1995) successfully described such a method for bifidobacteria and used it to trace them at the strain level in human faeces (McCartney *et al.*, 1996; Kimura *et al.*, 1997). Briefly, the endonucleases used were *Nar*I, *Pvu*II and *Bam*HI. After transfer of restriction fragments to a nylon membrane, a radioactively labelled 16S rDNA probe obtained from *B. infantis* ATCC 15697^T was used to select fragments of interest. Thereafter, marked restriction fragments were detected by autoradiography.

3. 4. Polymerase Chain Reaction (PCR) DNA Amplification Using Species-Specific Primers

These methods are designed to detect the presence or absence of well-characterised bifidobacteria. They are of a straight forward interpretation but only allow for a detection of bifidobacteria at the genus or at the species level, depending on the primers chosen.

Various species-specific pairs of primers have been designed targeting different genes such as 16S rRNA (Matsuki *et al.*, 1998: Matsuki *et al.*, 1999). Also, pairs of primers yielding amplification products of various sizes have been designed to be used in combination to speed up the identification and reduce the cost of the analysis (Mullié *et al.*, 2003).

Lately, part of our work has focused on this approach for the identification of species isolated from human sources. Up to this day, they represent a dozen of species, if *Scardovia inopinata* and *Parascardovia denticolens*, recently excluded for the genus *Bifidobacterium* (Jian and Dong, 2002), are considered. One of the main advantages of using species-specific primers is the simplification of interpretation work, as a single and well-characterised amplicon (theoretical pair bases size obtained from an *in silico* analysis) is produced and can easily be detected through agarose electrophoresis. Moreover, once validation of the primers used is performed on reference strains, there is no need for reference profiles obtained from well characterised strains to allow for clustering and identification of unknown strains, as it is the case with fingerprinting techniques such as ribotyping, PCR-DGGE, PCR-TGGE or rep-PCR.

However, when they are designed for monoplex amplifications (allowing for the detection of a single species per each PCR attempt), the use of species-specific primers remains tedious and, to some extent, expensive when large series of strains are to be identified. Therefore, the purpose of our work was to reduce the number of PCR attempts and, thus, hands-on time for the identification of human species of bifidobacteria. To achieve this goal, primers that would be used to amplify 16S rRNA gene fragments of various sizes were designed and clustered in a convenient series of 3 multiplex PCRs based on 12 species belonging or related to the genus *Bifidobacterium* so far isolated from human samples (Table 2). In each of the 3 groups, 3 to 4 pairs of species specific primers have been gathered, yielding amplified products of various sizes, each typical of a defined species (from 126 up to 873 base pairs) (Table 3).

Table 2. Species of the genus *Bifidobacterium* isolated from human sources.

Human source	Species
Faecal or intestinal samples	*B. adolescentis, B. angulatum, B. bifidum, B. breve, B. catenulatum, B. gallicum, B. longum/B. infantis, B. pseudocatenulatum*
Oral (caries) samples	B. dentium, S. inopinata, P. denticolens
Others (urine, blood…) samples	B. scardovii

Design of Species-Specific Primers Targeting 16S rRNA Gene

16S rDNA sequences of type strains (from human and non human sources) of the genus *Bifidobacterium* were retrieved from GenBank and aligned using Clustal W software. Based on this alignment, species-specific primers were designed by selecting 17 to 21 nucleotide-long sequences in the most variable regions detected. Our first aim was to group the species the more genetically related in a same PCR bunch. But, for some of them, the expected sizes of amplified products were too close to be discriminated on agarose gels. Therefore, closely related species had to be identified in separate bunches (e. g. *B. longum* / *B. infantis* and *B. catenulatum/B. pseudocatenulatum* continuum / *B. adolescentis*). Finally, group 1 gathered three species mostly found in infant faeces: *B. bifidum*, *B. breve* and *B. infantis*. Group 2 included *B. angulatum*, *B. catenulatum /B. pseudocatenulatum* continuum, *B. dentium* along with *B. longum*. Group 3 gathered *B. adolescentis*, *P. denticolens*, *S. inopinata* and *B. gallicum*. *B. scardovii* specific primers were also designed and have been employed successfully in the Group 3 multiplex when used instead of *P. denticolens* primers (Table 3). All primers were purchased from Oligo-express (Montreuil, France).

These primers were employed to identify strains of the genus *Bifidobacterium* genus isolated either from infant or adult faeces as well as for the detection bifidobacterial species directly from DNA extracted from faecal material.

Table 3. Examples of *Bifidobacterium* and related species-specific primers based on 16S rDNA sequences.

Target species		Sequence	Target site[a]	Expected size (bp)	Reference
Group 1					
B. bifidum	FWD[a]	CCA CAT GAT CGC ATG TGA TTG	183-203	277	Matsuki *et al.*, 1999
	REV[a]	CCG AAG GCT TGC TCC CAA A	460-442		Matsuki *et al.*, 1999
B. breve	FWD	GGG AGC AAG GCA CTT TGT GT	432-451	568	Mullié *et al.*, 2003
	REV	GAA ACC CCA TCT CTG GGA TC	1000-981		Mullié *et al.*, 2003
B. infantis	FWD[a]	TTC CAG TTG ATC GCA TGG TC	169-188	826	Matsuki *et al.*, 1999
	REV	GAA ACC CCA TCT CTG GGA TC	995-976		Mullié *et al.*, 2003
Group 2					
B. angulatum	FWD	ATC GGC TGG AGC TTG CTC C	60-78	126	Mullié *et al.*, 2003
	REV	CAT GCG ATG GAC TGG AGC AT	186-167		Mullié *et al.*, 2003
B. catenulatum.B. pseudocatenulatum continuum[b]	FWD[a]	CGG ATG CTC CGA CTC CT	176-192	285	Matsuki *et al.*, 1999
	REV[a]	CGA AGG CTT GCT CCC GAT	474-442		Matsuki *et al.*, 1999
B. dentium[c]	FWD[a]	ATC CCG GGG GTT CGC CT	72-89	387	Matsuki *et al.*, 1999
	REV[a]	GAA GGG CTT GCT CCC GA	473-443		Matsuki *et al.*, 1999
B. longum	FWD[a]	TTC CAG TTG ATC GCA TGG TC	181-200	831	Matsuki *et al.*, 1999
	REV[a]	GGG AAG CCG TAT CTC TAC GA	1028-1008		Matsuki *et al.*, 1999
Group 3					
B. adolescentis	FWD[a]	CTC CAG TTG GAT GCA TGT C	184-202	278	Matsuki *et al.*, 1999
	REV[a]	CGA AGG CTT GCT CCC AGT	462-445		Matsuki *et al.*, 1999
P. denticolens	FWD	GGA AAG CTT TTG CGC CAT GG	196-215	788	Mullié *et al.*, 2003
	REV	GGC GAT CGT CAA CAT GTC AA	984-965		Mullié *et al.*, 2003
B. scardovii	FWD	GTG ATG GCC TAC CGT GGC TT	227-246	873	Mullié *et al.*, 2003
	REV	CGT GAG TTC CCA TCC AAA ATG	1100-1081		Mullié *et al.*, 2003

Table 3. Continued

Target species		Sequence	Target site[a]	Expected size (bp)	Reference
B. gallicum	FWD[a]	TAA TAC CGG ATG TTC CGC TC	157-176	681	Matsuki *et al.*, 1999
	REV	ACA CAG GAC CCG GAA AAA GG	838-819		Mullié *et al.*, 2003
S. inopinata	FWD	AAC CTG CCG TGT AGT TGG GT	123-142	494	Mullié *et al.*, 2003
	REV	CCC ACC GTT AAG CAG TAA GC	617-598		Mullié *et al.*, 2003

[a]: Target region for hybridisation in the 16S rRNA molecule, numbered according to the homologous *E. coli* sequence (Noller and Woese, 1981).

[b]: pairs of primers tested but discarded because of a lack of specificity : for *B. catenulatum* FWD: GCA GCT TGC TGC CTG GTG A and REV: CGA AGG GAG ACC GTA TCT CT and for *B. pseudocatenulatum* FWD: ACG GCG ACG TGG AGC GG and REV: TAG ACG GCT CCA TCC ACA AAG

[c]: Specific primers tested but discarded because of a too weak amplification product under multiplex conditions:
FWD: TCG GGA GCA AG CCC TTC G and REV: AGG CCG TAT CTC TAC GGC.

DNA Isolation from Cultures

DNA isolation from pure strains was performed after an anaerobic growth of 24 to 48h in pre-reduced Brain Heart Infusion broth. Then, 1.5 ml of the broth was centrifuged at 4.500 g (5 min, 4°C). The bacterial pellet was retained and subjected to DNA extraction using Nucleospin Tissue© kit (Macherey-Nagel, Hoerdt, France) according to the supplier's instructions (support protocol for Gram-positive bacteria). Briefly, a first incubation with lysozyme (20 mg/ml) was carried out at 37°C for 1h. It was followed by a second incubation with proteinase K (30 mg/ml) at 56°C for 3h. RNase A (Sigma, Saint Quentin-Fallavier, France) was then added to the mixture to eliminate contaminating RNA (incubation for 5 min at room temperature). A further 10 min incubation at 70°C with a buffer containing chaotropic salts (B3 buffer, Nucleospin Tissue© kit) was performed to obtain a thorough lysis of the bacteria. The resulting DNA suspension was then purified on a mini-column provided in the kit and frozen at –20°C until further use.

DNA Isolation from Faecal Swabs

Faecal swabs (Portagerm Amies agar, Biomérieux, Marcy l'étoile, France) collected from babies were homogenised in 1:4 Ringer solution supplemented with 0.3% (W/V) cysteine hydrochloride. Dilutions were carried out in decimal steps up to 10^{-6} in osmosed water pre-treated with diethyl pyrocarbonate (DEPC). Centrifugation was then applied to 1.35 ml of each dilution (30.000g, 5 min, 4°C) and the pellet subjected to the same extraction procedure as the bacterial pellet from cultures.

PCR Amplification

The PCR reaction mixture was composed of:

- 5 µl of 10× PCR buffer II (Six pack Ampli Taq Gold DNA polymerase with buffer II, Applied Biosystem, Courtaboeuf, France)
- 2 mM $MgCl_2$ (Applied Biosystem)
- 400 µM of each dNTP (Applied Biosystem)
- 0.4 µM of each species-specific primer required (Oligo-express, Montreuil, France)
- 2.5 U of Taq DNA polymerase (Applied Biosystem)
- 10 µl of template DNA.
- DEPC pre-treated ultrapure water (to a final volume of 50 µl)

Amplification was carried out in a PTC-200 thermal cycler (PTC-200, MJ Research). The amplification consisted of one cycle at 94°C for 5 min, 35 cycles of 94°C for 30s, 59°C for 30 s, and 72°C for 1min to 2 min 42 s (elongation gradient of 3 s for each additional cycle), and a final elongation cycle of 72°C for 5 min. Amplified products were run on 1.5% agarose gel, stained with ethidium bromide and visualised under UV transillumination.

Validation of PCR Primers

The strains of the genus *Bifidobacterium* and related species used to validate the technique were obtained from the DSMZ (Deutsche Sammlung von Mikroorganismen und Zellkulturen, Germany), ATCC (American Type Culture Collection, United States of America) and BCCM^{T}/LMG (Belgian Co-ordinated Collection of Micro-organisms/Laboratorium voor Microbiologie of Gent, Belgium) and are listed in Table 4.

Table 4. Reference strains used for the validation of species-specific primers.

Species	Strain
Bifidobacterium adolescentis	DSM[a] 20083^{T}
	ATCC[b] 15705
Bifidobacterium angulatum	DSM 20098^{T}
	LMG[c] 11568
Bifidobacterium animalis	DSM 20104^{T}
Bifidobacterium bifidum	DSM 20456^{T}
	CFPL[d] AA22
Bifidobacterium breve	DSM 20213^{T}
	CFPL C7
	CFPL C50
Bifidobacterium catenulatum	DSM 20103^{T}
	LMG 18894
Parascardovia denticolens	DSM 10105^{T}
Bifidobacterium dentium	DSM 20436^{T}
Bifidobacterium gallicum	DSM 20093^{T}
Bifidobacterium infantis	DSM 20088
	ATCC 17930
Scardovia inopinata	DSM 10107^{T}
Bifidobacterium lactis	DSM 20104^{T}
Bifidobacterium longum	DSM 20219^{T}
	CFPL C40
	CFPL C117
Bifidobacterium pseudocatenulatum	DSM 20438^{T}
Bifidobacterium scardovii	LMG 21589^{T}
	LMG 21590
Bifidobacterium suis	DSM 20210^{T}
Gardnerella vaginalis	DSM 4944^{T}

[a]: DSM = Deutsche Sammlung von Mikroorganismen und Zellkulturen, Germany
[b]: ATCC = American Type Culture Collection, United States of America
[c]: BCCM^{T}/LMG = Belgian Co-ordinated Collection of Micro-organisms/Laboratorium voor Microbiologie of Gent, Belgium
[d]: CFPL = Collection de la Faculté de Pharmacie de Lille, France and previously identified using DNA-DNA hybridisation (Bahaka *et al.*, 1993)

Primers were first tested in monoplex PCR reactions to ensure their specificity. This was done using chromosomal DNA extracted from type strains of the genus *Bifidobacterium* and related species found in human samples as well as from *B. animalis*, *B. lactis*, *B. suis* and *Gardnerella vaginalis*, used as external controls. Every pair of primers gave amplified products of the expected size. Amplified products were sequenced (AbiPrism 377, Perkin-Elmer) and, when compared to Genbank, the resulting sequences matched those of their respective type strain. The only cross-reactions witnessed were positive PCR results for DNA of *B. suis* DSM 20210^T when amplified with either *B. infantis* or *B. longum* specific primers and for *B. catenulatum* DSM 20103^T when amplified with *B. angulatum* primers.

To check that there was no interference with the spfecies-specific primers of one of the selected groups, PCR were performed by pooling them and using chromosomal DNA from the type strains of the species they recognised. Except for *B. angulatum,* which showed a weaker amplification signal, the amplification reactions of chromosomal DNA of each type strain were not impaired by the presence of the other pairs of primers. Under multiplex conditions, *B. catenulatum* DSM 20103^T DNA gave a single amplified product of 126 bp, the sequence of which matched the *B. angulatum* DSM 20098^T sequence. To identify wild isolates of *B. angulatum* and *B. catenulatum*, we therefore had to confirm each strain yielding a 126 bp amplicon by a simplex PCR using the *B. catenulatum/B. pseudocatenulatum continuum* primers. If no amplification was obtained, the strain was confirmed as belonging to the *B. angulatum* species and if a 285 bp was observed, the strain was assigned to the *B. catenulatum* species.

Once validated on known strains, primers were then also tested for the identification of wild strains isolated from faecal samples of either infants (3 to 5 month-old) or adults (19 to 29 years). These strains had been allotted to the genus *Bifidobacterium* on the basis of their anaerobic requirement, Gram staining, cellular morphology, as well as on the detection of their fructose 6-phosphate phosphoketolase activity. All strains were kept frozen at -20°C on Rosenow broth (Sanofi Diagnostic Pasteur) until the identification process took place. They were then subcultured anaerobically on Columbia broth (Oxoid, Dardilly, France) at 37°C for 24 to 48 h. Bacterial DNA was afterwards extracted, purified and amplified according to the procedure stated above.

The results obtained on the 103 strains processed are summarised in Table 5. The most frequently isolated species in infants was *B. bifidum*, corroborating former observations (Requena *et al.*, 2002). It was followed by *B. longum*, *B. breve,* two species also frequently identified in infants using phenotypic or genotypic techniques (Table 5) (Biavati *et al.*, 1984; Matsuki *et al.*, 1999). *B. dentium* was never identified among infant isolates. In adults, the most frequently isolated species was *B. adolescentis* followed by *B. catenulatum/B. pseudocatenulatum* and *B. longum* (Table 5), also confirming previous reports using both phenotypic and genotypic methods (Matsuki *et al.*, 1999). *B. breve* and *B. infantis* were never identified in adult isolates. In both adults and infants, we never detected *P. denticolens* or *S. inopinata*. This could be related to their usual location in the upper part of the gastro-intestinal tract (i. e. dental caries) (Crociani *et al.*, 1996). Thus, the results obtained with this multiplex approach are in accordance with the previously described age-related distribution of bifidobacterial species, further confirming its reliability. Overall, this identification technique is one of the most comprehensive and easy to handle methods when human species are concerned.

Table 5. Identification of wild strains of the genus *Bifidobacterium* isolated from infant or adult faecal samples

Species[a]	Infants		Adults	
	No. of strains (n=82)	No. of carriers (n=21)	No. of strains (n=21)	No. of carriers (n=7)
B. adolescentis	4 (4.9)[b]	4 (19)[c]	8 (38.1)	5 (71.4)
B. angulatum	2 (2.4)	2 (9.5)	0 (0)	0 (0)
B. bifidum	30 (36.7)	15 (71.4)	3 (14.3)	2 (28.6)
B. breve	19 (23.2)	8 (38.1)	0 (0)	0 (0)
B. catenulatum/B. pseudocatenulatum	3 (3.6)	2 (9.5)	4 (19.0)	3 (42.9)
B. dentium	0 (0)	0 (0)	1 (4.8)	1 (14.3)
B. infantis	2 (2.4)	1 (4.8)	0 (0)	0 (0)
B. longum	22 (26.8)	10 (47.6)	5 (23.8)	2 (28.6)

[a]: *B. gallicum, B. scardovii, S. inopinata* and *P. denticolens* were never detected either from infant or adult samples. Therefore, they are not presented in this table.

[b]: percentage of the total number of strains.

[c]: percentage of the total number of infants or adults.

For some infant faecal samples (40 samples), direct PCR on DNA isolated from faecal suspension and its decimal dilutions (up to 10^{-6}) gave the same amplicons as those obtained with the isolated strains. For a few samples, additional species which were not detected in the isolates yielded by the cultural approach could even be detected in the faecal suspension and first dilutions. For example, in one infant sample, *B. longum* was detected by the sole direct amplification of DNA extracted from the faeces while cultural isolates were identified as belonging to *B. breve*, *B. bifidum* and *B. catenulatum/B. pseudocatenulatum* continuum.

3. 5. PCR-Denaturing Gel Gradient Electrophoresis (DGGE) or PCR-Temperature Gel Gradient Electrophoresis (TGGE)

This method was primarily used for ecological purposes. It consists in a PCR amplification with primers specific for the genus *Bifidobacterium* followed by a sequence-specific separation in temperature gradient gel electrophoresis (TGGE) or denaturing gradient gel electrophoresis (DGGE). The profiles thus obtained represent the prominent bacteria in the community and identification can be carried out through the comparison with a reference profile of known species and/or strains. PCR-DGGE does not require culture and isolation. A diversity of genus specific primers has been described for the genus *Bifidobacterium* and this technique has been applied to human faeces (Satokari *et al.*, 2001 ; Requena *et al.*, 2002) and to probiotic products as well (Temmerman *et al.*, 2003 ; Fasoli *et al.*, 2003).

Targeted genes used include the 16S rRNA encoding one. In the course of our studies, we were also able to select and test one of those pair of primers, the sequences of which are reported below:

- Forward primer: 5'- AGT CCA CGC CGT AAA CGG TGG A – 3'
- Reverse primer:5'- AGG GGC ATG ATG ATC TGA CGT CA –3'

As primers are genus-specific, one of the main advantages of this method is the detection of all species within this genus, theoretically even those which have not been described so far (that can not be done through amplification with species-specific primers). However, the principal drawback is that you need a reference ladder build with well characterised strains to be able to assign a band present in a certain position to a species. Alternatively, if the species is unknown or you do not possess a suitable reference strain, you can sequence the amplified 16S rDNA fragment and challenge on-line databases to see whether a strong homology can be found with a sequence already submitted. Yet, an identification led this way will be rather expensive.

3. 6. Terminal Restriction Fragment Length Polymorphism (T-RFLP)

This technique measures the size polymorphism of terminal restriction fragments from a PCR amplified marker (Marsh, 1999). As for PCR-DGGE and PCR-TGGE, it was designed for ecological purposes: it was meant to be a tool for the study of complex microbial communities. Classically, PCR amplification is performed with a non-specific pair a primers targeting conserved sequences on the bacterial genome. Therefore, conserved 16S rRNA sequences are good candidates for such a technique (as other conserved sequences on the genome would be). The forward or reverse primer is usually marked with a fluorescent component (such as 6-FAM: 6-carboxyfluorescein) to allow for the detection of amplified fragments. PCR products are purified and then digested with a variety of restriction enzymes such as *Hha*I, *Bfa*I, *Msp*I or *Rsa*I. The analysis of digested PCR products can be carried out by automated systems such as capillary electrophoresis. Indeed, only the terminal fragment can be detected, as it is the only one tagged (through the marked forward primer), hence explaining the full name of the method. In summary, this technique is a combination of at least two others described earlier in this chapter: PCR amplification and analyses of restriction fragment length polymorphism (just like in the ribotyping method).

Human bifidobacterial communities have already been described using T-RFLP (Sakamoto *et al.*, 2002) using complete 16S rDNA amplification with universal primers 27F and 1492R and specific bifidobacterial amplification with primers Bif164-f and Bif662-r (Table 6). As with classical PCR, when a defined restriction enzyme is chosen, the size of T-RLF can be predicted *in silico* from the sequences obtained in electronically available databases. Hayashi *et al.* (2004) showed that the usual forward primer used for T-RFLP analysis (27F) was not the most suitable for bifidobacteria and proposed two alternative forward primers given in table 6. The reverse primer remained unchanged (1492R).

Table 6. T-RFLP primers proposed for the genus *Bifidobacterium* (5'→3').

	Sequence	Reference
Universal T-RFLP primers		
27F	AGA GTT TGA TCC TGG CTC AG	Sakamoto *et al.*, 2003
35F	CCT GGC TCA GGA TGA ACG	Hayashi *et al.*, 2004
529F	ACG TGC CAG CAG CCG CGG	Hayashi *et al.*, 2004
1492R	GGT TAC CTT GTT ACG ACT T	Sakamoto *et al.*, 2003
Bifidobacteria specific T-RFLP primers		
Bif-164f	GGG TGG TAA TGC CGG ATG	Sakamoto *et al.*, 2003
Bif-662r	CCA CCG TTA CAC CGG GAA	Sakamoto *et al.*, 2003

4. Use of 16S rRNA Gene Sequence for the Description of New Species

The "Report of the Ad Hoc Committee on Reconciliation of Approaches to Bacterial Systematics" was published by Wayne *et al.* (1987). For these authors, the definition of a species would generally include strains with approximately 70% or greater DNA-DNA hybridisation and with 5°C or less ΔTm (ΔTm is a parameter which reflects the thermal stability of the heteroduplex of DNAs provided by two strains; it is reduced by approximately 1°C for each 1% of unpaired bases within the duplex). On the basis of 16S rRNA high degree of conservation within a species, some authors assumed that bacterial strains sharing less than 97% sequence homology for this molecule should be considered as belonging to different species. This threshold was proposed because species having 70% or greater DNA homology usually exhibit more than 97% 16S rRNA sequence homology. Indeed, in 1993, 14% of the species descriptions contained information inferring the phylogenic positions exclusively on the basis of 16S rRNA analysis (Stackebrandt and Goebel, 1994). However, the correlation plot of DNA similarity and 16S rRNA homology is not linear. Groups of organisms which share almost 100% 16S rRNA homology can display less than 70% DNA-DNA reassociation (Stackebrandt and Goebel, 1994). On the other hand, less than 97% sequence homology could be found in strains belonging to the same species, as defined by DNA-DNA hybridisation. In fact, while only analysing 16S rRNA gene, you are only looking at a little more than 1500 bp in the whole bacterial genome and as the repartition of nucleotide changes is not uniform, these might concentrate is the small part of the genome you are analysing. The resolution power of hybridisation is therefore much higher than that of sequence analysis. Stackebrandt and Goebel (1994) concluded that 16S rRNA sequence analysis was not the appropriate method to replace DNA reassociation for the delineation of species and mesure intraspecies relationships. However, it was not until 2002 that the second "Report of the Ad Hoc Committee for the re-evaluation of the species definition in Bacteriology" was published by Stackebrandt *et al.* (2002). The Committee made various recommendations regarding the species definition in the light of developments in methodologies available to systematists. The species definition of Wayne *et al.* (1987) remains the "Gold Standard" but investigators are encouraged to propose new species based upon other genomic methods or techniques provided that they can demonstrate that, within the taxa studied, there is a sufficient degree of congruence between the technique used and DNA-DNA reassociation. Moreover, this consensus states that an informative level of phylogenetic information would be obtained from a minimum of five genes encoding metabolic functions and under stabilising selection (also known as housekeeping genes) (Stackebrandt *et al.*, 2002). However, between the publication of these two papers, a burst in the description of new species (sometimes described on a single isolate) uniquely on the 16S rRNA sequence criterion occurred. This led Christensen *et al.* (2001) to question these uses and recommendations now strongly encourage to back up a species description on more than a single strain on the basis of the arguments these authors raised. More recently, Janda and Abbott (2002) also questioned the species definition and strain identification on the sole 16S rRNA sequence basis, especially when clinical isolates are concerned. In conclusion, the sole sequencing of 16S rRNA gene is not sufficient for the description of a new species. Nevertheless, its complete (>1300 nucleotides) sequence should be included in all new species description (Stackebrandt *et al.*, 2002).

5. Limits of 16S rRNA Uses for the Genus *Bifidobacterium*

One of the major concerns while working on the multiplex PCR procedure described in paragraph 3.4. was to distinguish species from the two existing continua in the genus *Bifidobacterium*: *B. longum*/*B. infantis* and *B. catenulatum*/*B. pseudocatenulatum*. Continua are indeed groups of strains divided in two or more species but with such DNA-DNA hybridisation data that the legitimacy of these entities at the species level can be questioned. For *B. longum*/*B.infantis* continuum, the design of specific pairs of primers could be achieved. However, *B. infantis* was seldom identified in stool samples, even in infants (2 strains out of 82 tested). As it is meant to be a commonly carried species in infants (Biavati *et al.*, 1984), this raised the point of the primers being too specific as regards the 16S rDNA sequence of the *B. infantis* type strain. This could lead to the absence of amplification of some *B. infantis* 16S rDNA strains. Morevover, the cross-reaction of *B. suis* DSM 20210^T with both *B. infantis* and *B. longum* primers as well as the proposal to unify this continuum questions the interest of a too great specificity in identification (Sakata *et al.*, 2002). Additionally, only two infant strains could not be identified using our technique. Sequencing of 16S rDNA enabled the classification of both of them as *B. longum*, further questioning the too strict specificity of the *B. longum* primers used. For the *B. catenulatum*/*B.pseudocatenulatum* continuum, we were able to design in silico specific primers for each species (Table 3). In practice, this specificity was not achieved so we had to exploit the formerly described primers for the continuum (Matsuki *et al.*, 1998). In addition, the amplification of *B. catenulatum* DSM 20103^T with the *B. angulatum* primers under multiplex conditions further exposed the limits of using only 16S rDNA sequences for identification purposes. Interestingly, Requena *et al.* (2002) had the same difficulty in differentiating *B. angulatum* and *B. catenulatum* using the transaldolase gene. Therefore, when it comes to very closely related species, the discriminative power of the sole 16S rRNA sequence is not sufficient to ascertain the belonging of one strain to one species or its close neighbour. Alternative markers, such as 16S to 23S internal transcribed spacer sequence proposed by Leblond-Bourget *et al.* (1993), might provide more information on inter- and/or intraspecific relationships.

Another problem encountered for the genus *Bifidobacterium* (but not specific to this genus) with 16S rRNA sequences is that some entries in the databases can be erroneous. This is indeed the case with Genbank sequences M58737 for *Bifidobacterium indicum* ATCC 25912^T and M58742 for *Bifidobacterium pseudolongum* ATCC 25526^T (Dong *et al.*, 2000). These problems become fewer as deposited sequences are more controlled but incorrect inputs should be removed when they are signaled to ease the work of online database users.

A final point that might be of interest in the use of specific 16S rRNA primers is their application in quantitative PCR (for the analysis of intestinal bifidobacteria, for example), as proposed by Matsuki *et al.* (2004). This technique uses the detection of fluorescent products generated during each PCR cycle to quantify the original number of copies of the PCR target sequence within the sample. The sooner the fluorescent product is detected, the higher the number of copies in the sample. However, application of this method with 16S rRNA gene as a target raises some questions. As seen in the beginning of this chapter, the number of *rrn* operon copies within the genus *Bifidobacterium* varies from one species to another (from 2 to 5 copies) and some of these copies are not even similar to the other ones within a same strain (see *B. adolescentis* example). Matsuki *et al.* (2004) employed other techniques to enumerate bifidobacteria (culture and FISH) to validate their real-time quantitative PCR and avoid this bias. However, it seems difficult to apply this technique routinely as such a validation should be performed almost every time, especially for samples containing complex populations of bifidobacteria as variations within the proportions of each species would influence the final results. Indeed, Farrelly *et al.* (1995) stated that the knowledge of genome size and the copy number of *rrn* genes were vital information for PCR amplification efficiency. For example, if the number of copy is high, the amount of enzyme (*Taq* polymerase) or substrates (primers and/or dNTPs) might become the limiting factor of PCR reaction. Hence, the amplification will not be linked in linear way with the number of *rrn* genes in the original sample. Therefore, the use of 16S rRNA as a target for real-time PCR quantification should be taken cautiously and further studies using comparative enumeration methods should be performed before it can be used routinely either as such or with corrective factors that remain to be determined.

References

Bahaka, D., Neut, C., Khattabi, A., Monget, D. and Gavini, F. (1993) Phenotypic and genotypic analyses of human strains belonging or related to *Bifidobacterium longum*, *Bifidobacterium infantis* and *Bifidobacterium breve*. *Int. J. Syst. Bacteriol.*, 43, 565-573.

Biavati, B., Castagnoli, P., Crociani, F. and Trovatelli, L. D. (1984) Species of the *Bifidobacterium* in the feces of infants. *Microbiologica*, 7, 341-345.

Bingen, E.H., Denamur, E. and Elion, J. (1994) Use of ribotyping in epidemiological surveillance of nosocomial outbreaks. *Clin. Microbiol. Rev.*, 7, 311-327.

Christensen, H., Bisgaard, M., Frederiksen, W., Mutters, R., Kuhnert, P. and Olsen, J.E. (2001) Is characterization of a single isolate sufficient for valid publication of a new genus or species ? Proposal to modify Recommandation 30b of the *Bacteriological Code* (1990 Revision). *Int. J. Syst. Evol. Microbiol.*, 51, 2221-2225.

Crociani, F., Biavati, B., Alessandrini, A., Chiarini, C. and Scardovi, V. (1996) *Bifidobacterium inopinatum* sp. nov. and *Bifidobacterium denticolens* sp. nov., two new species isolated from human dental caries. *Int. J. Syst. Bacteriol.*, 46, 564-571.

Dong, X., Xin, Y., Jian, W., Liu, X. and Ling, D. (2000) *Bifidobacterium thermacidophilum* sp. nov., isolated from an anaerobic digester. *Int. J. Syst. Evol. Microbiol.*, 50, 119-25.

Farrelly, V., Rainey, F.A. and Stackebrandt E. (1995) Effect of genome size and *rrn* gene copy number on PCR amplification of 16S rRNA genes from a mixture of bacterial species. *Appl. Environ. Microbiol.*, 61, 2798-2801.

Fasoli, S., Marzotto, M., Rizotti, L., Rossi, F., Dellaglio, F. and Torriani, S. (2003) Bacterial composition of commercial probiotic products as evaluated by PCR-DGGE analysis. *Int. J. Food Microbiol.,* 82, 59-70.

Green, R. and Noller, H.F. (1997) Ribosomes and translation. *Annu. Rev. Biochem.*, 66, 679-716.

Hayashi, H., Sakamoto, M. and Benno, Y. (2004) Evaluation of three different forward primers by terminal restriction length polymorphism analysis for determination of fecal *Bifidobacterium* spp. In healthy subjects. *Microbiol. Immunol.*, 48, 1-6.

Janda, J.M. and Abbott, S.L. (2002) Bacterial identification for publication: when is enough enough ? *J. Clin. Microbiol.,* 40, 1887-1891.

Jian, W. and Dong, X. (2002) Transfer of *Bifidobacterium inopinatum* and *Bifidobacterium denticolens* to *Scardovia inopinata* gen. nov., comb. nov., and *Parascardovia denticolens* gen. nov., comb. nov., respectively. *Int. J. Syst. Evol. Microbiol.,* 52, 809-812.

Kaufmann, P., Pfefferkorn, A., Teuber, M. and Meile, L. (1997) Identification and quantification of *Bifidobacterium* species isolated from food with genus-specific 16S rRNA-targeted probes by colony hybridization and PCR. *Appl. Environ. Microbiol.*, 63, 1268–1273.

Kimura, K., McCartney, A.L., McConnell, M.A. and Tannock, G.W. (1997) Analysis of fecal populations of bifidobacteria and lactobacilli and investigation of the immunological responses of their human hosts to the predominant strains. *Appl. Environ. Microbiol.* 63, 3394-3398.

Langendijk, P.S., Schut, F., Jansen, G.J., Raangs, G.C., Kamphuis, G.R., Wilkinson H.F., Wellings G.W. (1995) Quantitative fluorescent in situ hibridization of *Bifidobacterium* spp. with genus-specific 16S rDNA-targeted probes and its application in fecal samples. *Appl. Environ. Microbiol.,* 61, 3069-3075.

Leblond-Bourget, N., Philippe, H., Mangin, I., and Decaris, B. (1996) 16S rRNA and 16S to 23S internal transcribed spacer sequence analyses reveal inter- and intraspecific *Bifidobacterium* phylogeny. *Int. J. Syst. Bacteriol.,* 46, 102-111.

Marsh, T.L. (1999) Terminal restriction fragment length polymorphism (T-RFLP) : an emerging method for characterizing diversity among homologous populations of amplification products. *Curr. Opinion Microbiol.*, 2, 323-327.

McCartney, A.L. and Tannock, G. W. (1995) Ribotyping of *Bifidobacterium* strains using cells embedded in agarose plugs and a 16S rDNA probe. *Microbial Ecol. Health Dis.*, 8, 79-84.

McCartney, A.L., Wenzhi, W. and Tannock, G.W. (1996) Molecular analysis of the composition of the bifidobacterial and Lactobacillus microflora of humans. *Appl. Environ. Microbiol.* 62, 4608-4613.

Matsuki, T., Watanabe, K., Tanaka, R. and Oyaizu, H. (1998) Rapid identification of human intestinal bifidobacteria by 16S rRNA-targeted species- and group-specific primers. *FEMS Microbiol Lett.*, 167, 113-121.

Matsuki, T., Watanabe, K., Tanaka, R., Fukuda, M. and Oyaizu, H. (1999) Distribution of bifidobacterial species in human intestinal microflora examined with 16S rRNA-gene-targeted species-specific primers. *Appl. Environ. Microbiol.*, 65, 4506-4512.

Matsuki, T., Watanabe, K., Fujimoto, J., Kado, Y., Takada, T., Matsumoto, K. and Tanaka, R. (2004) Quantitative PCR with 16S rRNA-gene-targeted species-specific primers for analysis of human intestinal bifidobacteria. *Appl. Environ. Microbiol.*, 70, 167-173.

Mullié, C., Odou, M.F., Singer, E., Romond, M.B. and Izard, D. (2003) Multiplex PCR using 16S rRNA gene-targeted primers for the identification of bifidobacteria from human origin. *FEMS Microbiol. Lett.*, 222, 129-136.

Nebra, Y., Bonjoch, X. and Blanch, A.R. (2003) Use of *Bifidobacterium dentium* as an indicator of the origin of fecal water pollution. *Appl. Environ. Microbiol.*, 69, 2651-2656.

Noller, H.F. and Woese, C.R. (1981) Secondary structure of 16S ribosomal RNA. *Science*, 212, 403-411.

Olsen, G.J. and Woese, C.R. (1993) Ribosomal RNA: a key to phylogeny. *FASEB J.*, 7, 113-123.

Requena, T., Burton, J., Matsuki, T., Munro, K., Simon, M.A., Tanaka, R., Watanabe, K. and Tannock, G.W. (2002) Identification, detection, and enumeration of human *Bifidobacterium* species by PCR targeting the transaldolase gene. *Appl. Environ. Microbiol.*, 68, 2420-2427.

Sakata, S., Kitahara, M., Sakamoto, M., Hayashi, H., Fukuyama, M. and Benno, Y. (2002) Unification of *Bifidobacterium infantis* and *Bifidobacterium suis* as *Bifidobacterium longum*. *Int. J. Syst. Evol. Microbiol.*, 52, 1945-1951.

Sakamoto, M., Hayashi, H. and Benno, Y. (2003) Terminal restriction fragment length polymorphism analysis from human fecal microbiota and its application for analysis of complex bifidobacterial communities. *Microbiol. Immunol.*, 47, 133-142.

Satokari, R.M., Vaughan, E.E., Akkermans, A.D.L., Saarela, M. and de Vos, W.M. (2001) Bifidobacterial diversity in human feces detected by genus-specific PCR and denaturing gradient gel electrophoresis. *Appl. Environ. Microbiol.*, 67, 504-513.

Scardovi, V. and Trovatelli, L.D. (1965) The fructose-6-phosphate shunt as peculiar pattern of hexose degradation in the genus *Bifidobacterium*. *Ann. Microbiol.*, 15, 19-29.

Schell, M.A., Karmirantzou, M., Snel, B., Vilanova, D., Berger, B., Pessi, G., Zwahlen, M.C., Desiere, F., Bork, P., Delley, M., Pridmore, R.D. and Arigoni F. (2002) The genome sequence of *Bifidobacterium longum* reflects its adaptation to the human gastrointestinal tract. *Proc. Natl. Acad. Sci. U.S.A.*, 99, 14422-14427.

Stackebrandt, E. and Goebel, B.M. (1994) Taxonomic note: a place for DNA-DNA reassociation and 16S rRNA sequence analysis in the present species definition in bacteriology. *Int. J. Syst. Bacteriol.*, 44, 846-849.

Stackebrandt, E., Rainey, F.A. and Ward-Rainey, N. L. (1997) Proposal for a new hierarchic classification system, *Actinobacteria* classis nov. *Int. J. Syst. Bacteriol.*, 47, 479-491.

Stackebrandt, E., Frederiksen, W., Garrity, G.M., Grimont, P.A., Kampfer, P., Maiden, M.C., Nesme, X., Rossello-Mora, R., Swings, J., Truper, H.G., Vauterin, L., Ward, A.C. and Whitman, W.B. (2002) Report of the ad hoc committee for the re-evaluation of the species definition in bacteriology. *Int. J. Syst. Evol. Microbiol.*, 52, 1043-1047.

Temmerman, R., Scheirlinck, I., Huys, G. and Swings, J. (2003) Culture-independent analysis of probiotic products by denaturing gradient gel electrophoresis. *Appl. Environ. Microbiol.*, 69, 220-226.

Vaugien, L., Prevots, F. and Roques, C. (2002) Bifidobacteria identification based on 16S rRNA and pyruvate kinase partial gene sequence analysis. *Anaerobe*, 8, 341-344.

Wang, R.F., Beggs, M.L., Robertson, L.H. and Cerniglia, C.E. (2002) Design and evaluation of oligonucleotide-microarray method for the detection of human intestinal bacteria in fecal samples. *FEMS Microbiol. Lett.*, 213, 175-182.

Wayne, L.G., Brenner, D.J., Colwell, R.R., Grimont, P.A.D., Kandler, O., Krichevsky, M.I., Moore, L.H., Moore, W.E.C., Murray, R.G.E., Stackebrandt, E., Starr, M.P. and Trüper, H.G. (1987) International Committee on Systematic Bacteriology. Report of the ad hoc committee on reconciliation of approaches to bacterial systematics. *Int. J. Syst. Bacteriol.*, 37, 463-464.

Woese, C.R., Fox, G.E., Zablen, L., Uchida, T., Bonen, L., Pechman, K., Lewis, B.J. and Stahl, D. (1975) Conservation of primary structure in 16S ribosomal RNA. *Nature*, 254, 83-86.

Woese, C.R. (1987) Bacterial evolution. *Microbiol. Rev.*, 51, 221-271.

Yamamoto, T., Morotomi, M. and Tanaka, R. (1992) Species specific oligonuleotide probes for five *Bifidobacterium* species detected in human intestinal flora. *Appl. Environ. Microbiol.*, 58, 4076-4079.

In: Trends in RNA Research
Editor: Peter A. McNamara, pp. 277-301

ISBN: 1-59454-506-5

Chapter XII

RNA Amplification Methodologies

Shaoli Che[1,2] and Stephen D. Ginsberg[1,2,3]

[1]Center for Dementia Research, Nathan Kline Institute, and [2]Department of Psychiatry and [3]Department of Physiology and Neuroscience, New York University School of Medicine, Orangeburg, NY 10962

Abstract

RNA amplification is a series of somewhat complicated methods to amplify genetic signals from minute quantities of starting materials for complementary (cDNA) array analysis and other downstream genetic methodologies. Essentially, RNA amplification produces quantities of RNA through *in vitro* transcription (IVT). The present review will identify and provide historical perspective for each step in conventional and newly developed RNA amplification schemes, starting from isolating the RNAs to be amplified though the generation of double stranded (ds) cDNA template, to labeling the amplified RNA species for use in microarray analysis and other related experiments. A novel methodology has been developed in this laboratory to amplify RNA from minute amounts of starting material. Briefly, an efficient means of second strand (ss) cDNA synthesis using a sequence-specific "terminal continuation" (TC) method is demonstrated. An RNA synthesis promoter is attached to the 3' and/or 5' region of cDNA utilizing the TC mechanism. The orientation of amplified RNAs is 'antisense', or a novel 'sense' orientation. TC RNA amplification is utilized for many downstream applications including gene expression profiling, microarray analysis, and cDNA library/subtraction library construction. Biological samples harvested for this procedure can originate from a myriad of *in vivo* and *in vitro* tissue sources. Moreover, a variety of fixations can be employed, and tissues can be processed for histochemistry and/or immunocytochemistry prior to microdissection for TC RNA amplification, allowing for tremendous cell type and tissue specificity of downstream genetic applications. A likely mechanism of action is that the TC primer binds preferentially to GC-rich, CpG islands

[1] Address correspondence to: Stephen D. Ginsberg, Ph.D., Center for Dementia Research, Nathan Kline Institute, New York University School of Medicine, 140 Old Orangeburg Road, Orangeburg, NY 10962, 845-398-2170 (phone), 845-398-5422 (FAX) ,ginsberg@nki.rfmh.org

flanking 5' regions of DNA that contain promoter sequences. In summary, the present review will focus on the development and practical usage of RNA amplification technologies, including T7-based RNA amplification methodologies and the mechanism(s) of action of each amplification step.

Key words: RNA amplification, in vitro transcription, expression profiling, postmortem human brain, microarray

Introduction

Standard molecular biology techniques have allowed researchers to evaluate gene expression levels across a myriad of normative situations, experimental paradigms, and disease states. These methods include Southern analysis (DNA detection), Northern analysis (RNA detection), polymerase-chain reaction (PCR; DNA detection), reverse-transcriptase-PCR (RT-PCR; RNA detection), ribonuclease (RNase) protection assay, and *in situ* hybridization, among others. These industry accepted approaches typically quantitate the abundance of individual transcripts one at a time (or a few at a time). Recent developments in highthroughput genomic based technologies enable the assessment of dozens to hundreds to thousands of genes simultaneously in a coordinated fashion.

The potential to understand physiological processes and disease pathogenesis has expanded geometrically with these applications. Microarray analysis has emerged as an important and effective tool to assess transcript levels in a multitude of systems and paradigms. One drawback to these high throughput technologies is the need for significant amounts of high quality input sources of RNA for increased sensitivity and reproducibility of these tools. Whole organism and regional studies can generate significant input amounts of RNA species without any amplification procedures. Unfortunately, the quantity of RNA harvested from a single cell, estimated to be approximately 0.1-1.0 picograms, is not sufficient for standard RNA extraction procedures [50,73,79]. Therefore, methods have been developed to increase the amount of input starting material for downstream genetic analyses, including exponential PCR-based analyses and linear RNA amplification procedures. This chapter focuses on molecular and cellular processes that occur during RNA amplification procedures.

PCR Amplification

The amplification of genetic signals is a significant advance in the development of molecular biological techniques. Basic studies of genetic mechanisms and function, as well as clinical diagnoses, are greatly facilitated by the ease of detection and manipulation of amplified genetic materials. Amplification of genetic signals can be achieved at both DNA and RNA levels. Therefore, final amplified products are either DNA or RNA. A common method for DNA amplification is PCR [72]. Starting material for PCR reactions can originate from genomic DNA or complementary DNA (cDNA) reverse transcribed from RNA.

Essentially, *Thermus aquaticus* (*Taq*) DNA polymerase, a member of the DNA polymerase I family, is used in combination with oligonucleotide primer sets to amplify exponentially cDNA products through repeated cycles. PCR is an effective method to amplify a DNA template. However, PCR is an exponential, nonlinear amplification. Variation can occur in individual mRNA species of different lengths. Moreover, the efficiency of amplification depends on the GC content of the template. PCR-based methods tend to amplify abundant genes over rare genes and may distort quantitative relationships among gene populations [50,73]. Therefore, the final PCR products may not be proportional to the abundance of the starting material, potentially skewing comparisons of relative levels of gene expression. Real-time quantitative PCR (qPCR) can quantitate PCR product formation during each cycle of amplification and has lessened concerns regarding linear amplification. Other advantages of real-time qPCR include high throughput capabilities, the ability to simultaneously multiplex reactions, enhanced sensitivity, reduced inter assay variation, and lack of post PCR manipulations. The method allows for direct detection of PCR products during the exponential phase of the reaction, combining amplification and detection in a single step.

Two basic tenets enabled discovery and subsequent development of real-time qPCR. First, DNA polymerases including *Taq* polymerase possess 5' to 3' exonuclease activity [48]. Second, the principle of fluorescence resonance energy transfer (FRET) [11] can be applied to double labeled oligonucleotide probes (e.g., PCR and/or RT-PCR primers). In a two probe system (ideal for qPCR), FRET occurs when fluorescence energy of one primer is quenched by close proximity of the second primer. Thus fluorescence signal emission occurs only upon cleavage of the two primers (i.e., amplification of a specific PCR product during a PCR cycle). A variety of dye chemistries are currently being exploited in qPCR systems including hydrolysis probes, molecular beacons, and double stranded (ds) DNA binding dyes [10,55]. A prime example of a hydrolysis probe is the Taqman assay. In this real-time qPCR method, *Taq* polymerase enzyme cleaves a specific Taqman probe during the extension phase of the PCR. The probe is dual-labeled, with a reporter such as 6-carboxyfluorescein (FAM), at one end of the probe and a quencher dye, such as 6-carboxytetramethylrhodamine (TAMRA) at the other. As long as the probe is intact (in its free form), fluorescence emission of the reporter dye is absorbed by the quenching dye [43]. Upon nuclease degradation of the probe during PCR, separation of the reporter and quencher dyes occurs and FRET ceases, resulting in an increase of reporter fluorescence emission [11]. This process occurs in every cycle of PCR, and does not interfere with the exponential accumulation of amplified product. Molecular beacons are probes that form a stem-loop structure from a single stranded DNA molecule [8]. Similar to hydrolysis probes, a fluorophore is linked to one end of the molecule and a quencher is linked to the other end. Fluorescence is quenched when the probe is in a hairpin-like conformation due to the proximity between quencher and fluorophore allowing FRET. When the probe sequence in the loop anneals to a complementary nucleic acid target sequence, a conformational change occurs and FRET ceases, increasing the fluorescence emission [8]. Molecular beacons are particularly useful for identifying point mutations, as targets that differ by only a single nucleotide can be delineated. Molecular beacons also display significantly more sequence specificity than conventional oligonucleotide probes of equivalent length and can be employed to study novel nucleic acid-protein interactions [84]. DNA binding dyes such as SYBR green incorporate selectively into ds DNA. SYBR green emits undetectable fluorescence levels when it is in its free form. Robust fluorescence is emitted upon binding to ds DNA [55]. An advantage of utilizing SYBR green is that this dye chemistry can be implemented to assay virtually any target sequence with virtually any set of primers, making this application quite flexible and considerably less expensive than probe-based dye chemistries. However, sensitivity can be diminished due to the risk of amplifying nonspecific PCR products. Careful design of the primer pairs and assay optimization can alleviate the majority of nonspecific binding concerns.

Rationale for RNA Amplification

In contrast to PCR-based technologies, RNA amplification-based procedures increase RNA in a linear fashion by amplifying directly the initial cDNA template sequence. Resultant amplified products are representative of the original mRNA expression levels. RNA amplification is achieved through *in vitro* transcription (IVT). Essentially, these molecular manipulations are useful methods for complex gene profiling when limited starting material is available. RNA amplification is an advantageous procedure because it can preserve quantitative relationships of different genes. In addition to the linear amplification of small amounts of input RNA, recent studies also show that RNA amplification procedures can increase the sensitivity of gene expression profiling on microarray platforms even though a large quantity of starting material is available [28,74]. In the presence of a proper buffering system, nucleotide triphosphates (NTPs) and a bacteriophage transcription promoter sequence (e.g., T7, T3) the respective RNA polymerase will synthesize hundreds to thousands of high fidelity copies of the ds cDNA template [65,67]. In the subsequent sections of this chapter, we will illustrate how RNA amplification methods have been developed in our laboratory to perform expression profiling studies using fresh-frozen and fixed tissues within the central nervous system of animal models of neurodegeneration and postmortem human brains.

Input Sources of RNA: Regional Assessments

Sources of input RNA for RNA amplification can originate from a variety of *in vivo* and *in vitro* applications. Fresh, frozen, and fixed tissues are viable to varying degrees and are the subject of several reviews [2,89,90]. For example, laboratories routinely isolate both genomic DNA and total RNA from fixed paraffin-embedded tissues [18,58,76,77]. Moreover, genetic material preserved by fixation in human postmortem brain tissue is an excellent resource because fresh and/or frozen tissue is frequently not available, whereas archives of fixed human tissue exist in many institutions.

As stated above, a variety of tissues and cells can be used to extract mRNA for gene profiling experiments. With mRNA as starting material, it cannot be emphasized enough the importance of preservation of RNA integrity in tissues and cells. RNA species are particularly sensitive to degradation by ribonuclease (RNase). Because they play an important role in nucleic acid metabolism, RNases are found in virtually every cell type [25]. RNases degrade RNA species through endonuclease and exonuclease activity. RNases are quite stable and retain their activity over a broad pH range [6,25]. Thus, RNase-free precautions need to be taken prior to and during all RNA amplification procedures. All biological samples require prompt handling, either through rapid RNA extraction, flash freezing, or fixation to minimize the degradation of intracellular RNAs by RNases.

Much discussion involves identifying optimal methods to fix brain tissues for use with single cell microdissection and subsequent microarray analysis [34,39,89,90]. At present, no consensus protocol is currently utilized. However, several laboratories have demonstrated significant success mining array data using tissue samples from human brains fixed in both crosslinking fixatives such as 10% neutral-buffered formalin and 4% paraformaldehyde as well as precipitating fixatives such as 70% ethanol buffered with 150 mM sodium chloride [30,31,39]. Many variables, including antemortem characteristics, duration of fixation, and length of storage have been assessed [2,57,89,90]. One method that we have found to be useful in assessing RNA quality in tissue sections is acridine orange (AO) histofluorescence. AO is a fluorescent dye that intercalates selectively into nucleic acids [66,93] and has been used to detect RNA and DNA in normal and developing brain tissues [63,88]. Upon excitation with ultraviolet spectra, AO that intercalates into RNA emits an orange-red fluorescence, whereas AO that intercalates into DNA emits a yellowish-green fluorescence. For example, AO histofluorescence detects the presence of RNA species in normal neurons in human and animal (e.g., mouse, rat, and monkey) brains. In addition, RNA sequestration has been found in senile plaques and neurons that bear neurofibrillary tangles (NFTs) in Alzheimer's disease (AD) using AO histofluorescence [38,40].

Prior to discussing single cell microdissection, the merits of regional dissections are illustrated. Groups of related cells from discrete regions of brain (or other organs and tissues) can be readily dissected from paraffin-embedded tissue sections (e.g., 5-6 μm thick) or frozen tissue sections (e.g., 20-40 μm thick). Unstained sections can be utilized, but optimal cellular resolution occurs using sections prepared for immunocytochemistry or histochemistry [35]. Regional dissections can be performed using a stereomicroscope and a scalpel blade or micropunch. RNA is then extracted from the remaining tissue using chemical extraction (e.g., Trizol reagent, Invitrogen, Rockville, MD) or by magnetic extraction (e.g., KingFisher, ThermoLabsystems, Franklin, MA) [32]. An advantage of regional analyses is that limited RNA amplification is necessary to generate significant hybridization signal intensity for microarrays. Moreover, differences in specific regions may illustrate the importance of individual genes in one region as compared to another. The obvious disadvantage of regional dissection procedures is the lack of single cell resolution, as neurons, nonneuronal cells, vascular elements, and epithelial cells will be included in the dissection of brain. In summary, regional analysis is a powerful approach for the identification of transcripts that are enriched in a specific organ, region, lamina, or nuclei that differ from adjacent or connected regions, but will contain an admixture of different cell types that comprise the region.

Input Sources of RNA: Single Cell Assessments

One of the most important aspects of single cell or single population gene expression analysis is that different cell types can be discriminated based upon their molecular fingerprint. For example, neurons can be identified readily by selective expression of relevant proteins such as cholinergic basal forebrain neurons [39,71] and midbrain dopaminergic neurons [26,85] for single cell RNA amplification and subsequent microarray analysis. Moreover, cells that lack a distinct or selective signature phenotype can be analyzed using a variety of Nissl stains {e.g., cresyl violet (CV) and thionin} [35,52,60] for downstream genetic applications. We have utilized CV previously to identify CA1 pyramidal neurons in postmortem human tissue sections for downstream genetic applications including single cell array analysis and qPCR [35,39]. Discrimination of adjacent cell types is quite important because this enables the differentiation of neuronal cells from glia, vascular epithelia, and other nonneuronal cells within the brain. Thus, a goal is to assess pathological changes in gene expression within vulnerable populations while avoiding potential contamination from spared neuronal cell types and nonneuronal populations.

The study of optimally prepared (e.g. fixed, sectioned, and stained) tissues is a principal method for histological and histopathological investigation. Several histochemical staining techniques have been developed to increase contrast of specific regional, cellular, or subcellular tissue constituents to enable the visualization of distinct cellular structures and cell types, as most populations of cells are relatively translucent and difficult to discern in unstained preparations. Although histological stains are typically not specific to an individual cell type or protein, much information can be gleaned by utilizing classical histological preparations in conjunction with contemporary protein (*e.g.*, immunocytochemistry) and molecular biological methodologies. Several groups have used hematoxylin and eosin (H&E) staining on peripheral tissues in cancer-related paradigms in combination with microdissection and PCR-based strategies as well as cDNA microarray platforms using RNA amplification methods [3,44,61,87]. Moreover, our laboratory has demonstrated that several Nissl stains including CV, H&E, and thionin perform as well as immunocytochemistry in terms of hybridization signal intensity detection via microarray analysis [35]. In contrast, several dyes that bind to RNAs directly such as AO and silver stain do not perform well in combination with microdissection and subsequent microarray analysis [35]. In summary, a variety of immunocytochemical and histochemical reagents enable detection of regions and specific cell types for subsequent microaspiration and downstream genetic analyses following a requisite linear RNA amplification or PCR-based amplification of the input RNAs.

Accession of Single Cells and Pooled Populations of Individual Cells

Advances in molecular techniques and functional genomics approaches allow scientists to extract nucleic acids from tissue sources for precise molecular analysis. The efficacy of sophisticated genetic testing methods is dependent upon the purity and precision of the analyzed cell population(s). Bulk homogenizing tissue samples results in the mixture of several different cell types. Cell-specific molecular analysis demands accurate, non-destructive isolation of cell populations from optimally prepared tissue sections. Microdissection of individual cells is performed to enable downstream gene expression profiling using microarrays or by PCR-based technologies. Provided that procedures are performed on well prepared tissue sections and RNase free conditions are employed, both immunocytochemical and histochemical procedures can be utilized to identify specific cell(s) of interest. Several different methodologies have been used to aspirate individual cells or groups of cells including single cell microaspiration and laser capture microdissection (LCM).

Single cell microaspiration entails visualizing an individual cell (or cells) using an inverted microscope connected to a micromanipulator, vacuum source, and an imaging workstation on an air table. Electrophysiology rigs can also be modified to aspirate cells from fixed tissue sections with minor modifications. Hand-held and syringe-pump driven vacuum sources can also be utilized, however, they are difficult to control and may cause inadvertent damage to the tissue section. Tissue sections are not coverslipped or counterstained and are immersed in RNase-free buffer. Individual cells are carefully aspirated from the tissue section of interest and placed in microfuge tubes for subsequent RNA amplification. This methodology results in accurate dissection of the neurons of interest with minimal disruption of the surrounding neuropil. Microaspiration of individual cells can be performed on a variety of tissues and cell types including normal samples as well as tissues undergoing various forms of degeneration [37,39,41,46,47]. The advantage of utilizing a single cell microaspiration technique is the extremely high cellular (and potentially subcellular, compartmental, and/or dendritic) aspiration of single elements. Single cells can be used alone, or pooled with other cells for subsequent RNA amplification. Potential disadvantages include the relative difficulty of performing the aspirating technique, experimenter error, and the lengthy time allotment necessary to perform microaspiration, especially if multiple cells are to be used.

Development of high throughput microaspiration devices has enabled rapid accession of single cells and populations for downstream molecular and cellular analyses. Notably, LCM is a method for acquiring cells from a myriad of tissue sources using histological and immunocytochemical methods. LCM is a widely used technique that was developed originally at the NIH [7,24]. After locating cells of interest to be microdissected, a thermoplastic cap is placed over a target area. Pulsing infrared laser energy causes focal activation of thermoplastic film embedded in the cap to form a thin protrusion that bridges the gap between the cap and tissue section, thus adhering to the target cell(s). Raising the thermoplastic cap separates targeted cells, now attached to the film, from surrounding undisturbed tissue. Populations of cells attached to the cap are suitable for microscopic examination and downstream genetic analysis. This form of LCM is termed positive extraction (a method used by the PixCell IIe from Arcturus, Mountain View, CA). Negative extraction (or non-contact laser extraction) procedures employ a laser source to cut around the area of interest within a tissue section, and the microdissected material is catapulted into a microfuge tube (a method utilized by the P.A.L.M. system, P.A.L.M. Microlaser Technologies, Bernried, Germany). Both positive and negative extraction methods allow captured cells and their processes to be examined microscopically to confirm the identity and quality of isolated cell population(s). This quality control step ensures validity of the results obtained from downstream analysis. Single cells as well as dozens to hundreds of cells can be collected by LCM instrumentation. RNA, DNA, and protein extraction methods can be performed on microdissected cells [19,23,29,81,83]. However, LCM use is most noted for RNA extraction and subsequent application to microarrays [7,24,35,39,92].

An example of single cell expression profiling is our ongoing work studying the molecular fingerprint of CA1 pyramidal neurons in postmortem human brain and in the brains of animal models of neurodegeneration [34,35,39,41]. CA1 neurons are visualized by neurofilament immunocytochemistry or by H&Estaining on paraffin-embedded 6 μm-thick tissue sections. Individual CA1 neurons are obtained via LCM. Twenty five CA1 neurons per section are utilized for each RNA amplification and subsequent cDNA array analysis and/or qPCR assessment. This state-of-the-art paradigm enables an extensive, concurrent representation of hundreds to thousands of genes in brain germane to neurodegenerative cascades in vulnerable as well as relatively spared cell types.

Reproducibility of single cell expression profiling is a critical parameter that continues to be improved upon. Advances at the level of tissue dissection, RNA amplification, microarray platforms, and developing powerful statistical methods will ultimately lead to greater utility and flexibility of these technologies. Recent advances include the utilization of pooled populations of individual cell types [34,35,39] to reduce variability in expression levels yet maintain an expression profile for a single cell type. Moreover, the likelihood of attaining reproducible data is increased greatly by replicate array analysis of aliquots of the same amplified RNA sample.

RNA Amplification Strategies

One method of linear amplification termed amplified antisense RNA (aRNA) amplification, developed by Eberwine and colleagues [21,22,50] involves a T7 RNA polymerase based amplification procedure that enables quantitation of the relative abundance of gene expression levels from identified single cells (Fig. 1). The resultant amplified aRNA maintains a proportional representation of the size and complexity of the initial mRNAs [22,91]. Briefly, aRNA amplification entails the hybridization of a 66 basepair (bp) oligonucleotide primer consisting of 24 thymidine triphosphates (TTPs) and a T7 RNA polymerase promoter sequence {oligo d(T)T7)} to mRNAs and conversion to an mRNA-cDNA hybrid by reverse transcriptase (RT) [86,91]. Upon conversion of the mRNA-cDNA hybrid to ds cDNA, a functional T7 RNA polymerase promoter is formed. aRNA synthesis occurs with the addition of T7 RNA polymerase and NTPs. Candidate mRNA expression levels can be identified using high-density cDNA microarrays or oligonucleotide arrays by probing with fluorescent and/or biotin tagged NTPs incorporated into aRNA products.

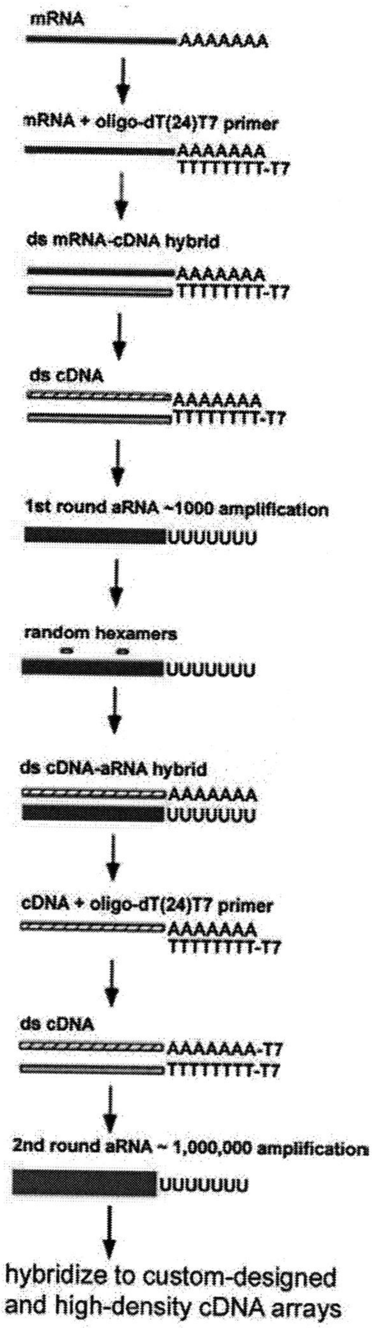

Figure 1. aRNA amplification scheme. An oligo-d(T)T7 primer is hybridized directly to poly(A)+ mRNAs. A ds mRNA-cDNA hybrid is formed by reverse transcribing the primed mRNAs with dNTPs and RT. The ds mRNA-cDNA hybrid is converted into ds cDNA forming a functional T7 RNA polymerase promoter. Following the removal of tertiary structures, first round of aRNA synthesis occurs using T7 RNA polymerase and NTPs. A second round of aRNA amplification begins by annealing random hexamers to the newly formed aRNA, and synthesizing a cDNA strand. The oligo(dT)-T7 primer is then reintroduced which binds to the poly(A)+ sequence on the newly synthesized cDNA strand, and a ds cDNA template is formed. Candidate mRNAs can be identified and quantitated using probes generated with radiolabeled (or fluorescently tagged) second round aRNA products. Adapted from reference [42].

Alternatively, radiolabeled NTPs can be incorporated into aRNA products and used for hybridization to membrane-based microarrays via reverse Northern blotting to known cDNA clones (Fig. 1). Each round of aRNA results in an approximate thousand fold amplification from the original amount of each polyadenylated {poly(A)+} mRNA in the sample [21,22]. Two rounds of are typically necessary to generate sufficient quantities of aRNA for subsequent downstream analyses such as cDNA microarrays. aRNA products are biased towards the 3' end of the transcript because of the initial priming at the poly(A)+ RNA tail [50,73]. This 3' bias exists for all amplified aRNA products and relative levels of gene expression can be compared [31,62,73]. However, when selecting cDNAs for expression analysis it is important to choose clones with 3' transcript representation. Another option is to assess multiple expressed sequence tagged cDNAs (ESTs) or oligonucleotides linked to the same gene [34,41,46]. Another caveat to the aRNA amplification procedure is that aRNA products tend not to be of full length [21,50,73]. Although aRNA is a laborious and often difficult procedure, we have generated successful results on microaspirated animal model and human tissues utilizing a wide variety of microarray platforms [37,41,46,47]. Modifications of the original aRNA procedure have been performed [61,94,95], and several kits that use aRNA technology to amplify small amounts of RNA are commercially available (e.g., MessageAmpII by Ambion, Austin, TX and RiboAmp by Arcturus), although single cell analysis is not necessarily tenable using these systems.

As stated above, aRNA is effective, yet can be a relatively difficult protocol to perform. A variety of strategies have been developed by independent laboratories to improve RNA amplification efficiency [9,49,64,94,96]. The principal obstacle is the problematic second strand (ss) cDNA synthesis. This impediment is not specific to the aRNA protocol. Rather, this issue is endemic to all current RNA amplification methods. Key factors to improving RNA amplification include increasing the efficiency of ss cDNA synthesis and allowing for flexibility in the placement of bacteriophage transcriptional promoter sequences.

A new RNA amplification procedure has been developed in our laboratory that utilizes a method of terminal continuation (TC) (Fig. 2). TC RNA amplification of genetic signals includes synthesizing fs cDNA complementary to the RNA template, subsequently generating ss cDNA complementary to the fs cDNA, and finally IVT using the ds cDNA as template [12]. Synthesis of the fs cDNA complementary to template mRNA entails the use of two oligonucleotide primers, a poly d(T) primer and a TC primer. The poly d(T) primer is similar to conventional primers that exploit the poly(A)+ sequence present on most mRNAs. The TC primer consists of an oligonucleotide sequence at the 5' terminus and a short span of three cytidine triphosphates (CTPs) at the 3' terminus. In this manner, ss cDNA synthesis can be initiated by annealing a second oligonucleotide primer complementary to the attached oligonucleotide [12]. By providing a known sequence at the 3' region of fs cDNA and a primer complementary to it, hairpin loops will not form, and ss cDNA synthesis can be performed with robust DNA polymerases, such as *Taq*, and the TC reaction is highly efficient. One round of amplification is sufficient for downstream genetic analyses [12,35]. Furthermore, TC RNA transcription can be driven using a promoter sequence attached to either the 3' or 5' oligonucleotide primers. Therefore, transcript orientation can be in an antisense orientation (similar to conventional aRNA methods) when the bacteriophage promoter sequence is placed on the poly d(T) primer or in a sense orientation when the promoter sequence is attached to the TC primer, depending upon the design of the experimental paradigm (Fig. 2).

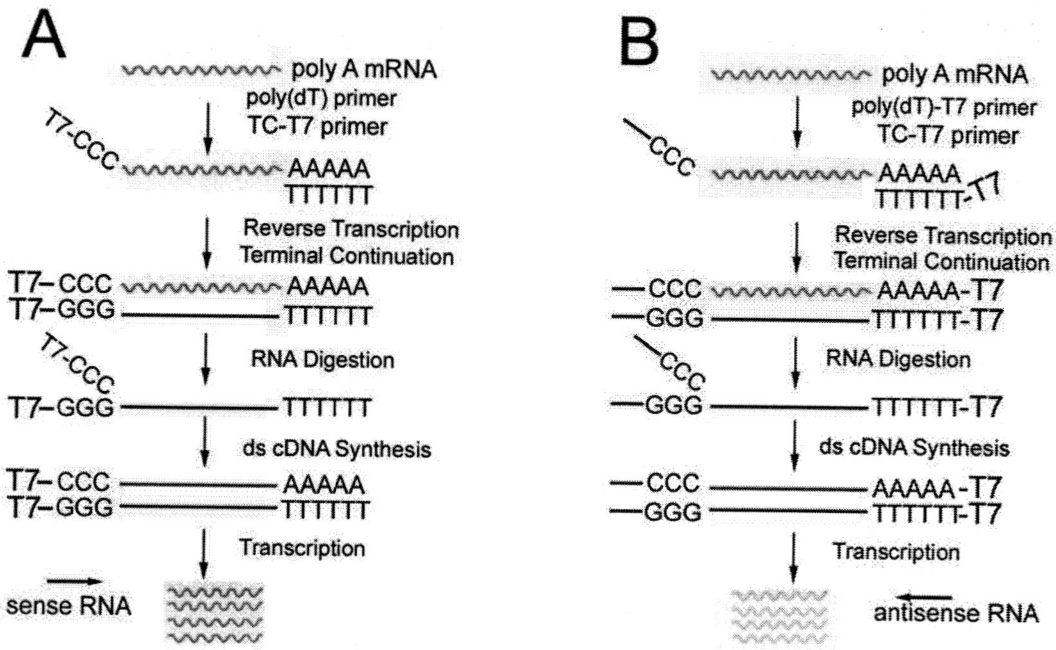

Figure 2. Overview and analysis of the TC RNA amplification method. A. A TC primer (containing a bacteriophage promoter sequence for sense orientation) and a poly d(T) primer are added to the mRNA population to be amplified (green rippled line). First strand (blue line) synthesis occurs as an mRNA-cDNA hybrid is formed after reverse transcription and terminal continuation of the oligonucleotide primers. Following RNase H digestion to remove the original mRNA template strand, second strand (red line) synthesis is performed using *Taq* polymerase. The resultant double stranded product is utilized as template for in vitro transcription, yielding high fidelity, linear RNA amplification of sense orientation (green rippled lines). B. Schematic similar to A, illustrating the TC RNA amplification procedure amplifying RNA in the antisense orientation (yellow rippled lines). Adapted from reference [12].

RNA Preparation

RNA amplification provides a powerful tool to overcome limitations in the amount of starting materials. However, these procedures do not diminish the importance of careful sample handling, since small quantities of RNAs are vulnerable to RNase degradation [6]. An RNase-free environment is absolutely critical for successful RNA amplification procedures and downstream genetic analyses. Long term storage of RNAs (especially amplified RNAs) is not recommended and repeated freeze-thaw should be avoided. Furthermore, dubious RNA handling can complicate any troubleshooting process when problems arise. A feasible break point in any RNA based experiment should be following the synthesis of either the fs or ss cDNA.

Efficient methodologies have been derived empirically to accommodate the expedient isolation of RNAs. In general, these methods yield cytoplasmic RNA, nuclear RNA, or mixtures of both. Isolation of RNA begins with cellular lysis. Two distinct buffers are typically used. One buffer system consists of harsh chaotropic agents, such as guanidinium, sodium dodecyl sulfate (SDS), urea, phenol, and/or chloroform, which disrupt the plasma membrane and subcellular organelles and simultaneously inactivate RNases [15-17]. A second buffer system solubilizes the plasma membrane using hypotonic lysis buffers such as Igepal CA-630 (a variant of the defunct NP-40) in the presence of RNase inhibitors [5,27]. Additionally, poly(A)+ RNA can be extracted using columns or resins that contain oligo d(T). Kits that enable extraction of total and poly(A)+ are available commercially, some of which are designed for small RNA input volumes.

Our laboratory routinely purifies total cellular RNA (from regional dissections as well as LCM preparations) with Trizol reagent (Invitrogen). Trizol contains guanidine isothiocyanate and phenol. Tissues or single cells are directly submerged in Trizol reagent in a microfuge tube or homogenizing vessel. A homogenizing step is applied when necessary. For example, frozen tissue pieces usually require gentle homogenizing whereas single cells need no more than a brief vortex to break cell membranes. For whole paraffin embedded tissue sections, proteinase K digestion is included to ensure that cell membranes are disrupted and that RNAs are released from the tissues prior to adding Trizol reagent [12,34]. After proteinase K digestion, Trizol LS reagent (Invitrogen) is added into the sample to extract RNAs. Trizol LS is similar to Trizol except that it is designed for liquid samples rather than tissue blocks. During sample homogenization and lysis, Trizol reagent inactivates RNases and therefore maintains the integrity of the RNA, while disrupting cells and dissolving cell components. Addition of chloroform followed by centrifugation separates the solution into an aqueous phase and an organic phase. RNA remains exclusively in the aqueous phase. After removal of the aqueous phase, the RNA is recovered by precipitation with isopropyl alcohol. RNAs purified from Trizol reagent are total cellular RNAs. mRNA can be affinity purified using oligo d(T) columns or beads. Kits are available to perform RNA (total and/or mRNA) extraction from *in vivo* and *in vitro* sources provided that sufficient quantity of total RNA is available. However, the yield of mRNA from a single cell may not withstand multiple purification steps. We have found that total RNA obtained from single cells, either via microaspiration or LCM, is more than adequate for aRNA or TC RNA amplification and the subsequent application as probe to microarrays [35,37,41].

First Strand cDNA Synthesis

fs cDNA synthesis is an RNA-dependent DNA synthesis performed in conjunction with a reverse transcription reaction. Moloney murine leukemia virus reverse transcriptase (MMLV-RT), and to a lesser degree, avian myeloblastosis reverse transcriptase (AMV-RT) is widely used for the synthesis of cDNA. In addition to having RNA dependent DNA polymerase activity, unmodified retroviral RT possesses an undesirable RNase H activity that degrades RNA strands in an RNA-DNA hybrid [4,69]. RTs, like many other DNA polymerases, also possess terminal nucleotidyl transferase (TdT) activity [14], a non-template directed transferase adding nucleotides to the 3' end of the first strand cDNA. RNAse H activity is thought to be partially responsible for the limitation on efficient synthesis of long cDNAs, and it is reasonable to assume that longer cDNA products lead to longer RNA probes.

Reverse transcriptase that lacks RNAse H activity (RNAse H- RT) is commercially available (e.g., SuperScript III from Invitrogen). Elimination of RNAse H activity also reduces the TdT activity of RT. For example, SuperScript III has decreased TdT activity (and reduced RNAse H activity) compared to its predecessor to Superscript II [75]. Our laboratory has utilized SuperScript III routinely for TC RNA amplification and has generated excellent hybridization signal intensity on microarray platforms from these probes [12,34-36,39].

Second Strand cDNA Synthesis

Synthesis of ss cDNA is required to generate a ds cDNA template as part of an RNA amplification procedure. Traditionally, ss synthesis has been a problematic and inefficient step. Two procedures utilized for ss cDNA synthesis include self-priming and replacement synthesis. Self-priming uses a hairpin at the 3' end of fs cDNA to self-prime ss cDNA synthesis. However, the loop formed at the end must be removed using S1 nuclease digestion in a poorly controlled reaction and invariably leads to the loss of the 5' signal [12,79]. In addition, self-priming is performed with the Klenow fragment of *E. coli* DNA polymerase I, an enzyme with relatively low processivity, further decreasing the efficiency of the method. The second procedure currently used for ss cDNA synthesis, termed replacement synthesis, avoids S1 nuclease digestion and has been used for RNA amplification. Replacement synthesis employs multiple enzymes to synthesize ss cDNA including RNase H, DNA polymerase I, and T4 DNA ligase [45,61]. After the fs cDNA is synthesized, it is annealed to the template RNA strand. Replacement synthesis uses the activity of RNase H to produce multiple nicks in the RNA strand. These nicks then serve as initiation sites for ss cDNA synthesis by DNA polymerase I. Through this process the majority of RNA fragments are replaced by DNA. ds cDNA molecules are finally treated with T4 DNA ligase to fill in any remaining nicks in the ss cDNA. However, replacement synthesis suffers from low efficiency, likely caused by the multiple enzymatic steps involved.

Table 1. Representative oligonucleotide sequences utilized for the poly d(T) and TC primers for the TC RNA amplification method.

Antisense RNA Orientation	
poly d(T)-T7 primer (66 bp):	5'- AAA CGA CGG CCA GTG AAT TGT AAT ACG ACT CAC TAT AGG CGC TTT TTT TTT TTT TTT TTT TTT TTT -3'
TC primer (17 bp):	5'- TAT CAA CGC AGA GTC CC -3'
Sense RNA Orientation	
poly d(T) primer (18 bp):	5'- TTT TTT TTT TTT TTT TTT -3'
TC-T7 primer (51 bp):	5'- AAA CGA CGG CCA GTG AAT TGT AAT ACG ACT CAC TAT AGG CGC GAG AGC CCC -3'

To increase the sensitivity and reproducibility of the ss cDNA synthesis, we developed a novel TC RNA amplification method. TC method increases the efficiency of ss cDNA synthesis, effectively enabling more abundant and longer length cDNA templates to be available for IVT. Specifically, the TC method requires presence of two primers: the TC primer in the RT reaction in addition to a poly d(T) primer (Table I). A two primer system allows the use of more robust DNA polymerases, such as *Taq* polymerase, for ss cDNA synthesis (even though this is not a PCR-based application). *Taq* polymerase has superb processivity as well as thermal stability.

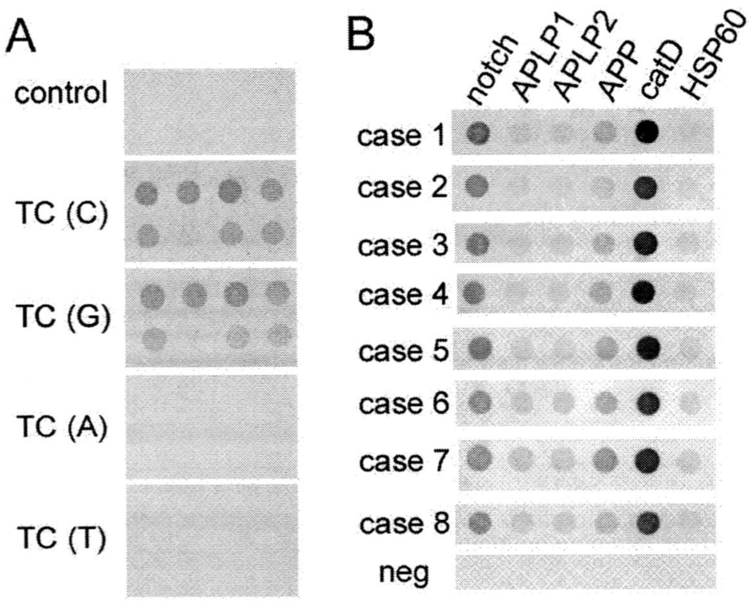

Figure 3. Representative custom-designed cDNA arrays illustrating utility and reproducibility of the TC RNA amplification procedure. A. cDNA array analysis of postmortem hippocampus following TC RNA amplification with TC primers of differing nucleotide composition. Representative custom-designed cDNA arrays illustrating relative expression levels of eight transcripts {column 1, clones 1-4, left to right; cyclic AMP response element binding protein (CREB), cyclin D1 (cycD1), immediate early gene zif268, nuclear factor-kappa B (NFKb); column 2, clones 5-8, left to right; neural cell adhesion molecule (NCAM), cyclooxygenase-2 (COX-2), and galanin receptor 1 and 2 (GalR1, GalR2)}. Strong levels of hybridization signal intensity are observed for the eight transcripts using TC primers containing CTP or GTP at the 3' terminus, respectively. In contrast, a paucity of hybridization signal intensity using TC primers containing ATP or TTP at the 3' terminus, respectively. B. Single cell cDNA analysis of CA1 pyramidal neurons using custom-designed cDNA arrays and TC RNA amplification. Representative arrays illustrate a wide dynamic range of hybridization signal intensities for eight human postmortem cases. The negative control (neg) is a single CA1 pyramidal neuron from the first case that does not have the primers necessary for TC RNA amplification. In addition, a moderate variation of gene level expression across the eight human cases is also observed, indicating the utility of using postmortem human samples for normative and neuropathological investigations. Key: notch, amyloid-ß precursor-like proteins 1and 2 (APLP1, APLP2), amyloid-ß precursor protein (APP), cathepsin D (catD), and heat shock protein 60 (HSP60).

Moreover, *Taq* polymerase synthesizes high fidelity DNA at high speed and at a high temperature that effectively reduces background that tends to manifest as nonspecific hybridization on arrays. We recommend to use "hot start" for the ss cDNA synthesis, e.g. adding in *Taq* polymerase at the denaturing step when the reaction temperature is at 95 °C

[12,70]. A TC primer contains at least two or three dCTPs or dGTPs (CTPs and GTPs work as well; dATPs, ATPs, and TTPs do not perform well [12]; Fig. 3) at the 3' end, which in turn can anneal to GC-rich regions at 5'-end of mRNAs. In this manner, a TC primer acts as an extended template for RT to continue. This results in a very long fs cDNA consisting of the poly d(T) primer, a sequence complementary to the original mRNA, and a sequence complementary to the TC primer present [12].

The mechanism of TC primer annealing to mRNA is not certain. A theory is that the 3' region of a TC primer anneals to a stretch of CTPs added nonspecifically onto mRNA through the TdT activity of RT. However, this hypothesis contradicts with our results. As mentioned earlier, a thermal stable RT, SuperScript III, has decreased RNase-H and TdT activity [75]. Use of this enzyme should decrease TC efficiency if TdT function was requisite for the mechanism of TC activity. In contrast, excellent probes are generated using the TC RNA amplification methodology in concert with SuperScript III [12,35,36,39]. Consistent these observations, TdT activity of RT appears more likely to add extra nucleotides such as adenosine triphosphates (ATPs) or thymidine triphosphates (TTPs) to mRNA rather than guanosine triphosphates (GTPs) or CTPs [14]. This suggests that GTPs and CTPs added nonspecifically on mRNA by RT are not requisite to anneal TC primer to provide an extended template. A mechanism for the TC primer to anneal preferentially to 5' regions of transcripts is being investigated. Based upon sequence analysis of TC RNA amplified products from human and mouse brains, we hypothesize that the TC primer, with its span of CTPs (GTPs work as well [34]) anneals preferentially within CpG islands. CpG islands are nonmethylated GC-rich regions of the genome that tend to include the 5' end of genes and are found at a significantly less frequency (CpGs are 25% less frequent than predicted) throughout the rest of the genome [1,20]. Estimates suggest that upwards of 60% of all human genes are located near CpG islands [1,20,56]. It is apparent that the TC primer has to base pair with the complementary CTPs or GTPs at the termination site of the RT reaction in order to provide a short template for DNA synthesis to continue. Replacement of the CTPs/GTPs with ATPs or TTPs vastly diminishes the TC reaction [34] (Fig. 3). In addition, random base pairing of ATPs and TTPs with complementary TTPs and ATPs in mRNAs may interrupt a proper RT process that is essential for generation of the fs cDNA [12,34].

In Vitro Transcription

IVT is a process of RNA synthesis facilitated by an RNA polymerase under the direction of a ds DNA template coupled to a bacteriophage transcription promoter sequence. Contrary to DNA synthesis, RNA synthesis does not need a primer, rather a promoter sequence is required. To initiate RNA synthesis, DNA-dependent RNA polymerase must bind to a specific promoter DNA sequence that lies upstream of the start site for transcription, separate the two strands of the DNA in the vicinity of the start site to form a transcription bubble, and initiate RNA synthesis using the coding strand of the downstream DNA as a template [53,80,82]. Common RNA polymerases used for in vitro transcription reactions include T7, T3, and SP6 polymerases, named for the bacteriophages from which they were cloned. These aforementioned enzymes are single subunit RNA polymerases capable of transcribing complete genes without requiring additional proteins [13]. Among the three enzymes, T7 and T3 RNA polymerases have compatible efficacy in transcribing RNAs, whereas SP6 is somewhat weaker. A specific and distinct bacteriophage RNA synthesis promoter is required for each polymerase termed the T7, T3, and SP6 bacteriophage transcription promoter, respectively (Table II).

Although consensus sequences of all three promoters presumably contain a minimum sequence for activation (e.g., from −17 to +1; Table II), this may not be enough sequence to initiate efficient RNA synthesis. For example, levels of IVT are virtually undetectable when a T7 promoter of the minimum consensus sequence composition is attached to ds cDNA template [31]. In contrast, a T7 promoter of approximately 50 nucleotides initiated a robust IVT of both antisense RNA and sense RNA from TC RNA amplification reactions [12,34]. Furthermore, decreased efficiency of IVT is observed when a bacteriophage transcription promoter without a leader sequence is attached to a PCR-generated template [59]. Interestingly, these authors also found that a SP6 promoter could hardly induce any IVT from a PCR-generated minigene construct, although the same SP6 promoter could induce RNA synthesis within a plasmid preparation [59].

Table II. Depicted are representative bacteriophage transcription promoter sequences. The underlined sequences are consensus sequences of the respective promoters. Specificity of each promoter has been confirmed functionally. Notably, when a DNA vector contained all three phage promoters in tandem, each RNA polymerase was found to bind its own promoter and initiate transcription from a specific starting point without cross talk to other sequences (see *www.ambion.com* for further experimental details).

T7	+1
*TAATACGACTCACTATA**G**G*GAGA	
T3	+1
*AATTAACCCTCACTAAAGG*GAGA	
SP6	+1
*ATTTAGGTGACACTATAGA*AGNG	

To create a template for IVT, a bacteriophage transcription promoter can be attached to a ds cDNA through a primer containing promoter sequence. Conventional aRNA methodologies attach an RNA synthesis promoter through the poly d(T) primer in the fs synthesis step [50,73]. In this manner, a primer composed of an RNA synthesis promoter at its 5' end and a poly d(T) sequence at its 3' end is used to prime the RT reaction. A ds primer is formed through the ss cDNA synthesis. This process results in an antisense orientation of the synthesized RNA products [21,22]. In the TC RNA amplification method, a promoter sequence can be attached either through a poly d(T) primer in the same manner as the aRNA method, or through the novel TC primer at the 5' region of mRNA [12]. As the result, the poly d(T) promoter initiates aRNA in the antisense direction whereas the TC promoter drives RNA transcription in the sense orientation [12]. Direction of the RNA product is important due to 3' bias of conventional aRNA amplification and when RNA probes are used to hybridize single stranded oligonucleotide cDNA arrays. Thus, a potential advantage of using sense TC RNA amplification is the preservation of the 5' mRNA signal where biologically relevant protein coding sequences tend to reside. Synthesized sense TC amplified RNA can also be used as a template for *in vitro* protein translations and downstream proteomic applications.

TC RNA Amplification

TC RNA amplification offers high sensitivity, flexibility, and throughput capabilities for downstream genetic analyses. Following TC RNA amplification, a large proportion of genes can be assessed quantitatively as evidenced by bioanalysis and microarray analysis in mouse and human postmortem brain tissues [12,32-36,39]. Robust linear amplification is consistently observed. Amplification efficiency of approximately 2500-3000 fold is demonstrated with commercially available purified mRNAs and approximately 1000-1500 fold amplification is found after one round using biological samples of RNA extracted from a variety of brain sources [12]. Results indicate a high degree of expression level similarity for high, moderate, and low expressed genes using the TC RNA amplification method. The threshold of detection of genes with low hybridization signal intensity is also greatly increased, as many genes that are at the limit of detection using conventional aRNA can be readily observed with the TC method [12]. Importantly, increased sensitivity appears greatest for genes with relatively low abundance (Fig 3). Moreover, background hybridization is significantly attenuated when using TC RNA amplification [12,35,36].

Data collected via RNA amplification and microarray analysis does not allow absolute quantitation of mRNA levels, but generates an expression profile of the relative changes in mRNA levels [12,35,36,39,46,47,71]. Relative changes in individual mRNAs are analyzed by univariate statistics {e.g., analysis of variance (ANOVA) with post-hoc Newmann-Keuls test} for individual comparisons [34,35,37,39,41,42]. Differential expression greater than approximately two-fold is accepted conventionally as relevant for further examination [31,39,46,47,68,71]. Differentially expressed genes can be clustered into functional protein categories for multivariate coordinate gene expression analyses [30,39,54,78]. Computational analysis is critical for optimal use of microarrays due to the enormous volume of data that is generated from a single probe. Additionally, access to relational databases is desirable, especially when evaluating hundreds of ESTs that may/may not be linked to genes (and subsequent proteins) of known function.

Strategies for cDNA Microarray Verification

Independent verification of individual gene level changes discovered by microarray analysis by alternate techniques is of critical importance. However, many current techniques are fraught with methodological considerations that have to be appreciated when applying the respective assay(s). For example, PCR-based analyses are exponential, rather than linear analyses. Therefore, following the initial cDNA synthesis step by RT, subsequent amplification by PCR is exponential. Unless careful quantitative RT-PCR is performed, the PCR amplification step can induce significant cell to cell variability [9,50]. Moreover, to attain the necessary level of sensitivity for messages of low abundance a second round of PCR with internal nested primers is performed. The disadvantage of this approach is the high risk of contamination and the small number of genes that can be analyzed simultaneously. In contrast, real-time qPCR lessens concerns regarding linear amplification, as direct detection of PCR products occurs during the exponential phase of the reaction, combining amplification and detection in a single step. *In situ* hybridization histochemistry is an extremely useful technique, especially when applied to well-processed brain sections, allowing for the localization and estimation of abundance of specific mRNA transcripts within single cells [51]. Quantitation of reaction product density in the case of non-radioactive probes such as digoxigenin, or densitometry of silver grains from radiolabeled probes, can be applied. Due to several technical considerations including limited sensitivity and relative inability to assess multiple mRNAs simultaneously, *in situ* hybridization is best suited for regional assessments of gene expression patterns, or as a companion technique to PCR or linear RNA amplification based single cell analyses. Further, it is highly desirable to supplement array data with measurements of protein expression by immunoblotting, immunocytochemical, ELISA, or proteomic techniques. In terms of protein assessments, single cell resolution can be attained in tissue sections using immunocytochemical methods. However, immunoblotting and proteomic-based procedures at present are extremely difficult to perform (at least *in vivo*)on single cells and must rely on tissue homogenate preparations. Thus, a combination of multidisciplinary approaches is ideal for verification of gene expression level alterations, with the explicit knowledge that the sum of the evaluations may be more informative and reflect the actual biology of the system than an individual method.

Conclusions

The combination of single cell microdissection, TC RNA amplification, and cDNA microarrays enables high resolution, high throughput expression profiling of hundreds to thousands of genes simultaneously from a single cell or population. The next level of understanding of cellular and molecular mechanisms underlying normative function and the pathophysiology of disease lies in the ability to combine these aforementioned technologies with appropriate models to recapitulate the structure and connectivity of these systems *in vivo* and *in vitro*. For example, there is a preponderance of evidence that complex biological processes are not governed by the action of a single isolated gene, but rather by the coordinate interactions of numerous genes, whose increased or diminished expression may play a role in normative function under physiological conditions and may contribute to the mechanisms underlying disease states under pathological conditions. In the absence of understanding coordinate gene expression patterns, studies of single genes have generally failed to provide a unified picture of gene expression during disease pathogenesis. Single cell profiling techniques coupled with cDNA microarray platforms have the potential to quantify simultaneous expression levels of numerous genes in a given cell, thereby allowing for previously unobserved gene interactions to become more evident. A goal of modern molecular and cellular neuroscience is to bridge basic mechanisms with translational studies to help develop novel agents for pharmacotherapeutic interventions. A prime example lies in late-onset, progressive neurodegenerative disorders where current therapies are not optimal and often based upon treating end-stage symptoms of the disease process. Experimental paradigms such RNA molecular fingerprinting studies may help develop novel agents that target specific circuits or systems that are affected selectively and/or specifically during precise stages in disease pathogenesis. Therefore, this reduces the potential problems of drug interactions and unwanted side effects. To reiterate, pharmacotherapeutic interventions developed under the auspices of an expression profiling paradigm differ from current strategies, whereby a drug is delivered that effects the body in a global pattern, with relatively poor organ, regional, and/or cellular specificity. Thus, one of the true goals of the next generation of pharmacotherapeutics is to target regions and cell types that are most affected in specific disease states. The brain is an obvious site for single cell RNA exploratory studies, due to the plethora of cell types and intricate connectivity of regions. However, RNA amplification procedures, notably the TC RNA amplification methodology, have much broader applications. Virtually any *in vivo* or *in vitro* paradigm can be employed for TC RNA amplification. Disciplines that will benefit from this technology include, but are not restricted to, cancer biology, development, drug discovery, and a myriad applications where input sources of RNA are limited and/or cell-type specific.

Acknowledgements

We thank Ms. Joanna M. Eanniello, Ms. Irina Elarova, and Mr. Marc D. Ruben for expert technical assistance Support for this project (SDG) comes from the NINDS (NS43939, NS48447), NIA (AG10668, AG14449, AG17617, AG09466), NCI (CA94520), and Alzheimer's Association. We also express our appreciation to the families of the patients studied here who made this research possible.

References

[1] Antequera, F. and Bird, A., Number of CpG islands and genes in human and mouse, *Proc Natl Acad Sci U S A*, 90 (1993) 11995-11999.

[2] Bahn, S., Augood, S.J., Ryan, M., Standaert, D.G., Starkey, M. and Emson, P.C., Gene expression profiling in the post-mortem human brain--no cause for dismay, *J Chem Neuroanat*, 22 (2001) 79-94.

[3] Becker, I., Becker, K.-F., Röhrl, M.H., Minkus, G., Schütze, K. and Höfler, H., Single-cell mutation analysis of tumors from stained histologic slides, *Lab Invest*, 75 (1996) 801-807.

[4] Berger, S.L., Wallace, D.M., Puskas, R.S. and Eschenfeldt, W.H., Reverse transcriptase and its associated ribonuclease H: interplay of two enzyme activities controls the yield of single-stranded complementary deoxyribonucleic acid, *Biochemistry*, 22 (1983) 2365-2372.

[5] Besse, S. and Puvion-Dutilleul, F., Anchorage of adenoviral RNAs to clusters of interchromatin granules, *Gene Expr*, 5 (1995) 79-92.

[6] Blumberg, D.D., Creating a ribonuclease-free environment, *Methods Enzymol*, 152 (1987) 20-24.

[7] Bonner, R.F., Emmert-Buck, M., Cole, K., Pohida, T., Chuaqui, R., Goldstein, S. and Liotta, L.A., Laser capture microdissection: molecular analysis of tissue, *Science*, 278 (1997) 1481-1483.

[8] Bonnet, G., Tyagi, S., Libchaber, A. and Kramer, F.R., Thermodynamic basis of the enhanced specificity of structured DNA probes, *Proc Natl Acad Sci U S A*, 96 (1999) 6171-6176.

[9] Brail, L.H., Jang, A., Billia, F., Iscove, N.N., Klamut, H.J. and Hill, R.P., Gene expression in individual cells: analysis using global single cell reverse transcription polymerase chain reaction (GSC RT-PCR), *Mutat Res*, 406 (1999) 45-54.

[10] Bustin, S.A., Quantification of mRNA using real-time reverse transcription PCR (RT-PCR): trends and problems, *J Mol Endocrinol*, 29 (2002) 23-39.

[11] Cardullo, R.A., Agrawal, S., Flores, C., Zamecnik, P.C. and Wolf, D.E., Detection of nucleic acid hybridization by nonradiative fluorescence resonance energy transfer, *Proc Natl Acad Sci U S A*, 85 (1988) 8790-8794.

[12] Che, S. and Ginsberg, S.D., Amplification of transcripts using terminal continuation, *Lab Invest.*, 84 (2004) 131-137.

[13] Cheetham, G.M. and Steitz, T.A., Insights into transcription: structure and function of single-subunit DNA-dependent RNA polymerases, *Current Opinion in Structural Biology*, 10 (2000) 117-123.

[14] Chen, D. and Patton, J.T., Reverse transcriptase adds nontemplated nucleotides to cDNAs during 5'-RACE and primer extension, *Biotechniques*, 30 (2001) 574-582.

[15] Chirgwin, J.M., Przybyla, A.E., MacDonald, R.J. and Rutter, W.J., Isolation of biologically active ribonucleic acid from sources enriched in ribonuclease, *Biochemistry*, 18 (1979) 5294-5299.

[16] Chomczynski, P., A reagent for the single-step simultaneous isolation of RNA, DNA and proteins from cell and tissue samples, *Biotechniques*, 15 (1993)532-537.

[17] Chomczynski, P. and Sacchi, N., Single-step method of RNA isolation by acid guanidinium thiocyanate-phenol-chloroform extraction, *Anal Biochem*, 162 (1987) 156-159.

[18] Coombs, N.J., Gough, A.C. and Primrose, J.N., Optimisation of DNA and RNA extraction from archival formalin-fixed tissue, *Nucleic Acids Res*, 27 (1999) e12.

[19] Craven, R.A., Totty, N., Harnden, P., Selby, P.J. and Banks, R.E., Laser capture microdissection and two-dimensional polyacrylamide gel electrophoresis: evaluation of tissue preparation and sample limitations, *Am J Pathol*, 160 (2002) 815-822.

[20] Cross, S.H., Clark, V.H. and Bird, A.P., Isolation of CpG islands from large genomic clones, *Nucleic Acids Res*, 27 (1999) 2099-2107.

[21] Eberwine, J., Kacharmina, J.E., Andrews, C., Miyashiro, K., McIntosh, T., Becker, K., Barrett, T., Hinkle, D., Dent, G. and Marciano, P., mRNA expression analysis of tissue sections and single cells, *J Neurosci*, 21 (2001) 8310-8314.

[22] Eberwine, J., Yeh, H., Miyashiro, K., Cao, Y., Nair, S., Finnell, R., Zettel, M. and Coleman, P., Analysis of gene expression in single live neurons, *Proc Natl Acad Sci USA*, 89 (1992) 3010-3014.

[23] Ehrig, T., Abdulkadir, S.A., Dintzis, S.M., Milbrandt, J. and Watson, M.A., Quantitative amplification of genomic DNA from histological tissue sections after staining with nuclear dyes and laser capture microdissection, *J Mol Diagn*, 3 (2001) 22-25.

[24] Emmert-Buck, M.R., Bonner, R.F., Smith, P.D., Chuaqui, R.F., Zhuang, Z., Goldstein, S.R., Weiss, R.A. and Liotta, L.A., Laser capture microdissection, *Science*, 274 (1996) 998-1001.

[25] Farrell Jr, R.E., *RNA Methodologies, Second Edition*, Academic Press, San Diego, 1998.

[26] Fasulo, W.H. and Hemby, S.E., Time-dependent changes in gene expression profiles of midbrain dopamine neurons following haloperidol administration, *J Neurochem*, 87 (2003) 205-219.

[27] Favaloro, J., Treisman, R. and Kamen, R., Transcription maps of polyoma virus-specific RNA: analysis by two-dimensional nuclease S1 gel mapping, *Methods Enzymol*, 65 (1980) 718-749.

[28] Feldman, A.L., Costouros, N.G., Wang, E., Qian, M., Marincola, F.M., Alexander, H.R. and Libutti, S.K., Advantages of mRNA amplification for microarray analysis, *Biotechniques*, 33 (2002) 906-914.

[29] Fend, F., Emmert-Buck, M.R., Chuaqui, R., Cole, K., Lee, J., Liotta, L.A. and Raffeld, M., Immuno-LCM: laser capture microdissection of immunostained frozen sections for mRNA analysis, *Am J Pathol*, 154 (1999) 61-66.

[30] Galvin, J.E. and Ginsberg, S.D., Expression profiling and pharmacotherapeutic development in the central nervous system, *Alzheimer Dis Assoc Disord*, 18 (2004) 264-269.Ginsberg, S.D., Gene expression profiling using single cell microdissection combined with cDNA microarrays. In D.H. Geschwind (Ed.), *DNA Microarrays: The New Frontier in Gene Discovery and Gene Expression Analysis.*, Society for Neuroscience Press, Washington, 2001, pp. 61-70.

[31] Ginsberg, S.D., High quality mRNA extraction with KingFisher for molecular fingerprinting, *Focus*, 1 (2002) 5.

[32] Ginsberg, S.D., Glutamatergic neurotransmission expression profiling in the mouse hippocampus after perforant-path transection, Am J Geriatr Psychiatry, 13 (2005) 1052-1061.Ginsberg, S.D. and Che, S., RNA amplification in brain tissues, *Neurochem Res*, 27 (2002) 981-992.

[33] Ginsberg, S.D. and Che, S., Combined histochemical staining, RNA amplification, regional, and single cell analysis within the hippocampus, *Lab Invest*, 84 (2004) 952-962.

[34] Ginsberg, S.D. and Che, S., Expression profile analysis within the human hippocampus: comparison of CA1 and CA3 pyramidal neurons, *J Comp Neurol,* 487 (2005) 107-188.Ginsberg, S.D., Crino, P.B., Hemby, S.E., Weingarten, J.A., Lee, V.M.-Y., Eberwine, J.H. and Trojanowski, J.Q., Predominance of neuronal mRNAs in individual Alzheimer's disease senile plaques, *Ann Neurol*, 45 (1999) 174-181.

[35] Ginsberg, S.D., Crino, P.B., Lee, V.M.-Y., Eberwine, J.H. and Trojanowski, J.Q., Sequestration of RNA in Alzheimer's disease neurofibrillary tangles and senile plaques, *Ann Neurol*, 41 (1997) 200-209.

[36] Ginsberg, S.D., Elarova, I., Ruben, M., Tan, F., Counts, S.E., Eberwine, J.H., Trojanowski, J.Q., Hemby, S.E., Mufson, E.J. and Che, S., Single cell gene expression analysis: implications for neurodegenerative and neuropsychiatric disorders, *Neurochem Res*, 29 (2004) 1054-1065.

[37] Ginsberg, S.D., Galvin, J.E., Chiu, T.-S., Lee, V.M.-Y., Masliah, E. and Trojanowski, J.Q., RNA sequestration to pathological lesions of neurodegenerative disorders, *Acta Neuropathol*, 96 (1998) 487-494.

[38] Ginsberg, S.D., Hemby, S.E., Lee, V.M.-Y., Eberwine, J.H. and Trojanowski, J.Q., Expression profile of transcripts in Alzheimer's disease tangle-bearing CA1 neurons, *Ann Neurol*, 48 (2000) 77-87.

[39] Ginsberg, S.D., Schmidt, M.L., Crino, P.B., Eberwine, J.H., Lee, V.M.-Y. and Trojanowski, J.Q., Molecular pathology of Alzheimer's disease and related disorders. In A. Peters and J.H. Morrison (Eds.), *Cerebral Cortex, vol. 14. Neurodegenerative and Age-related Changes in Structure and Function of Cerebral Cortex*, Kluwer Academic/Plenum, New York, 1999, pp. 603-653.

[40] Giulietti, A., Overbergh, L., Valckx, D., Decallonne, B., Bouillon, R. and Mathieu, C., An Overview of Real-Time Quantitative PCR: Applications to Quantify Cytokine Gene Expression, *Methods*, 25 (2001) 386-401.

[41] Goldsworthy, S.M., Stockton, P.S., Trempus, C.S., Foley, J.F. and Maronpot, R.R., Effects of fixation on RNA extraction and amplification from laser capture microdissected tissue, *Mol Carcinog*, 25 (1999) 86-91.

[42] Gubler, U. and Hoffman, B.J., A simple and very efficient method for generating cDNA libraries, *Gene*, 25 (1983) 263-269.

[43] Hemby, S.E., Ginsberg, S.D., Brunk, B., Arnold, S.E., Trojanowski, J.Q. and Eberwine, J.H., Gene expression profile for schizophrenia: discrete neuron transcription patterns in the entorhinal cortex, *Arch Gen Psychiat*, 59 (2002) 631-640.

[44] Hemby, S.E., Trojanowski, J.Q. and Ginsberg, S.D., Neuron-specific age-related decreases in dopamine receptor subtype mRNAs, *J Comp Neurol*, 456 (2003) 176-183.

[45] Holland, P.M., Abramson, R.D., Watson, R. and Gelfand, D.H., Detection of specific polymerase chain reaction product by utilizing the 5'----3' exonuclease activity of Thermus aquaticus DNA polymerase, *Proc Natl Acad Sci U S A*, 88 (1991) 7276-7280.

[46] Iscove, N.N., Barbara, M., Gu, M., Gibson, M., Modi, C. and Winegarden, N., Representation is faithfully preserved in global cDNA amplified exponentially from sub-picogram quantities of mRNA, *Nat Biotechnol*, 20 (2002) 940-943.

[47] Kacharmina, J.E., Crino, P.B. and Eberwine, J., Preparation of cDNA from single cells and subcellular regions, *Methods Enzymol*, 303 (1999) 3-18.

[48] Kadkol, S.S., Gage, W.R. and Pasternack, G.R., In situ hybridization-theory and practice, *Mol Diagn 1999*, 4 (1999) 169-183.

[49] Kamme, F., Salunga, R., Yu, J., Tran, D.T., Zhu, J., Luo, L., Bittner, A., Guo, H.Q., Miller, N., Wan, J. and Erlander, M., Single-cell microarray analysis in hippocampus CA1: demonstration and validation of cellular heterogeneity, *J Neurosci*, 23 (2003) 3607-3615.

[50] Kochetkov, S.N., Rusakova, E.E. and Tunitskaya, V.L., Recent studies of T7 RNA polymerase mechanism, *FEBS Lett*, 440 (1998) 264-267.

[51] Kotlyar, M., Fuhrman, S., Ableson, A. and Somogyi, R., Spearman correlation identifies statistically significant gene expression clusters in spinal cord development and injury, *Neurochem Res*, 27 (2002) 1133-1140.

[52] Kricka, L.J., Stains, labels and detection strategies for nucleic acids assays, *Ann Clin Biochem*, 39 (2002) 114-129.

[53] Krieg, A.M., Now I know my CpGs, *Trends Microbiol*, 9 (2001) 249-252.

[54] Leonard, S., Logel, J., Luthman, D., Casanova, M., Kirch, D. and Freedman, R., Biological stability of mRNA isolated from human postmortem brain collections, *Biol Psychiatry*, 33 (1993) 456-466.

[55] Lewis, F., Maughan, N.J., Smith, V., Hillan, K. and Quirke, P., Unlocking the archive--gene expression in paraffin-embedded tissue, *J Pathol*, 195 (2001) 66-71.

[56] Logel, J., Dill, D. and Leonard, S., Synthesis of cRNA probes from PCR-generated DNA, *Biotechniques*, 13 (1992) 604-610.

[57] Luo, L., Salunga, R.C., Guo, H., Bittner, A., Joy, K.C., Galindo, J.E., Xiao, H., Rogers, K.E., Wan, J.S., Jackson, M.R. and Erlander, M.G., Gene expression profiles of laser-captured adjacent neuronal subtypes, *Nat Med*, 5 (1999) 117-122.

[58] Luzzi, V., Holtschlag, V. and Watson, M.A., Expression profiling of ductal carcinoma in situ by laser capture microdissection and high-density oligonucleotide arrays, *Am J Pathol*, 158 (2001) 2005-2010.

[59] Madison, R.D. and Robinson, G.A., lRNA internal standards quantify sensitivity and amplification efficiency of mammalian gene expression profiling, *Biotechniques*, 25 (1998) 504-514.

[60] Mai, J.K., Schmidt-Kastner, R. and Tefett, H.-B., Use of acridine orange for histologic analysis of the central nervous system, *J Histochem Cytochem*, 32 (1984) 97-104.

[61] Matz, M., Shagin, D., Bogdanova, E., Britanova, O., Lukyanov, S., Diatchenko, L. and Chenchik, A., Amplification of cDNA ends based on template-switching effect and step-out PCR, *Nucleic Acids Res*, 27 (1999) 1558-1560.

[62] Melton, D.A., Krieg, P.A., Rebagliati, M.R., Maniatis, T., Zinn, K. and Green, M.R., Efficient in vitro synthesis of biologically active RNA and RNA hybridization probes from plasmids containing a bacteriophage SP6 promoter, *Nucleic Acids Res*, 12 (1984) 7035-7056.

[63] Mikel, U.V. and Becker Jr., R.L., A comparative study of quantitative stains for DNA in image cytometry, *Analyt Quant Cytol Histol*, 13 (1991) 253-260.

[64] Milligan, J.F., Groebe, D.R., Witherell, G.W. and Uhlenbeck, O.C., Oligoribonucleotide synthesis using T7 RNA polymerase and synthetic DNA templates, *Nucleic Acids Res*, 15 (1987) 8783-8798.

[65] Mirnics, K., Middleton, F.A., Marquez, A., Lewis, D.A. and Levitt, P., Molecular characterization of schizophrenia viewed by microarray analysis of gene expression in prefrontal cortex, *Neuron*, 28 (2000) 53-67.

[66] Moelling, K., Reverse transcriptase and RNase H: present in a murine virus and in both subunits of an avian virus, *Cold Spring Harb Symp Quant Biol*, 39 (1975) 969-973.

[67] Moretti, T., Koons, B. and Budowle, B., Enhancement of PCR amplification yield and specificity using AmpliTaq Gold DNA polymerase, *Biotechniques*, 25 (1998) 716-722.

[68] Mufson, E.J., Counts, S.E. and Ginsberg, S.D., Single cell gene expression profiles of nucleus basalis cholinergic neurons in Alzheimer's disease, *Neurochem Res*, 27 (2002) 1035-1048.

[69] Mullis, K.B., The unusual origin of the polymerase chain reaction, *Sci Am*, 262 (1990) 56-65.

[70] Phillips, J. and Eberwine, J.H., Antisense RNA Amplification: A Linear Amplification Method for Analyzing the mRNA Population from Single Living Cells, *Methods Enzymol Suppl*, 10 (1996) 283-288.

[71] Polacek, D.C., Passerini, A.G., Shi, C., Francesco, N.M., Manduchi, E., Grant, G.R., Powell, S., Bischof, H., Winkler, H., Stoeckert, C.J., Jr. and Davies, P.F., Fidelity and enhanced sensitivity of differential transcription profiles following linear amplification of nanogram amounts of endothelial mRNA, *Physiol Genomics*, 13 (2003) 147-156.

[72] Potter, J., Zheng, W. and Lee, J., Thermal stability and cDNA synthesis capability of SuperScript III reverse transcriptase, *Focus*, 25 (2003) 19-24.

[73] Qin, Y., Heine, V.M., Karst, H., Lucassen, P.J. and Joels, M., Gene expression patterns in rat dentate granule cells: comparison between fresh and fixed tissue, *J Neurosci Methods*, 131 (2003) 205-211.

[74] Rupp, G.M. and Locker, J., Purification and analysis of RNA from paraffin-embedded tissues, *Biotechniques*, 6 (1988) 56-60.

[75] Ryan, M.M., Huffaker, S.J., Webster, M.J., Wayland, M., Freeman, T. and Bahn, S., Application and optimization of microarray technologies for human postmortem brain studies, *Biol Psychiatry*, 55 (2004) 329-336.

[76] Sambrook, J., Fritsch, E. and Maniatis, T., *Molecular cloning: a laboratory manual. Third edition*, Cold Spring Harbor Laboratory Press, Cold Spring Harbor, 2001.

[77] Severinov, K., T7 RNA polymerase transcription complex: what you see is not what you get, *Proc Natl Acad Sci U S A*, 98 (2001) 5-7.

[78] Simone, N.L., Remaley, A.T., Charboneau, L., Petricoin, E.F., 3rd, Glickman, J.W., Emmert-Buck, M.R., Fleisher, T.A. and Liotta, L.A., Sensitive immunoassay of tissue cell proteins procured by laser capture microdissection, *Am J Pathol*, 156 (2000) 445-452.

[79] Skinner, G.M., Baumann, C.G., Quinn, D.M., Molloy, J.E. and Hoggett, J.G., Promoter binding, initiation, and elongation by bacteriophage T7 RNA polymerase. A single-molecule view of the transcription cycle, *J Biol Chem*, 279 (2004) 3239-3244.

[80] Suarez-Quian, C.A., Goldstein, S.R., Pohida, T., Smith, P.D., Peterson, J.I., Wellner, E., Ghany, M. and Bonner, R.F., Laser capture microdissection of single cells from complex tissues, *Biotechniques*, 26 (1999) 328-335.

[81] Tan, W., Fang, X., Li, J. and Liu, X., Molecular beacons: a novel DNA probe for nucleic acid and protein studies, *Chemistry*, 6 (2000) 1107-1111.

[82] Tang, W.X., Fasulo, W.H., Mash, D.C. and Hemby, S.E., Molecular profiling of midbrain dopamine regions in cocaine overdose victims, *J Neurochem*, 85 (2003) 911-924.

[83] Tecott, L.H., Barchas, J.D. and Eberwine, J.H., In situ transcription: specific synthesis of complementary DNA in fixed tissue sections, *Science*, 240 (1988) 1661-1664.

[84] To, M.D., Done, S.J., Redston, M. and Andrulis, I.L., Analysis of mRNA from microdissected frozen tissue sections without RNA isolation, *Am J Pathol*, 153 (1998) 47-51.

[85] Topaloglu, H. and Sarnat, H.B., Acridine orange-RNA fluorescence maturing neurons in the perinatal rat brain, *Anat Rec*, 224 (1989) 88-93.

[86] Van Deerlin, V.M.D., Ginsberg, S.D., Lee, V.M.-Y. and Trojanowski, J.Q., Fixed post mortem brain tissue for mRNA expression analysis in neurodegenerative diseases. In D.H. Geschwind (Ed.), *DNA Microarrays: The New Frontier in Gene Discovery and Gene Expression Analysis*, Society for Neuroscience, Washington, D.C. 2000, pp. 118-128.

[87] Van Deerlin, V.M.D., Ginsberg, S.D., Lee, V.M.-Y. and Trojanowski, J.Q., The use of fixed human post mortem brain tissue to study mRNA expression in neurodegenerative diseases: applications of microdissection and mRNA amplification. In D.H. Geschwind and J.P. Gregg (Eds.), *Microarrays for the Neurosciences: An Essential Guide*, MIT Press, Boston, 2002, pp. 201-235.

[88] VanGelder, R., von Zastrow, M., Yool, A., Dement, W., Barchas, J. and Eberwine, J., Amplified RNA (aRNA) synthesized from limited quantities of heterogeneous cDNA, *Proc Natl Acad Sci USA*, 87 (1990) 1663-1667.

[89] Vincent, V.A., DeVoss, J.J., Ryan, H.S. and Murphy, G.M., Jr., Analysis of neuronal gene expression with laser capture microdissection, *J Neurosci Res*, 69 (2002) 578-586.

[90] von Bertalanffy, L. and Bickis, I., Identification of cytoplasmic basophilia (ribonucleic acid) by fluorescence microscopy, *J Histochem Cytochem*, 4 (1956) 481-493.

[91] Wang, E., Miller, L.D., Ohnmacht, G.A., Liu, E.T. and Marincola, F.M., High-fidelity mRNA amplification for gene profiling, *Nat Biotechnol*, 18 (2000) 457-459.

[92] Xiang, C.C., Chen, M., Ma, L., Phan, Q.N., Inman, J.M., Kozhich, O.A. and Brownstein, M.J., A new strategy to amplify degraded RNA from small tissue samples for microarray studies, *Nucleic Acids Res*, 31 (2003) E53.

[93] Zhumabayeva, B., Diatchenko, L., Chenchik, A. and Siebert, P.D., Use of SMART-generated cDNA for gene expression studies in multiple human tumors, *Biotechniques*, 30 (2001) 158-163.

Index

A

D

E

F

G

H

I

J

K

L

M

N

R

S

T

U

V

W

Y